W0228078

S. K. Nigam D. C. H. McBrien
T. F. Slater (Eds.)

Eicosanoids,
Lipid Peroxidation
and Cancer

With 126 Figures and 58 Tables

Springer-Verlag Berlin Heidelberg GmbH

Santosh K. Nigam, Ph.D., M.B., B.Ch.
Frauenklinik, Universitäts-Klinikum Steglitz
Hindenburgdamm 30, 1000 Berlin 45, FRG

David C. H. McBrien, Ph.D.
Department of Biology and Biochemistry, Brunel University
Uxbridge, Middlesex UB8 3PH, U.K.

Trevor F. Slater, Ph.D., D.Sc., M.D. (Hon.)
Department of Biology and Biochemistry, Brunel University
Uxbridge, Middlesex UB8 3PH, U.K.

ISBN 978-3-540-18932-9   ISBN 978-3-642-73424-3 (eBook)
DOI 10.1007/978-3-642-73424-3

Library of Congress Cataloging-in-Publication Data
Eicosanoids, lipid peroxidation, and cancer / S. Nigam, D.C.H. McBrien, T.F. Slater
(eds.). p. cm. Includes bibliographies and index. ISBN 978-3-540-18932-9 . 1. Carcinogenesis. 2. Eicosanoic acid–Derivatives–Metabolism. 3. Lipids–Peroxidation. I. Nigam, S.
(Santosh) II. McBrien, D.C.H. III. Slater, T.F. (Trevor Frank), 1931- . [DNLM: 1. Eicosanoic Acids–metabolism. 2. Lipid Peroxides–metabolism. 3. Neoplasms–chemically induced. QZ 202 E34] RC268.5.E33 1988 616.99'4071–dc19 88–38646

This work is subject to copyright. All rights are reserved, whether the whole or part of the
material is concerned, specifically the rights of translation, reprinting, re-use of illustra-
tions, recitation, broadcasting, reproduction on microfilms or in other ways, and storage
in data banks. Duplication of this publication or parts thereof is only permitted under the
provisions of the German Copyright Law of September 9, 1965, in its version of June 24,
1985, and a copyright fee must always be paid. Violations fall under the prosecution act
of the German Copyright Law.

© Springer-Verlag Berlin Heidelberg 1988
Originally published by Springer-Verlag Berlin Heidelberg New York in 1988
Softcover reprint of the hardcover 1st edition 1988

The use of registered names, trademarks, etc. in the publication does not imply, even in
the absence of a specific statement, that such names are exempt from the relevant protective laws and regulations and therefore free for general use.
Product Liability: The publisher can give no guarantee for information about drug dos-
age and application thereof contained in this book. In every individual case the respective user must check its accuracy by consulting other pharmaceutical literature.

Typesetting and printing: Zechnersche Buchdruckerei Speyer
Bookbinding: J. Schäffer, Grünstadt
2121/3140-543210 – Printed on acid-free paper

Preface

This book records the proceedings of the 5th International Symposium on cancer topics organised in collaboration with the Association for International Cancer Research, a cancer charity based in the United Kingdom. The Symposium was held at the Klinikum Steglitz, Free University of West Berlin, Germany, between 1 and 3 October 1987. The Organising and Scientific Committees are deeply grateful to the President of the Free University, Professor Heckelmann, for allowing us to meet in the Klinikum Steglitz and for the University's generous financial support.

It was a special pleasure to have a positive and generous input from the West Berlin Senate. Their support was crucial in making the Symposium a scientific success.

The Symposium received financial support from a number of sponsors, and we are indeed happy to acknowledge our gratitude to them: Behringwerke AG, Berthold AG, Boehringer Mannheim GmbH, the British Council, Deutsche Abbott GmbH, Deutsche Bank AG, Du Pont De Nemours GmbH, Knoll AG, Krewelwerke GmbH, Nunc GmbH, Schering AG, Schwarz GmbH and Varian GmbH.

The Symposium was the most ambitious so far mounted by the Association for International Cancer Research. The organisation and planning were carried through by an Organising Committee including Professor J. Hammerstein, Professor K.D. Asmus, Barbara Steiger and Renate Nigam. The Scientific Committee included Dr. D.C.H. McBrien, Professor T. Dormandy, Professor K.R. Rees, Professor M.U. Dianzani and Professor H. Esterbauer. We thank these colleagues very much for their dedication, hard work and expert input.

Proceedings of previous symposia of the Association for International Cancer Research were on "Free radicals, lipid peroxidation and cancer" (1982), "Cancer of the uterine cervix, biochemical and clinical aspects" (1984), "Biochemical mechanisms of platinum antitumour drugs" (1986), and "Protective agents in cancer" (1987).

The Symposium held in the Klinikum Steglitz was a very successful meeting in terms of exchange of new scientific information relating to lipid peroxidation, eicosanoids and cancer. This book is a valuable and up to date account of those discussions. We anticipate that all who read it will find much of interest, and a substantial amount of new data.

S.K. Nigam (Chairman, Organising Committee)
T.F. Slater (Chairman, Scientific Committee)

Contents

Lipid Peroxidation and Cancer I

Lipid Peroxidation and Cancer II

Biochemical Aspects of Gynaecological and Liver Cancer

List of Speakers

Dr. E. Albano
Dr. C. Benedetto
Prof. I. L. Bonta
Prof. P. Braquet
Dr. K. H. Cheeseman
Prof. M. Comporti
Prof. T. A. Connors
Prof. T. L. Dormandy
Prof. H. Esterbauer
Prof. M. L. Foegh
Dr. H. de Groot
Prof. S. Hammarström
Prof. K. V. Honn
Dr. J. S. Hurst
Prof. B. A. Jakschik
Prof. H. Kappus
Prof. L. Levine
Prof. U. M. Marinari

Dr. A. Morgan
Prof. S. Muszbek
Dr. S. K. Nigam
Priv.-Doz. Dr. G. Nöhammer
Prof. G. E. Plante
Prof. G. Poli
Dr. P. Principe
Prof. P. W. Ramwell
Dr. W. Rojanapo
Dr. R. J. Schaur
Dr. R. Seifert
Prof. H. Sies
Prof. T. F. Slater
Dr. D. J. Spargo
Prof. A. Stier
Prof. C. Thomson
Prof. V. Ullrich

*Arachidonic Acid Metabolism
and Tumour Promotion*

Thromboxane A$_2$ and Prostacyclin in Tumorigenesis

V. Ullrich, M. Hecker, R. Nüsing and T. Rosenbach

Faculty of Biology, University of Konstanz, P.O. Box 5560, 7750 Konstanz, FRG

1 Introduction

A tumour cell arises from a normal cell by changes in the processes of cell regulation, growth and differentiation. The complexity of such regulatory mechanisms, however, makes it difficult to understand the pathogenesis of malignant growth.

It is of no surprise that derivatives of arachidonic acid show many-fold relations to carcinogenesis since eicosanoids are involved in transducing information between cells and thereby control growth and development. Unfortunately, our knowledge of eicosanoids and their actions is far from complete and not more advanced than our understanding of cancer. Some caution, therefore, is appropriate when eicosanoids and cancer are discussed and this certainly also applies to possible roles of thromboxane A$_2$ and prostacyclin in tumorigenesis.

Among the prostaglandins these two representatives probably exhibit the most interesting physiological effects. Prostacyclin (PGI$_2$) is a potent vasodilator and anti-aggregatory compound, whereas thromboxane A$_2$ (TXA$_2$) has opposite effects. This antagonistic principle in a Yin-Yang mode also seems to apply to their action on tumour cell growth, since it has been reported that PGI$_2$ inhibits tumour growth, whereas TXA$_2$ accelerates it (Honn and Meyer 1981; Honn et al. 1987). Inhibition of TXA$_2$ biosynthesis also has a retarding effect. Therefore, the PGI$_2$/TXA$_2$ ratio has been proposed as either a diagnostic or prognostic parameter for tumour growth.

In this contribution we have attempted to compile our knowledge of the biochemistry and physiology of TXA$_2$ and PGI$_2$ in order to obtain a mechanistic background for such assumptions. Special emphasis will be put on the biosynthesis of both prostaglandins.

Nigam et al. (Eds.), Eicosanoids,
Lipid Peroxidation and Cancer
© Springer-Verlag Berlin Heidelberg 1988

2 The Biochemistry of PGI$_2$ and TXA$_2$ Formation

All prostaglandins are derived from the cyclo-oxygenase product PGH$_2$ which is formed in a complex series of oxygenation steps from the precursor arachidonic acid. In a recent study (Hecker et al. 1987a) we have confirmed and extended an earlier proposed mechanism; this involves the steps illustrated in Fig. 1.

As a novel side-product we have isolated, from incubations with purified cyclo-oxygenase, 11(R)-HPETE which obviously arises from an intermediate with the wrong stereochemistry. About 5% of the reaction ends up in this product. Similarly, 15-HPETE is formed from the first radical intermediate in comparable yield, indicating that the isomerisation of the double bond can also occur with carbon 14. This 15-HPETE and the subsequent 15-HPETE formation is aspirin-sensitive in contrast to the 15-lipoxygenase reaction found in eosinophilic PMN.

The conversion of PGH$_2$ to PGI$_2$ and TXA$_2$ by the respective enzymes prostacyclin and thromboxane synthase is an isomerisation reaction but it involves formation of a new carbon-oxygen bond and, therefore, a cleavage of the endope-

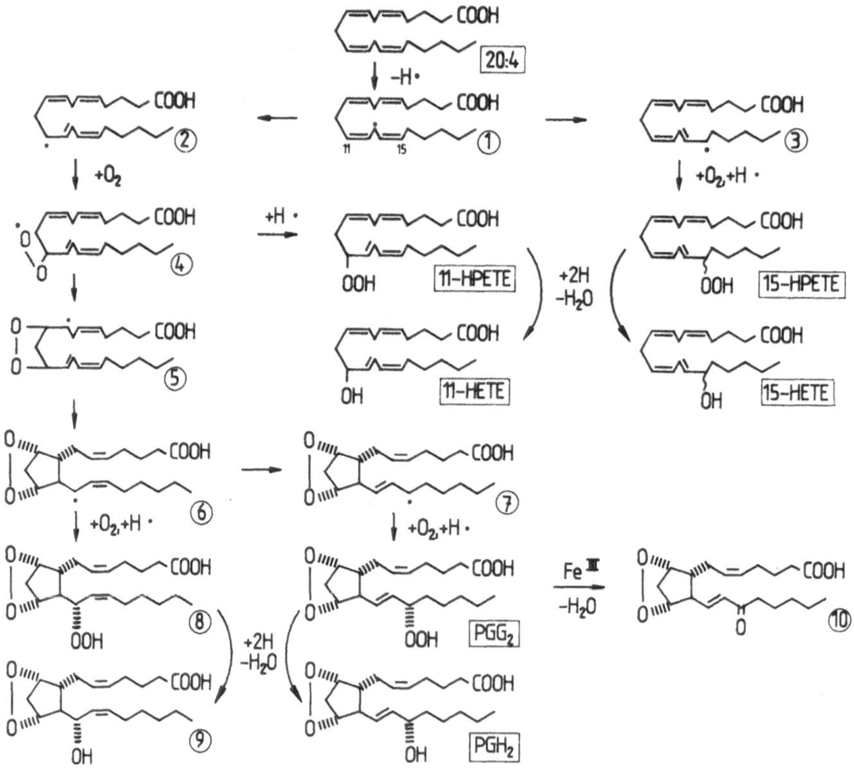

Fig. 1. Reaction scheme of the cyclooxygenase-catalyzed metabolism of arachidonic acid. ⑧, 13(S)-hydroxy-5(Z), 14(Z)-PGG$_2$; ⑨, 13(S)-hydroxy-5(Z), 14(Z)-PGH$_2$; ⑩, 15-keto-PGH$_2$. (Hecker et al. 1987a)

roxide bond under concomitant activation of oxygen is required. We have postulated from analogy with cytochrome P450 mediated hydroxylations a hemethiolate active centre in both enzymes and could verify this assumption by the isolation of both enzymes as homogenous hemoproteins of the P450 type. In their spectral properties both enzymes are almost identical (cf. Table 1). (Ullrich and Graf 1984; Haurand and Ullrich 1985).

We have presented evidence that both enzymes bind their substrate PGH$_2$ with the 9-oxygen atom to the ferric heme (Ullrich et al. 1981; Hecker et al. 1987b) so that a different folding and binding of the side chains at the cyclopentane ring must be responsible for the two different products PGI$_2$ and TXA$_2$. The different stereochemistry in the binding of the substrate is also supported by the variety of pyridine or imidazole-based thromboxane synthase inhibitors (Hecker et al. 1986) which are not acting on PGI$_2$-synthase and, therefore, can be used to influence the biosynthesis of both eicosanoids independently. The similarities and differences between both synthases is illustrated by the data in Table 1.

In accordance with the different functions of PGI$_2$ and TXA$_2$ the localisation of both enzymes is within different cells. Prostacyclin synthase occurs mainly in endothelial and smooth muscle cells from which it was isolated (De-Witt and Smith 1983). Thromboxane synthase is abundant in platelets but by employing antisera and monoclonal antibodies we have also verified its presence in tissue macrophages from liver and lung. Monocytes contain little activity and gave no immunohistochemical staining with the antibodies. However, after migration into tissues the activity appears rapidly (Nüsing, unpublished results). A knowledge of the source of TXA$_2$ within a given organ is essential for an evaluation of its action. since TXA$_2$ has a half-lifetime of 30s which certainly does not allow a systemic action. This may be different for PGI$_2$ which has a half-lifetime of about 3 min but even here no systemic activity has been shown.

Neither prostacyclin nor thromboxane synthase seem to be regulated by feedback control, phosphorylation or calcium. Since for cyclooxygenase this has also not yet been reported, the release of arachidonic by the action of phospholipase

Table 1. Properties of purified prostacyclin and thromboxane synthase

	PGI$_2$ synthase (from bovine or pig aorta)	TXA$_2$ synthase (from human platelets)
M_r	55 000 (53 000)	58 800
Opt. abs. (ox.)	417, 532, 568	418, 537, 570
Opt. abs. (red. + CO)	424, 451, 545	424, 450, 545
EPR-spectrum	1.90, 2.01, 2.25, 2.46	1.90, 1.92, 2.04, 2.25
(g values at)		2.41, 2.46
Substrate	PGH$_2$	PGH$_2$
K_M (µM)	5	22
Product (s)	PGI$_2$	TXA$_2$, HHT + MDA
Mol. act. (min^{-1})	150–200	3000–5500

Opt. abs. (ox.) optical absolute spectrum (oxidized); (red. + Co), reduced with sodium dithionite in the presence of carbon monoxide: mol. act., molecular activity. (From Ullrich and Graf 1984; Haurand and Ullrich 1985)

A_2 may be the major determinant in PGI_2 and TXA_2 formation. Although phospholipase A_2 seems to be regulated mainly by the concentration of intracellular calcium, other activating or inhibiting protein factors may also be important (Flower and Blackwell 1979). However, since these do not operate as short term regulators (within 1 min) and since the increase of intracellular calcium seems to follow the extent of receptor activation, the release of PGI_2 and TXA_2 from the synthesizing cells may be a direct function of the activation of receptors which control the calcium level in these cells.

3 Action of PGI_2 and TXA_2 on Target Cells

A physiological action of prostaglandins implies the existence of receptors on target cells. Unlike steroid hormones the corresponding receptors have been exclusively found on the cell surface and this is also true for PGI_2- and TXA_2-receptors. Receptors for PGI_2 or TXA_2 or both have been found mainly on platelets, endothelial cells and smooth muscle cells of various arteries and veins (Needleman et al. 1986).

After occupancy of the receptor, the message is transduced into the cell and causes the cell to respond. So far, all PGI_2-receptors have been found to evoke cAMP production indicating that PGI_2-receptors only couple to the adenylate-cyclase system (Needleman et al. 1986). In contrast, platelet TXA_2-receptors couple to phospholipase C and thus cause the PI-response (Siess et al. 1983). Of course, both signalling systems exert different actions in different cells, but in general an increase in cAMP switches the response of most cells to quiescence and anabolism concomitant with a decrease in the cytosolic calcium concentration, wheres the PI-response activates the cells, leads to catabolic reactions and is accompanied by an increase in intracellular calcium concentration as well as in intracellular pH. The antagonistic effects of both compounds are essentially due to the coupling of the corresponding receptors to the two different second messenger systems.

PGI_2 and TXA_2 are autacoids and act within a given organ or tissue. This virtually precludes exact measurements of their effective concentrations, since these are steady states determined by the rate of synthesis. transport to the target cell within the intercellular space. non-enzymic degradation and enzymic metabolism as well as receptor concentration, which, again, is a steady state from formation and internalisation. For most purposes one can use stable receptor agonists and antagonists to study the action of the physiological molecules PGI_2 and TXA_2 (Coleman et al. 1984).

4 PGI$_2$ and TXA$_2$ in Tumorigenesis

It is obvious that according to the above-mentioned mechanism of action both prostaglandins can only exert their effect via the corresponding receptors. A cell type without receptor will certainly not respond. If a receptor for PGI$_2$ or TXA$_2$ is available we can now ask what effect on tumour formation can be predicted according to our present knowledge. Since it is already accepted that PGI$_2$ is ineffective in tumorigenesis or even prevents it, we can first discuss possible mechanisms of TXA$_2$ that can lead to tumour growth. This may occur at three different stages.

4.1 Metastasis

From an already existing tumour detached cells may reach the circulation and by adherence may initiate metastasis. In the case of an efficient elimination by the immune system the circulating cancer cell may disappear before adherence, but, if the process of adherence is enhanced, a higher risk of metastasis will arise. Tumour cells are known to respond to TXA$_2$ by increased adherence. They become "sticky" like platelets or granulocytes (Honn et al. 1987). PGI$_2$ causes the opposite effect so that after surgical removal of a tumour the metastatic risk can be diminished by synthetic and stable PGI$_2$-analogues or by inhibition of TXA$_2$ formation. One could even do both and in addition block the TXA$_2$-receptor.

4.2 Tumour Initiation

There are no indications that TXA$_2$ itself can chemically modify DNA and thus lead to chemical mutagenesis or carcinogenesis. However, it has been long known that concomitant with TXA$_2$ formation 12(S)-hydroxy-5,8,10-heptadeca-trienoic acid (HHT) together with malondialdehyde (MDA) is released from activated platelets. We have confirmed this for the isolated thromboxane synthase and have established a clear 1:1:1 relationship between all three products. Since changes of neither temperature, pH nor substrate concentration influenced this ratio, the two additional products must be intrinsic and coupled to the mechanism of the isomerisation reaction (Hecker et al. 1987c).

Malondialdehyde is a bifunctional reactive aldehyde with a mutagenic and carcinogenic potential (Shamberger et al. 1974). Although the larger portion of malondialdehyde formed in the body probably is a product of lipid peroxidation. 20%–30% may be derived from the thromboxane synthase reaction since after application of thromboxane synthase inhibitors or aspirin the serum levels decrease by about this percentage (Violi et al. 1985).

However, the carcinogenic and mutagenic properties of MDA remain controversial because highly purified MDA is only a weak mutagen in Salmonella strains and lacks initiating or promoting properties when applied to mouse skin

(Marnett 1985). The carcinogenicity described in the early assay appears to be due to side products during preparation of MDA. like β-alkoxy-acrolein, a compound which is 30 times more mutagenic than MDA (Marnett and Tuttle 1980). In summary, many questions remain to be answered regarding a direct or indirect involvement of thromboxane synthase metabolites and PGI_2 in tumour initiation or promotion.

4.3 Tumour Promotion

According to the current view on tumorigenesis the promotion phase for an already existing genetic defect or for the presence of an oncogen seems to be equally or even more important for the incidence of a tumour. Although ill-defined, the stage of tumour promotion coincides with a proliferation and growth response of cells for which the induction of ornithine decarboxylase is a suitable indicator (Weeks et al. 1984). Since it is known that activators of protein kinase C. like phorbolesters which mimic the physiological stimulator diacyl-glycerol (Castagna et al. 1982) are inducers of ornithine decarboxylase and tumour promotors, one can postulate that during the PI-response tumour promotion can take place. Studies concerning the role of TXA_2 in tumour promotion are hampered by its instability in aqueous solution. One has to employ either stable TXA_2-mimetica, like 11,9-epoxymethano-PGH_2 (U46619) or draw conclusions from the effects of inhibitors. U46619 but also TXB_2 increased the proliferation of B16 amelanotic melanoma cells in a dose-dependent manner, whereas PGI_2 decreased it. Similarly, the thromboxane synthase inhibitors U51605 and U54701 prevented cell proliferation in vitro and also decreased DNA-synthesis (Honn and Meyer 1981). One important parameter for these cellular events may be the cAMP-level, but intracellular calcium, pH and protein kinase C activity together with cGMP levels may be other factors of not yet fully understood significance in the process of proliferation and tumour promotion.

5 Significance of the PGI_2/TXA_2 Ratio

It is difficult to imagine a short lasting pulse of TXA_2 as a tumour promoting event. Rather, a longer lasting shift in the cellular second messenger systems should stimulate the genetic response of the cell causing proliferation to meet the conditions of new physiologic or pathophysiologic conditions. Therefore, the PGI_2/TXA_2 ratio has been proposed to be a better indicator for cellular activity. Changes in this ratio towards smaller values are found e.g. in diabetes or arteriosclerosis (Bunting et al. 1983). The latter obviously brings about damages in the vascular endothelium and thus diminishes PGI_2 production. Also an increased "oxidative stress" which lowers the antioxidant potential of the tissues can lead to a drop in intravascular PGI_2/TXA_2 ratio since prostacyclin synthase is rather sensitive towards oxidizing conditions, whereas thromboxane synthase is not (Ham et al. 1979). This is surprising in view of the structural similarities of both enzymes.

To improve the PGI$_2$/TXA$_2$ ratio variations in the diet regarding the pattern of polyunsaturated fatty acids are recommended. With a higher portion of eicosapentaenoic acid, as in fish oil, the resulting prostacyclin metabolite PGI$_3$ still has PGI$_2$-like activity, whereas TXA$_3$ fails to induce platelet aggregation (Needleman et al. 1979). Improvement of the body's antioxidant potential might also be a useful therapy in order to prevent prostacyclin synthase from oxidative inactivation.

As a criticism against the unreflected use of the PGI$_2$/TXA$_2$ ratio one should keep in mind that the corresponding data are collected from blood and not from the tissues and that the major source of TXA$_2$ is platelets which in case of activation may contribute enormously to the TXA$_2$ levels in the vascular systems without being significant for the tissue TXA$_2$ concentrations.

6 Conclusions

It is abvious that the experimentally observed correlation of a lowered PGI$_2$/TXA$_2$ ratio with an increased tumour incidence can be biochemically supported. Most likely an increased cell activation by a lowered adenylcyclase activity or an increased PI-response will favour the conditions of tumour promotion. There is also a clear mechanistic basis for increased metastasis, but the possible tumour initiation by malondialdehyde requires further experimental support.

It is crucial to show that initiated cells have PGI$_2$- and TXA$_2$-receptors in order to respond to an altered PGI$_2$/TXA$_2$ ratio. This is known for most cells, although quantitative data are lacking. there is also scant knowledge on the actual concentration of both agonists in tissues. Blood levels may not be very relevant, since the tumorigenic potential of TXA$_2$ and the antagonistic effect of PGI$_2$ may depend very much on the locus of formation.

Nevertheless, it seems desirable to monitor the PGI$_2$/TXA$_2$ ratio since it is also an indicator for arteriosclerotic risk or for changes in the immune system.

Acknowledgments. T.R. thanks the Deutsche Forschungsgemeinschaft for a training stipend (Ro 695/1-1).

References

Bunting S, Moncada S, Vane JR (1983) The prostacyclin-thromboxane A$_2$ balance. Pathophysiological and therapeutic implications. Br Med Bull 39:271–276

Castagna M, Takai Y, Kaibuchi K, Sano K, Kikkawa U, Nishizuka Y (1982) Direct activation of calcium-activated, phospholipid-dependent protein kinase by tumor promoting phorbol esters. J Biol Chem 257:7847–7851

Coleman RA, Humphrey PPA, Kennedy I, Lumley P (1984) Prostanoid receptors – the development of a working classification. Trends Pharmacol Sci 7:303–306

DeWitt DL, Smith WL (1983) Purification of prostacyclin synthase from bovine aorta by immunoaffinity chromatography. J Biol Chem 258:3285–3293

Flower RJ, Blackwell GJ (1979) Antiinflammmatory steroids induce biosynthesis of a phospholipase A_2 inhibitor which prevents prostaglandin generation. Nature 278:456–459

Ham EA, Egan RW, Soderman DD, Gale PH, Kuehl FA jr (1979) Peroxidase dependent deactivation of prostacyclin synthethase. J Biol Chem 254:2191–2194

Haurand M, Ullrich V (1985) Isolation and purification of thromboxane synthase from human platelets as a cytochrome P-450 enzyme. J Biol Chem 260:15059–15067

Hecker M, Haurand M, Ullrich V, Terao S (1986) Spectral studies on structure activity relationships of thromboxane synthase inhibitors. Eur J Biochem 157:217–223

Hecker M, Ullrich V, Fischer C, Meese CO (1987a) Identification of novel arachidonic acid metabolites formed by prostaglandin H synthase. Eur J Biochem 169:113–123

Hecker M, Baader WJ, Weber P, Ullrich V (1987b) Thromboxane synthase catalyzes hydroxylations of prostaglandin H_2-analogs in the presence of iodosylbenzene. Eur J Biochem 169:563–569

Hecker M, Haurand M, Ullrich V, Diczfalusy U, Hammarström S (1987c) Products, kinetics, and substrate specifity of homogeneous thromboxane synthase from human platelets: Development of a novel enzyme assay. Arch Biochem Biophys 254:124–135

Honn KV, Meyer J (1981) Thromboxanes and prostacyclins: Positive and negative modulators of tumor growth. Biochem Biophys Res 102:1122–1129

Honn KV, Menter DG, Steinert BW, Taylor JD, Onoda JM, Sloane BF (1987) Analysis of platelet, tumor cell and endothelial cell interactions in vivo and in vitro. In: Garaci E, Paoletti R, Santoro G (eds) Prostaglandins and cancer research. Springer Berlin Heidelberg New York, pp 172–184

Marnett LJ (1985) Arachidonic acid metabolism and tumor initiation. In: Marnett LJ (ed) Arachidonic acid metabolism and tumor initiation. Nijhoff, Boston, pp 39–42

Marnett LJ, Tuttle MA (1980) Comparison of the mutagenicities of malondialdehyde and the side products formed during its chemical synthesis. Cancer Res 40:276–282

Needleman P, Raz A, Minkes MS, Ferrendelli JA, Sprecher H (1979) Triene prostaglandins: Prostacyclin and thromboxane biosynthesis and unique biological properties. Proc Natl Acad Sci USA 76:944–948

Needleman P, Turk J, Jakschik BA, Morrison AR, Lefkowith JB (1986) Arachidonic acid metabolism. Annu Rev Biochem 55:69–102

Shamberger RJ, Adreone TL, Willis CE (1974) Antioxidants and cancer. IV. Initiating activity of malondialdehyde as a carcinogen. HJNCI 53:1771–1774

Siess W, Cuatrecasas P, Lapetina EG (1983) A role for cyclooxygenase products in the formation of phosphatidic acid in stimulated human platelets. Differential mechanisms of action of thrombin and collagen. J Biol Chem 258:4683–4686

Ullrich V, Graf H (1984) Prostacyclin and thromboxane synthase as P-450 enzymes. Trends Pharmacol Sci 7:352–355

Ullrich V, Castle L, Weber P (1981) Spectral evidence for the cytochrome P450 nature of prostacyclin synthethase. Biochem Pharmacol 30:2033–2036

Violi F, Ghiselli A, Alessendri C, Iuliano L, Cordovar C, Balsano F (1985) Relationship between platelet cyclooxygenase pathway and plasma malondialdehyde-like material. Lipids 20:322–324

Weeks CE, Slaga TJ, Boutwell RK (1984) The role of polyamines in tumor promotion. In: Slaga TJ (ed) Tumor promotion and skin carcinogenesis. CRC Boca Raton, pp 127–143

Tumour Promoters and Prostaglandin Production

L. Levine

Department of Biochemistry, Brandeis University, Waltham, Massachusetts 02254, USA

1 Introduction

Our knowledge of prostaglandin biosynthesis came originally from the studies by Bergström et al. (1964) and van Dorp et al. (1964). During the past 20 years, Samuelsson and his colleagues have done much to describe the biochemical pathways for the biosynthesis and metabolism of the arachidonic acid metabolic products (Samuelsson 1976, 1985; Hamberg and Samuelsson 1974).

Arachidonic acid does not exist free in cells but is found esterified in the form of phospholipids, steroidesters and triglycerides. The phospholipids, found primarily in the cellular membranes, are the richest source. Increased arachidonic acid metabolism is thought to reflect, at least in part, increased de-esterification from these cellular phospholipids. Thus, de-esterification reactions are part of the sequence of events involved in generation of the prostaglandins. A scheme that includes several de-esterification reactions in mammalian cells is shown in Fig. 1 (Levine et al. 1987). In this scheme, reaction ① represents the generation of diacylglycerol by an endogenous phosphatidylcholine- hydrolyzing phospholipase C (PLC_{PC}). Sequential hydrolysis of the diacylglycerol by a diacylglycerol lipase and possibly a monoacylglycerol lipase ② liberates arachidonic acid. The diacylglycerol formed by ① activates a protein kinase C ③. Diacylglycerol also is generated by several agonist-receptor interactions and growth factors by phosphoinositide hydrolysis (PLC_{PI}) ④; the diacylglycerol formed by ④ also activates the protein kinase C ③. The inositol trisphosphate generated in ④ increases free cytosolic Ca^{2+} ⑤ which also stimulates protein kinase C activity. The activated protein kinase C in this model has two functions: (a) it *may* regulate the endogenous PLC_{PC} activity by phosphorylation of the enzyme or it *may* inactivate, by phosphorylation, an inhibitor of the PLC_{PC} ⑥; and (b) it can stimulate phospholipase A_2 activity ⑨. Diacylglycerol kinase phosphorylates the diacylglycerol to form phosphatidic acid ⑦, which in addition to phosphatidylcholine and phosphatidylethanolamine, may be the substrate for phospholipase A_2 activity ⑧. The increase in free cytosolic Ca^{2+} ⑤ stimulates phospholipase A_2 activity and also *may* regulate the activity of the PLC_{PC} ⑥.

Nigam et al. (Eds.), Eicosanoids,
Lipid Peroxidation and Cancer
© Springer-Verlag Berlin Heidelberg 1988

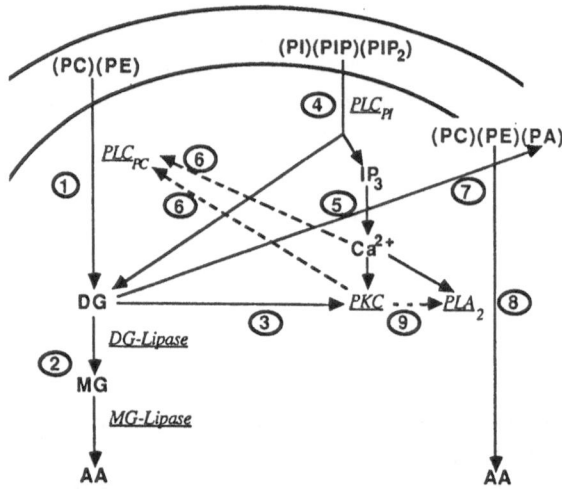

Fig. 1. Proposed reactions that lead directly and/or indirectly to liberation of arachidonic acid from phospholipids. Reaction ① represents the hydrolysis of phosphatidylcholine (*PC*) or phosphatidylethanolamine (*PE*) by phospholipase C (PLC$_{PC}$). Reaction ② describes the liberation of arachidonic acid (*AA*) from diacylglycerol (*DG*), the product of ①, after hydrolysis by a DG-lipase and/or hydrolysis of monoacylglycerol (*MG*) by a MG-lipase. Reaction ③ represents the activation of protein-kinase C (*PKC*) by the DG. Reaction ④ represents the hydrolysis of phosphatidylinositol or polyphosphatidylinositols (*PI*)(*PIP*)(*PIP$_2$*), by a PLC$_{PI}$ to generate DG and inositol-trisphosphate (*IP$_3$*); the latter increases free cytosolic Ca $^{2+}$ ⑤. Both the increased free Ca^{2+} and PKC regulate ⑥ the activity of the PLC$_{PC}$ in ①. Reaction ⑦ represents the phosphorylation of DG by a DG-kinase to form phosphatidic acid (*PA*) which may be a substrate for phospholipase A$_2$ (*PLA$_2$*) ⑧. PKC and the increased free Ca^{2+} also stimulate PLA$_2$ activity ⑨

2 Stimulation of Arachidonic Acid Metabolism by Tumour Promoters

Most tumour promoters stimulate arachidonic acid metabolism in a dose-dependent manner (Levine et al. 1984). The concentrations of several tumour promoters that increase PGI$_2$ production (measured serologically as its stable-hydrolytic product 6-keto-PGF$_{1\alpha}$) 2 fold after 20 hours of incubation with rat liver cells are shown in Table 1. Palytoxin, a cytolysin present in several *Palythoa* species is most potent – ≈ 400 times more potent than okadaic acid, a cytotoxic polyether present in some species of sponges, and ≈ 1000 times more potent than the phorbol esters, teleocidin and aplysiatoxin. The latter three tumour promoters, classified as TPA-type tumour promoters (Fujiki and Sugimura 1988), are found in the seeds of the plant species *Croton tiglium,* the mycelia of *Streptomyces mediocidicus,* and in the blue-green alga, *Lyngbya majuscula,* respectively. Iodoacetic acid, anthralin and benzoyl peroxide, are much less potent and benzo(e)pyrene and saccharine are inactive at the levels tested. Palytoxin stimulates arachidonic acid metabolism in cells other than rat liver cells. A list of the cells that have responded to palytoxin with increased prostaglandin production

Table 1. Effect of tumour promoters on 6-keto-PGF$_{1\alpha}$ production by rat liver (C-9) cells

Reagent	Concentrations required for 2-fold stimulation[a] (nM)
Palytoxin	0.007
Okadaic acid	3
TPA	6.8
Teleocidin	12
Aplysiatoxin	16
PDD	24
Iodoacetic acid	1500
Anthralin	7000
Benzoyl peroxide	10000
4α-PDD	b
Benzo(e)pyrene	c
Saccharine	d

[a] C-9 cells (5×10^5 cells/35-mm dish) were incubated in 1 ml MEM containing various levels of reagent for 20 h at 37°C. Data were calculated from dose-response curves
[b] 0 stimulation at 5 μM.
[c] 0 stimulation at 60 μM.
[d] 0 stimulation at 8 mM.

is shown in Table 2 (Ohuchi et al. 1985; Lazzaro et al. 1987; Levine, unpublished).

Several of the tumour promoters vary qualitatively and quantitatively in their capacity to induce ornithine decarboxylase (ODC), to irritate skin, to induce HL-60 cell adhesion, to inhibit [³H]TPA binding, and to activate protein kinase C (Table 3) (Fujiki and Sugimura 1988; Suganuma et al. 1988). Based on these activities, several of the tumour promoters were originally calssified as TPA-Types and non-TPA types (Fujiki et al. 1984). Palytoxin (Fujiki et al. 1984), okadaic acid (Suganuma et al. 1988) and thapsigargin (Fujiki and Sugimura 1988), the latter found in the root of *Thapsia garganica,* are non-TPA type tumour promoters (Fujiki and Sugimura 1988). Structures of some of the tumour promoters studied with respect to arachidonic acid metabolism are shown in Fig. 2 (Fujiki and Sugimura 1988; Suganuma et al. 1988).

Table 2. Cells in which arachidonic acid metabolism is stimulated by palytoxin to produce cyclo-oxygenase products

1. Rat liver cells
2. Bovine aorta smooth muscle cells
3. Bovine aorta endothelial cells
4. Porcine aorta endothelial cells
5. Rat keratinocytes
6. Squirrel monkey aorta smooth muscle cells
7. Rat peritoneal macrophages
8. Mouse calvaria: osteoclasts and/or osteoblasts

Table 3. Effects of several tumour promoters on irritation of mouse ear, induction of ornithine decarboxylase (ODC), inhibition of [³H]TPA binding, induction of HL-60 adhesion, activation of protein kinase C activity, stimulation of arachidonic acid metabolism and promotion of tumours in mouse skin

	Irritation mouse ear	Induction of ODC	Induction of HL-60 cell adhesion	Inhibition of specific [³H]-TPA binding	Activation of protein kinase C in vitro	Stimulation of arachidonic acid metabolism	Promotion of tumours
TPA	+	+	+	+	+	+	+
Teoleocidin	+	+	+	+	+	+	+++
Aplysiatoxin	+	+	+	+	+	+	++++
Palytoxin	+	−	−	−	−	+	++
Okadaic acid	+	+	−	−	−	+	++
Thapsigargin	+	−	−	−	−	+	+

(From Fujiki and Sugimura 1988; Suganuma et al. 1988)

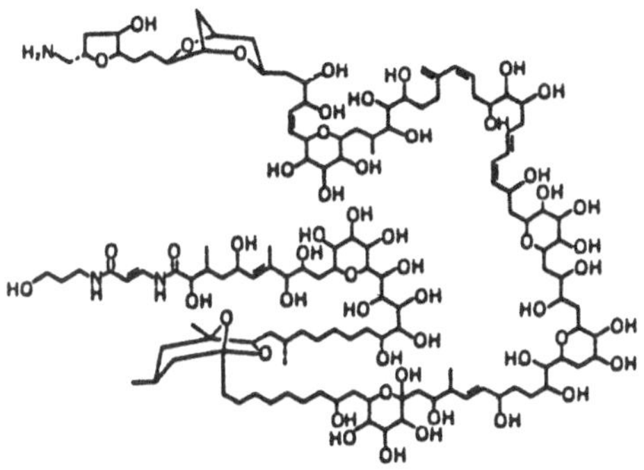

Palytoxin

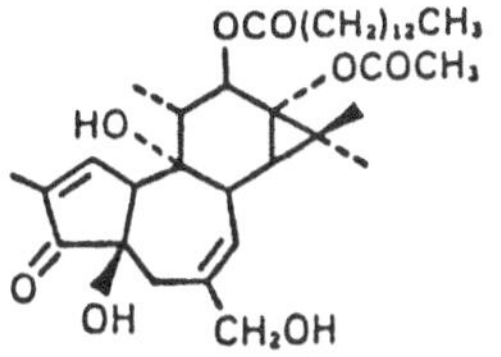

TPA

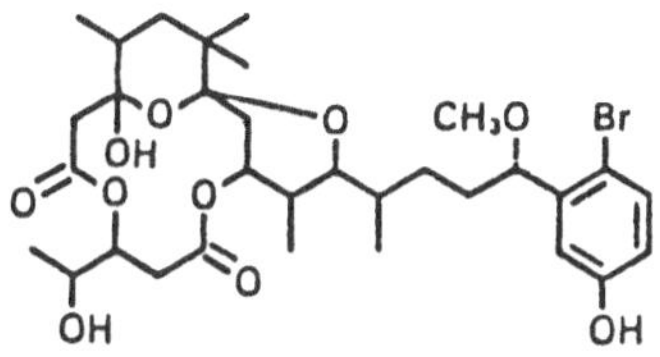

Aplysiatoxin

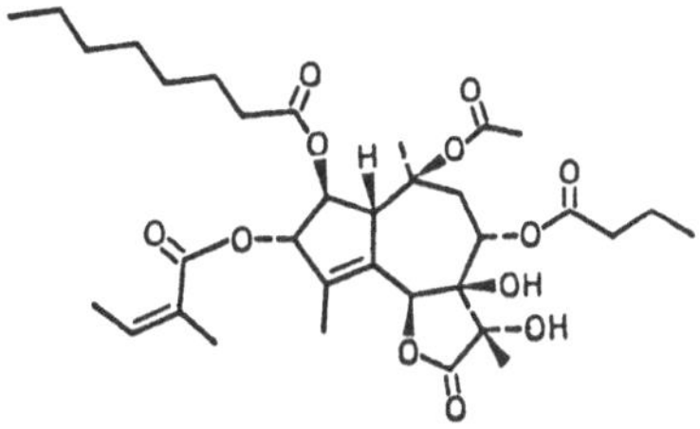

Thapsigargin

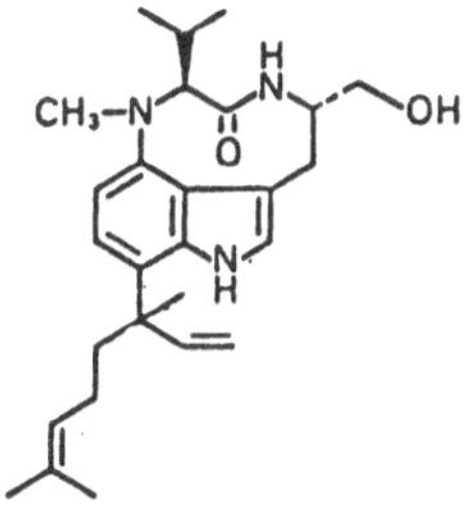

Teleocidin A

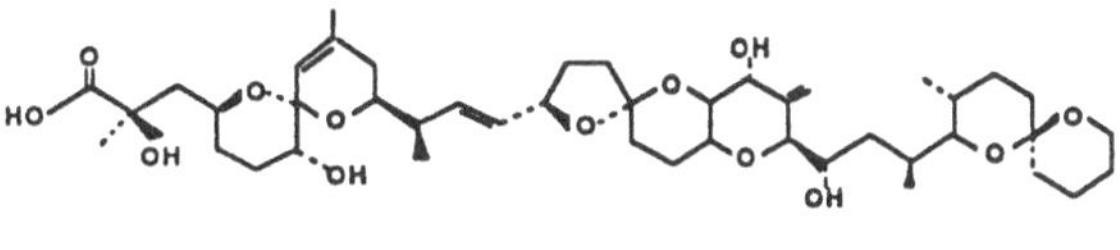

Okadaic Acid

Fig. 2. Structures of some tumour promoters

The time course of stimulation of arachidonic acid metabolism varies among the tumour promoters tested. For example, whereas the TPA-type tumour promoters stimulate 6-keto-PGF$_{1\alpha}$ production without a clearly defined lag, the lags in 6-keto-PGF$_{1\alpha}$ production when the liver cells are treated with the non-TPA-type tumour promoters, palytoxin and okadaic acid are pronounced (Fig. 3). The time course of 6-keto-PGF$_{1\alpha}$ production by the rat liver cells during incubation with anthralin and benzoyl peroxide is also accompanied by a pronounced lag period (data not shown).

Palytoxin's mechanism of stimulation appears to contain a pathway common to that of the TPA-type tumour promoters. For example, stimulation of arachidonic acid metabolism by palytoxin is synergistic with that of TPA, aplysiatoxin and teleocidin, whereas stimulation of arachidonic acid metabolism by mixtures of the TPA-type tumour promoters is not (Table 4) (Levine and Fujiki 1985). Palytoxin's stimulation of arachidonic acid metabolism differs from that of TPA in several ways. For example, the combintion of palytoxin and 1-oleoyl-2-acetyl-

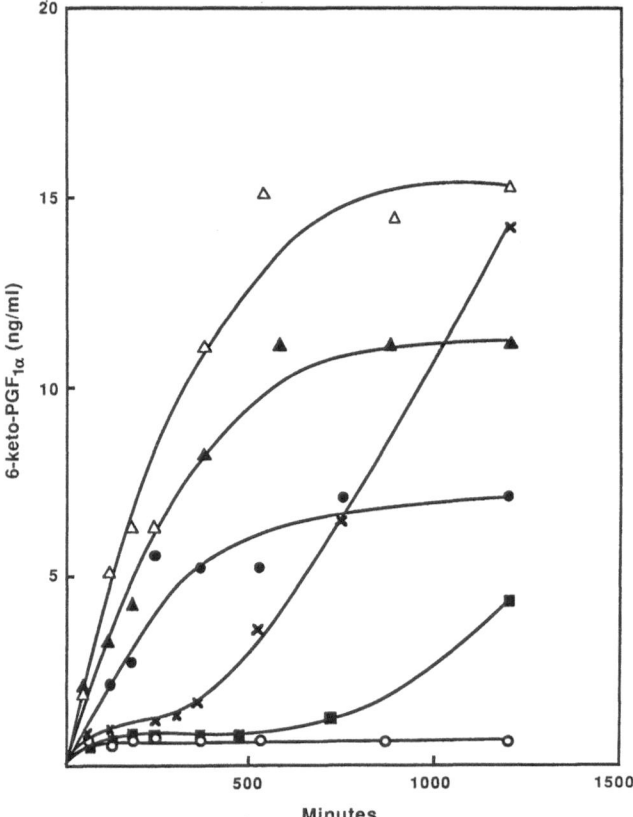

Fig. 3. Time course of 6-keto- PGF$_{1\alpha}$ production in rat liver cells by aplysiatoxin, 15 nM ($\triangle$); teleocidin, 23 nM ($\blacktriangle$); TPA, 16 nM ($\bullet$); palytoxin, 19 pM (**X**); okadaic acid, 60 nM ($\blacksquare$); MEM ($\bigcirc$)

Table 4. Stimulation of arachidonic acid metabolism by palytoxin and TPA in the presence of aplysiatoxin and teleocidin, and by palytoxin plus TPA

Tumour promoter	6-keto-PGF$_{1\alpha}$ ng/ml[a]	PGF$_{2\alpha}$ ng/ml[a]
MEM	0.05[b]	<0.01[c]
TPA (55 nM)	0.20	<0.01
Palytoxin (0.015 nM)	0.21	<0.01
Aplysiatoxin (50 nM)	0.17	<0.01
Teleocidin (12 nM)	0.16	<0.01
Palytoxin (0.015 nM)+TPA (55 nM)	1.40	0.028
Palytoxin (0.015 nM)+aplysiatoxin (50 nM)	2.20	0.051
Palytoxin (0.015 nM)+teleocidin (12 nM)	1.40	0.044
TPA (55 nM)+aplysiatoxin (50 nM)	0.20	<0.01
TPA (55 nM)+teleocidin (12 nM)	0.19	<0.01

[a] C-9 cells (5×10^5/35-mm dish) were incubated with 1 ml MEM for 20 h at 37°C.
[b] Data represent the means of at least three dishes. Analyses of each culture dish agreed within 20%.
[c] Data represent a single analysis by radioimmunoassay of a pool of three dishes.
(From Levine and Fujiki 1985)

glycerol (OAG), but not TPA and OAG, synergistically stimulates production of PGI$_2$ and release of arachidonic acid from rat liver cells (Fig. 4) (Levine et al. 1986a). In addition, PGI$_2$ production by combinations of palytoxin and of re-

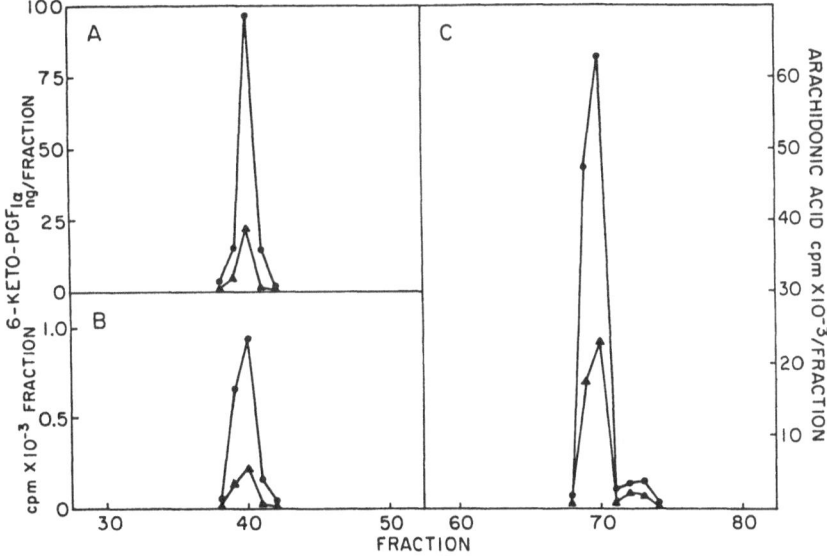

Fig. 4. HPLC analyses of released, radiolabelled compounds (**B, C**) from [³H]arachidonic acid labelled cells; (●) treated with palytoxin (37 pM) and OAG (62 μM) and (▲) with palytoxin (37 pM) for 8 h. Cell density was 8 · 10⁵/60-mm dish. A RIA analyses. (From Levine et al. 1986b)

combinant epidermal growth factor, palytoxin and recombinant transforming growth factor-α, palytoxin and recombinant insulin growth factor 1, palytoxin and insulin, palytoxin and interleukin-1 and palytoxin and 1,2-dioctanoyl-*sn*-glycerol is strikingly synergistic while combinations of TPA and these same factors are much less synergistic (Table 5) (Levine et al. 1986b; Xiao and Levine 1986; Levine, unpublished). Synergistic stimulation of arachidonic acid metabolism by these factors, all of which affect phorphorylation reactions, may be explained most simply by the scheme shown in Fig. 5. In this scheme, an agonist binds to its specific receptor in the plasma membrane, initiating a series of enzymic reactions, a, b, c, d, ... i, and generating products, Ⓐ, Ⓑ, Ⓒ, Ⓓ, ... Ⓘ, leading to de-esterification and arachidonic acid metabolism. Other stimulants of arachidonic acid metabolism, e.g. phorbol ester tumour promoters, growth factors, palytoxin, also stimulate arachidonic acid metabolism by a complex series of reactions (a′, b′, c′, d′, ... i′). If an activity, or factor generated by reactions a′, b′, c′, d′, ... i′, yields products identical to Ⓐ, Ⓑ, Ⓒ, Ⓓ, ... Ⓘ or factors identical to those generated by a, b, c, d, ... i, amplification of arachidonic acid metabolism will result.

Table 5. Synergistic stimulation of arachidonic acid metabolism of several biologically active compounds with palytoxin and TPA

Compound	Palytoxin	TPA
Recombinant epidermal growth factor	+ + + +	+
Recombinant transforming growth factor-α	+ + + +	+
Recombinant insulin growth factor 1	+ + + +	+
Insulin	+ + + +	+
Recombinant interleukin-1 α and β	+ + + +	+
Transforming growth factor-β	+ +	−
Recombinant tumour necrosis factor α	+ + + +	−
1,2,dioctanoyl-*sn*-glycerol	+ + + +	−

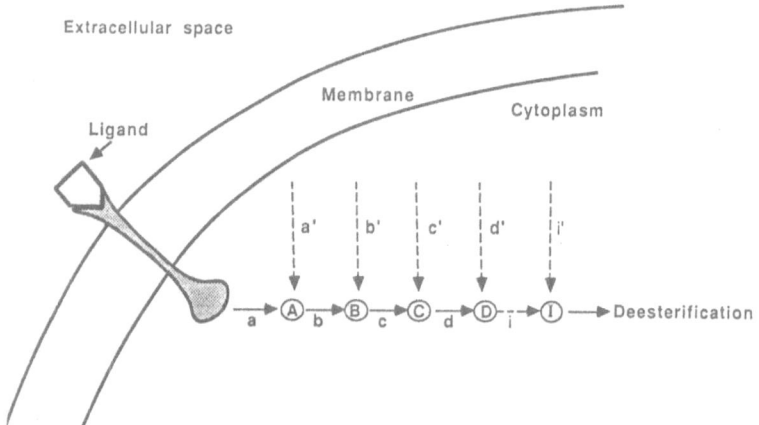

Fig. 5. Hypothetical scheme for synergism in arachidonic acid metabolism

3 Conclusions

Increased arachidonic acid metabolism per se may not be causally related to tumour promotion. Whereas treatment of eukaryotic cells with most tumour promoters tested leads to increased arachidonic acid metabolism (Tables 1, 3), not all stimulators of arachidonic acid metabolism, e.g. thrombin, serum, bradykinin, vasopressin, phenothiazines, serotonin, norepinephrine (Hong et al. 1976, 1977; Hong and Levine 1976; Levine and Moskowitz 1979; Rigas et al. 1981; Coughlin et al. 1981; Uglesity et al. 1983), are tumour promoters. The tumour promoters may interfere with regulatory mechanisms of signal transmission and intracellular communication and, concomitantly, alter the lipid composition of the membranes. The latter events probably will increase de-esterification and alter arachidonic acid metabolism. The tumour promoters probably affect the order of phosphorylations and consequently dephosphorylations of the regulatory molecules, which results in a "tumour-promoting" pattern of intracell communication. It is the unique "tumour promoting" pattern of phosphorylating and dephosphorylating events that is responsible for tumour promotion. It should be stressed that the parameters used to characterize the tumour promoters (Fujiki and Sugimura 1988; Suganuma et al. 1988) (Table 3) do not include kinase activities other than protein kinase C. Nor can the possibility that protein kinase C is active *intracellularly* after treatment with palytoxin, okadaic acid or thapsigargin be ruled out. There may be multiple endogenous factors, e.g. epidermal growth factor, transforming growth factors, as well as multiple environmental factors that can lead to this unique "tumour promoting" pattern.

Several compounds have been reported to inhibit tumour formation in the multi-stage model of chemical carcinogenesis (Wattenberg 1985; Fujiki and Sugimura 1988). It is likely that these inhibitors block, at diverse stages, formation of the unique "tumour-promoting" pattern of phosphorylations and dephosphorylations. If an endogenous protein that prevents formation of this unique "tumour promoting" pattern can be cloned and overexpressed, tumour promotion could be prevented.

Acknowledgements. This work was supported by Grant GM 27256 from the National Institutes of Health. L. L. is an American Cancer Society Research Professor of Biochemistry (Award PRP-21).

I am grateful to Dr. Hirota Fujiki, National Cancer Center Research Institute, Chuo-ku, Tokyo 104, Japan, for his permission to refer to his unpublished work and for our several collaborative studies with palytoxin. In addition, I wish to thank Nancy Worth, Xiaoying Chen, Jeffrey A. Bessette, Norma Henricksen, and Constantia Petrou for their technical assistance and Inez Zimmerman for preparation of the manuscript.

Publication No. 1666, Department of Biochemistry, Brandeis University, Waltham, Massachusetts 02254.

References

Bergström S, Danielsson H, Samuelsson B (1964) The enzymatic formation of prostaglandin E_2 from arachidonic acid. Biochim Biophys Acta 90:207–210

Coughlin SR, Moskowitz MA, Antoniades HN, Levine L (1981) Serotonin receptor-mediated stimulation of bovine smooth muscle cell prostacyclin synthesis and its modulation by platelet-derived growth factor. Proc Natl. Acad Sci USA 78:7134–7138

Fujiki H, Sugimura T (1987) New classes of tumor promoters: teleocidin, aplysiatoxin and palytoxin. Adv Cancer Res 49:223–264

Fujiki H, Suganuma M, Tahira T, Yoshioka A, Nakayasu M Endo Y, Shudo K, et al. (1984) Nakahara memorial lecture: new classes of tumor promoters: teleocidin, aplysiatoxin and palytoxin. In: Fujiki H, Hecker E, Moore RE, Sugimura T, Weinstein IB (eds) Cellular interactions of environmental tumor promoters. Japan Scientific Societies, Tokyo/VNU Science, Utrecht, pp 37–45

Hamberg M, Samuelsson B (1974) Prostaglandin endoperoxides. Novel transformations of arachidonic acid in human platelets. Proc Natl Acid Sci USA 71:3400–3404

Hong SL, Levine L (1976) Stimulation of prostaglandin synthesis by bradykinin and thrombin and their mechanisms of action on MC5-5 fibroblasts. J Biol Chem 251:5814–5816

Hong SL, Polsky-Cynkin R, Levine L (1976) Stimulation of prostaglandin biosynthesis by vasoactive substances in methylcholanthrene-transformed mouse BALB/3T3. J Biol Chem 251:776–780

Hong SL, Polsky-Cynkin R, Levine L (1977) Effects of serum on prostaglandin production by cells in culture. In: Silver MJ, Smith JB, Kokcis JJ (eds) Prostaglandins in hematology. Spectrum, New York, pp 103–120

Lazzaro M, Tashjian AH Jr, Fujiki H, Levine L (1987) Palytoxin: an extraordinarily potent stimulator of prostaglandin production and bone resorption in cultured mouse calvariae. Endocrinology 120:1338–1345

Levine L, Moskowitz MA (1979) α- and β-adrenergic stimulation of arachidonic acid metabolism in cells in culture. Proc Natl Acad Sci USA 76:6632–6636

Levine L, Fujiki H (1985) Stimulation of arachidonic acid metabolism by different types of tumor promoters. Carcinogenesis 6:1631–1634

Levine L, Goldstein SM, Snoek GT, Rigas A (1984) Arachidonic acid metabolism by cells in culture: effects of tumor promoters. In: Thaler-Dao H, de Paulet AC, Paoletti R (eds) Icosanoids and cancer. Raven, New York, pp 115–125

Levine L, Xiao D-M, Fujiki H (1986a) A combination of palytoxin with 1-oleoyl-2-acetyl-glycerol (OAG) or insulin or interleukin-1 synergistically stimulates arachidonic acid metabolism, but combinations of 12-O-tetradecanoylphorbol-13-acetate (TPA)-type tumor promoters with OAG do not. Carcinogenesis 7:99–103

Levine L, Xiao D-M, Fujiki H (1986b) Combinations of palytoxin or 12-O-tetradecanoylphorbol-13-acetate and recombinant insulin growth factor-1 or insulin synergistically stimulate prostaglandin production in cultured rat liver cells and squirrel monkey aorta smooth muscle cells. Prostaglandins 31:669–681

Levine L, Xiao D-M, Little C (1987) Increased arachidonic acid metabolites from cells in culture after treatment with the phosphatidylcholine-hydrolyzing phospholipase C from Bacillus cereus. Prostaglandins 34:633–642

Ohuchi K, Watanabe M, Yoshizawa K, Tsurufuji S, Fujiki H, Suganuma M, Sugimura T, Levine L (1985) Stimulation of prostaglandin E_2 production by 12-O-tetradecanoylphorbol-13-acetate (TPA)-type and non-TPA tumor promoters in macrophages and its inhibition by cacloheximide. Biochim Biophys Acta 834:42–47

Rigas VA, van Vunakis H, Levine L (1981) The effect of phenothiazines and their metabolites on prostaglandin production by rat basophilic leukemia cells in culture. Prostaglandins Med 7:183–198

Samuelsson B (1976) New trends in prostaglandin research. Prostaglandin Thromboxane Res 1:1–6

Samuelsson B (1985) Leukotrienes and related compounds. Adv Prostaglandin Thromboxane Leukotriene Res 15:1–9

Suganuma M, Fujiki H, Suguri H, Yoshizawa S, Hirota M, Nakayasu M, Ojika M, et al. (1988) Okadaic acid: a new non-12-O-tetradecanoylphorbol-13-acetate type tumor promoter. Proc Natl Acad Sci USA, (in press)

Uglesity A, Kreisberg JI, Levine L (1983) Stimulation of arachidonic acid metabolism in rat kidney mesangial cells by bradykinin, antidiuretic hormone and their analogues. Prostaglandins Leukotrienes Med 10:83–93

Van Dorp DA, Beerthuis RK, Nugterin DH, Vonkeman H (1964) The biosynthesis of prostaglandins. Biochim Biophys Acta 90:204–207

Wattenberg LW (1985) Chemoprevention of cancer. Cancer Res 45:1–8

Xiao D-M, Levine L (1986) Stimulation of arachidonic acid metabolism: differences in potencies of recombinant human interleukin-1α and interleukin-1β on two cell types. Prostaglandins 32:709–718

Eicosanoid Regulation of Macrophage In Vitro Cytostatic Function Towards Tumour Cells

I. L. Bonta and G. R. Elliott

Department of Pharmacology, Faculty of Medicine, Erasmus University Rotterdam, P.O. Box 1738, 3000 DR Rotterdam, The Netherlands

1 Introduction

Macrophage functions are, at least in part, controlled by metabolites of the cyclo-oxygenase and 5-lipoxygenase pathways of arachidonic acid. For example macrophage lysosomal enzyme secretion is stimulated by the 5-lipoxygenase product leukotriene (LT)C_4 and inhibited by the cyclo-oxagenase metabolic prostaglandin (PG)E_2. Further, the stimulatory action of LTC$_4$ is enhanced by inhibitors of cyclo-oxygenase (Schenkelaars anf Bonta 1986). Also the discharge of the polypeptide interleukin-1 (IL-1) is regulated by eicosanoids. Inhibition of the biosynthesis of PGE$_2$ and exposure to LTB$_4$ or LTD$_4$ can augment the IL-1 production by monocytes (Rola-Pleszczynski and Lemaire 1985). Thus it appears that macrophages are stimulated by 5-lipoxygenase leukotrienes and inhibited by PGE$_2$. In order to test the application of this concept to other than secretory functions, we investigated the influence of inhibitors of eicosanoid production on the cytostatic activity of macrophages towards a tumour cell line.

2 Macrophage Cytostasis and Inhibitors of Eicosanoid Biosynthesis

The antitumour potential of quiescent macrophages is very low. The antitumour effect, however, is a characterisitic function of the activated macrophage (Adams and Hamilton 1984). This antitumour function is expressed by the macrophage either as a cytostatic effect, i.e. inhibition of growth of tumour cells or as a cytotoxic effect, i.e. destruction of the tumour cells. For assessing the cytostatic function of the macrophage on tumour cells in co-culture the inhibition of [^{3}H]-thymidine incorporation in the tumour cells is a readily useful parameter, particularly because the thymidine uptake capacity of macrophages is very low in comparison with the rate of uptake by the tumour cells. In our experience measurements of the [^{3}H-]thymidine uptake of the two kinds of cells in separate culture disclosed such a marked difference (macrophage 811 ± 67 cpm, MOPC-315 tumour cells 13.858 ± 873 cpm) that under conditions of co-culture with the tumour cells it is justifiable to assume that the contribution of macrophages to the

Nigam et al. (Eds.), Eicosanoids,
Lipid Peroxidation and Cancer
© Springer-Verlag Berlin Heidelberg 1988

cpm count is negligible. We used a co-culture consisting of resident macrophages purified from a peritoneal cell wash from BALB/c mice, and of syngeneic MOPC-315 tumour cells. The method was essentially that which has been published extensively elsewhere (Ophir et al. 1987). Briefly, a co-culture of 10^5 macrophages and 10^3 tumour cells together with the test substances of the respective vehicles was incubated for 24h. Thereafter [^{3}H-]thymidine was added for another 16h incubation. That was terminated by harvesting the cells on glass fibre filter discs which were punched out to feed into a beta counter to assess the cellular uptake of the labelled thymidine.

The thymidine uptake of separate MOPC-315 cells was not appreciably influenced by indomethacin, whereas this inhibitor of cyclo-oxygenase exerted a stimulatory influence on the thymidine incorporation of macrophages (Ophir et al. 1987). However, under the conditions of co-culture of the two cell populations indomethacin caused the thymidine uptake to decline (Fig. 1). It is reason-

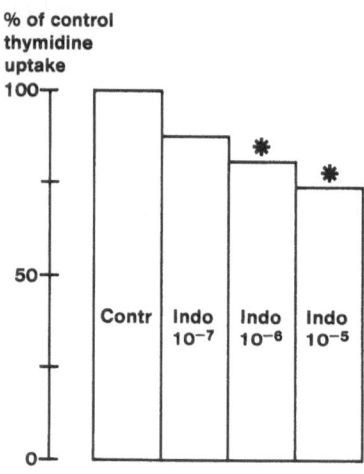

Fig. 1. Indomethacin *(Indo)* enhances macrophage cytostasis against MOPC-315 tumour cells in co-culture. Reduced rate of thymidine uptake reflects a decrease in the metabolism of tumour cells which are damaged by adjacent macrophages. *Asterisks, p <0.05*

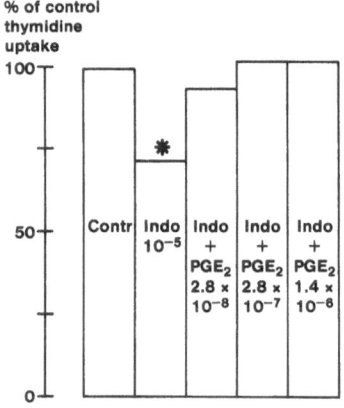

Fig. 2. PGE$_2$ abolishes the enhancing effect of indomethacin *(Indo)* on macrophage cytostasis against MOPC-315 tumour cells. *Asterisk, p <0.05*

able to assume that, following exposure to indomethacin the macrophages exerted a cytostatic influence on the tumour cells, reflected as a decrease in the metabolism and a reduced rate of [^{3}H-]thymidine incorporation. The effect of indomethacin was abolished by adding PGE_2 to the system (Fig. 2). These results clearly show that after removal of endogenous PGE_2 the macrophages are activated to such an extent as to cause damage to co-cultured tumour cells. Inhibition of the lipoxygenase pathway by nordihydroguaiaretic acid (NDGA) per se failed to affect the thymidine uptake in the system, but the effect of indomethacin was abolished by NDGA (Fig. 3). It would appear that the impairment of the lipoxygenase pathway leads to suppression of the cytostatic function of macrophages that have been activated by removal of endogenous PGE_2. This observation is complementary to the earlier finding which showed that the 5-lipoxygenase product LTD_4 additively enhances the effect of indomethacin on the tumour cell damaging activity of macrophages (Ophir et al. 1987). Suppression of cytotoxicity has been shown with natural killer cells following the impairment of the lipoxygenase pathway in such cells (Rossi et al. 1985).

3 Macrophage Secretory Functions and Cytostasis

Tumour cell-damaging activity has been shown with supernatants of activated blood monocytes (Matthews 1981). This activity has been attributed to the presence of a soluble material called tumour necrosis factor (TNF). At present there is more than one TNF identified. The substance called TNF_α is secreted by macrophages and natural killer cells (Palladino and Finkle 1986), TNF_α is not the only macrophage-derived cytokine which can damage tumour cells. The growth of some tumour cell lines in vitro has been shown to be directly inhibited by IL-1 (Onozaki et al. 1985). Furthermore, besides the two cytokines, other secretory products, for example arginase or some of the lysosomal enzymes, may be in-

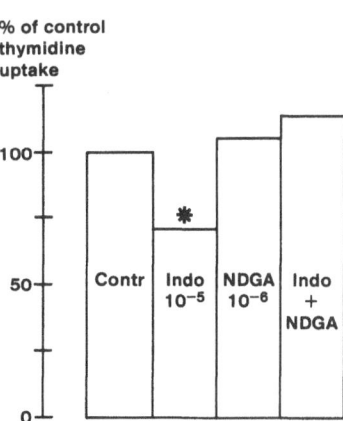

Fig. 3. Nordihydroguaiaretic acid *(NDGA)* abolishes the enhancing effect of indomethacin *(Indo)* on macrophage cytostasis against MOPC-315 tumour cells. *Asterisk, $p < 0.05$*

volved in the antitumour potential of macrophages. Eicosanoids are implicated in regulating the discharge of lysosomal enzymes (Schenkelaars and Bonta 1986) and IL-1 (Rola-Pleszczynski and Lemaire 1985). However, with other cytostatic or cytotoxic substances which are secreted by macrophage the possible regulatory role of eicosanoids has not been investigated as yet. It also needs to be remarked that, although the involvement of eicosanoids has been reported for the discharge of IL-1 (Rola-Pleszczynski and Lemaire 1985), more recently the existence of more than one IL-1 has been proposed (Oppenheim et al. 1986).

4 Summary and Closing Remarks

In the course of examining the cytostatic activity of macrophages towards a syngeneic tumour cell line, we obtained evidence, reported here, that impairment of the cyclo-oxygenase pathway, particularly the removal of endogenous PGE_2, enhances this function. Conversely, inhibiting the biosynthesis of 5-lipoxygenase products leads to reduction of the antitumour effects of macrophages in vitro. The present results amplify the earlier finding which showed that LTD_4 additively reinforces the enhancing effect of indomethacin on the cytostatic activity of macrophages. The emergent picture clearly shows that cytostasis by macrophages in vitro is modulated by a delicate balance of the biosynthesis of cyclo-oxygenase and lipoxygenase metabolites of arachidonic acid. Removal of the inhibitory PGE_2 facilitates macrophage cytostatic activation by 5-lipoxygenase products (e. g. LTD_4). Also the secretory activities of macrophages are stimulated by leukotrienes of the 5-lipoxygenase pathway and inhibited by PGE_2. Furthermore, the stimulatory effect of leukotrienes on macrophage secretory function can be reinforced by inhibitors of cyclo-oxygenase. Accordingly, it appears that eicosanoids and inhibitors of eicosanoid biosynthesis modulate the cytostatic function of macrophages in a way which resembles their effect on secretory functions of the cells. Indeed, the two functions may be interrelated because the tumour cell-damaging function of macrophages is, at least to a considerable extent, mediated through secretion of cytostatic or cytotoxic substances. However, cell-to-cell contact is another mechanism by which macrophages may damage adjacent tumour cells. The co-culture method as applied in the present study does not allow discrimination between cytostasis by secretion or by cell-to-cell contact. Therefore, at present it is not possible to exclude the possibility that eicosanoid regulation of macrophage antitumour function in vitro is valid for both mechanisms.

Acknowledgement. Financial support was obtained from the Netherlands Cancer Foundation (Koningin Wilhelmina Fonds).

References

Adams DO, Hamilton TA (1984) The cell biology of macrophage activation. Annu Rev Immunol 2:283–318

Matthews N (1981) Production of an anti-tumor cytotoxin by human monocytes. Immunology 44:135–142

Onozaki K, Matsushima K, Aggarwal BB, Oppenheim JJ (1985) Human interleukin 1 is a cytocidal factor for several tumor cell lines. J Immunol 131:3962–3968

Ophir R, Ben-Efraim S, Bonta IL (1987) Leukotriene D$_4$ and indomethacin enhance additively the macrophage cytostatic activity in vitro towards MOPC-315 tumor cells. Int J Tissue React 9:189–194

Oppenheim JJ, Kovacs EJ, Matsushima K, Durum SK (1986) There is more than one interleumin-1. Immunol Today 7:45–55

Palladino MA, Finkle BS (1986) Immunopharmacology of tumor necrosing factors. Trends Pharmaco Sci 7:388–389

Rola-Pleszczynski M, Lemaire I (1985) Leukotrines augment interleukin I production by human monocytes. J Immunol 135:3958–3961

Rossi P, Lindgren JA, Kullman C, Jondal M (1985) Products of the lipoxygenase pathway in human natural killer cell cytotoxicity. Cell Immunol 93:1–8

Schenkelaars BJ, Bonta IL (1986) Cyclooxygenase inhibitors promote the leukotriene C$_4$ induces release of beta-glucoronidase from rat peritoneal macrophages: prostaglandine E$_2$ suppresses. Int J Immunopharmacol 8:305–311

Role of Tumour Cell Eicosanoids and Membrane Glycoproteins IRGPIb and IRGPIIb/IIIa in Metastasis

K.V. Honn, I.M. Grossi, H. Chopra, B.W. Steinert, J.M. Onoda, K.K. Nelson and J.D. Taylor

Departments of Radiation Oncology and Biological Sciences, Wayne State University; and Gershenson Radiation Oncology Center, Harper Grace Hospitals, Detroit, Michigan 48202, USA

1 Eicosanoid Metabolism

Eicosanoids are a group of oxygenated arachidonic acid metabolites which include prostaglandins, thromboxanes, leukotrienes, lipoxins, and various hydroperoxy and hydroxy fatty acids. Eicosanoids are implicated in diverse cellular functions such as chemotaxis, proliferation, cell-cell signaling etc. The first committed step in the biosynthesis of each of the eicosanoids, the incorporation of molecular oxygen into polyunsaturated fatty acids, is catalyzed by one of a group of enzymes calles fatty acid oxygenases. This group includes the cyclooxygenase (COX) of prostaglandin endoperoxide synthase (PGH) and various lipoxygenases (LOX).

2 Role of Eicosanoid Metabolites in Tumour Cell Metastasis

2.1 Cyclooxygenase Pathway

Due to limitations in space this review will only deal with the role of eicosanoids in tumour cell-platelet-endothelial cell interactions. Tumour cells have been reported to posses a platelet activating material and procoagulant activity (Gasic et al. 1978; Tohgo et al. 1986; Tanaka et al. 1986), which has recently been purified (Cavanaugh et al. 1985, and in press). Gasic et al. (1968) was the first to provide direct experimental evidence for the role of platelets in tumour cell metastasis. Other laboratories have confirmed the in vitro aggregation of platelets by both animal and human tumour cells and a correlation exists between platelet aggregating ability, tumour colony forming ability and metastatic ability in vivo. However, exceptions have been reported (Estrada and Nicolsin 1984). The mechanism of platelet facilitated metastasis is unknown, although several hypotheses have been set forth. Honn et al. (1984) have suggested that platelets sta-

Nigam et al. (Eds.), Eicosanoids,
Lipid Peroxidation and Cancer
© Springer-Verlag Berlin Heidelberg 1988

bilize the attachment of tumour cells to endothelial cells and subendothelial matrix. Evidence for this mechanism has been demonstrated by Menter et al. (1987a,b). The participation of platelets in tumour cell arrest has suggested that some form of antiplatelet therapy may be efficacious in reducing tumour cell metastasis. Numerous agents have been used, often with conflicting results (Honn et al. 1987a). Honn et al. (1981, 1983) first proposed the hypothesis that eicosanoids produced by the platelets, the tumour cell and the vessel wall were key determinents in the interaction of those three cell types. They proposed that prostacyclin (PGI_2) or PGI_2 stimulating agents may limit tumour cell metastasis (Honn et al. 1983), whereas thromboxane A_2 (TXA_2) would promote tumour cell metastasis (Honn 1983).

Prostacyclin inhibits tumour cell induced platelet aggregation (TCIPA) in vitro (Menter et al. 1984; 1987c). In addition, PGI_2 inhibits the platelet release reaction during TCIPA (Menter et al. 1987d) and platelet adhesion to the tumour cell surface (Honn et al. 1985; Menter et al. 1987c). Tumour cells undergo specific morphological changes in their plasma membrane in response to platelet adhesion (Honn et al. 1985; Menter et al. 1987e). Therefore, tumour cells undergo a discrete and specific response to platelets (Honn et al. 1985; Menter et al. 1987c, e). This response can be inhibited in vitro by exogenous PGI_2 (Honn et al. 1985; Menter et al. 1987c). Thromboxane A_2 is the principal COX metabolite of platelets and a potent inducer of platelet aggregation. This led Honn et al. (1984) to suggest the use of thromboxane synthase inhibitors (TXSI) as antimetastatic agents. Several TXSI were found to reduce both experimental and spontaneous metastasis in vivo (Honn, 1983). However, TXSI did not inhibit TCIPA in vitro (Honn et al. 1984). This lack of an in vitro TXSI effect may be due to endoperoxide PGH_2 (an intermediate in the biosynthesis of TXA_2) which interacts with the TXA_2 receptor resulting in TCIPA. These results suggest that a TXA_2 receptor antagonist should inhibit TCIPA, an effect which has been reported by Mehta et al. (1986). Honn et al. (1987b) have recently reported that the inhibition of TCIPA by COX inhibitors is dependent upon the agonist strength (i.e., number and thrombogenicity of tumour cells). At low agonist strength, TCIPA is reversible and completely inhibited by COX inhibitors (i.e., indomethacin; Honn et al. 1987b). However, at high agonist strength COX inhibitors are ineffective (Honn et al. 1987b). Lipoxygenase inhibitors alone are ineffective regardless of agonist strength. However, at high agonist strength the combination of a COX and a LOX inhibitor completely inhibited TCIPA (Honn et al. 1987b and Steinert et al., submitted).

Steinert et al. (submitted) have demonstrated that the platelet COX component necessary for TCIPA is TXA_2. The combination of a LOX inhibitor (i.e., quercetin) and a TXSI (i.e., CGS 14854) resulted in complete inhibtion of TCIPA (Fig. 1) whereas neither agent was effective alone. Similar results were observed using a LOX inhibitor and a TXA_2 receptor antagonist (Steinert et al., submitted).

Mehta and coworkers (1983, 1984) determined in vitro that the PGI_2 biosynthetic capability of vessels removed from patients with osteogenic sarcoma was less than that of vessels removed from patients without cancer. In addition, the plasma PGI_2 concentrations of patients with bone tumours were less than

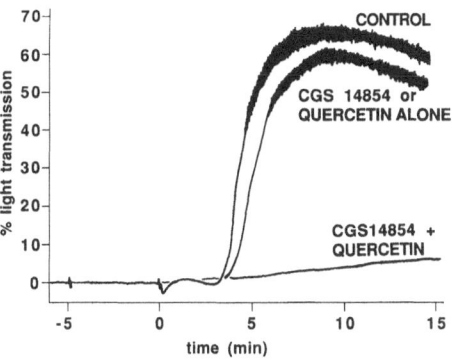

Fig. 1. Effect of TXSI (CGS 14854) and LOX (quercetin) inhibitors on Walker 256 carcinosarcoma-induced platelet aggregation. Aggregation of homologous platelet-rich plasma was induced by 5×10^5 tumour cells. Platelets were preincubated with either or both inhibitor (50 μM) 5 min prior to initiation of aggregation

normal controls. Plasma TXB_2 levels were unchanged in patients with bone cancer. Serneri et al. (1986) examined eicosanoid metabolism by arterial wall samples removed from patients with renal cell carcinoma. They observed a significant decrease in prostacyclin production by these vessels when compared to vessels removed from control patients undergoing nephrectomies. Interestingly, in the study by Serneri et al. (1986) the reduction in PGI_2 biosynthesis was selective for PGI_2; as other COX products (i. e., PGE_2) produced by the vessel wass were unchanged between cancer patients and normal controls.

2.2 Lipoxygenase Pathway

In addition to the above mentioned studies by Honn et al. (1987b) and Steinert et al. (submitted) scant information exists for a role of the LOX pathway in tumour cell-platelet-endothelial cell interactions; however, several interesting observations are worth noting. Honn et al. (1982) has reported significant anti-metastatic effects in mice in response to nafazatrom. Similar results were reported by Drago and Al-Mondhry (1984). Nafazatrom is a pyrazolone derivative with an interesting biochemical mechanism of action. Marnett et al. (1984) have reported that nafazatrom serves as a cooxygenation substrate for prostaglandin endoperoxide synthase and protects PGI_2 synthase from free radical inactivation resulting in increased PGI_2 production. In addition, Honn and Dunn (1982) have reported that nafazatrom is a potent LOX inhibitor. Therefore, the antimetastic effects observed with nafazatrom could be due either to its enhancement of PGI_2 synthesis or its LOX inhibitory activity. These two effects are not mutually exclusive. Serneri et al. (1986), found a concommitant increase in vessel wall synthesis of 12-hydroxyeicosatetranoic acid (12-HETE) in addition to the decrease in vessel wall PGI_2 synthesis in cancer patients. 12-HETE is a potent inhibitor of PGI_2 synthase (Hadjiagapiow and Spector 1986). On the other hand, nafazatrom is a potent inhibitor of 12-lipoxygenase (Honn and Dunn 1982). These results suggest that increased 12-lipoxygenase activity in the vessel wall could produce a condition which would favor metastasis. Maniglia et al. (1986) have examined the effects of nafazatrom on tumour cell degradation of extracellular matrix. They found a significant inhibition of matrix degradation in the presence of nafazatrom. These authors did not correlate their nafazatrom effects with an inhibition of tumour cell LOX activity. However, Sloane et al. (1982) have reported that nafazatrom inhibits the release of the cysteine proteinase, cathepsin B, from tumour cells. These authors speculate that release of proteinases from the surface of tumour cells is mediated by tumour LOX products (Sloane and Honn 1985).

3 Platelet Adhesion Glycoproteins

Of the numerous glycoproteins found on the surface of platelets, several function in aggregation and adhesion. GPIb is the principle sialoglycoprotein of the platelet membrane (Nurden et al. 1986). GPIIb/IIIa is a calcium dependent heterodimer complex in resting and activated human platelets (Coller et al. 1983). Both GPIb and IIb/IIIa are transmembrane proteins (Fitzgerald and Phillips 1988). In the resting platelet, both glycoproteins have been localized on the platelet surface and to internal vacuolar structures. GPIb serves as the receptor for thrombin and also as a receptor for von Willebrand factor (Peterson et al. 1987). Platelet activation transforms the GPIIb/IIIa complex into a receptor for fibrinogen, fibronectin and von Willebrand factor (Nurden et al. 1986; George et al. 1984; Plow et al. 1986). The binding of fibrinogen is a requisite for platelet aggregation, while the binding of fibronectin and von Willebrand factor may be necessary for platelets to adhere and spread on subendothelial surfaces (Plow et al. 1986). Menter et al. (1987 a, b) recently provided evidence that the platelet membrane and platelet cytoskeleton where the causal factors in platelet enhanced tumour cell adhesion to subendothelial matrix, whereas no role in tumour cell adhesion was found for components if platelet α granules and dense granules. These authors suggested that the platelet plasma membrane components involved in facilitation of tumour cell adhesion were GPIb and GPIIb/IIa. Recently, Grossi et al. (1987 a, b) reported an inhibition of TCIPA and platelet facilitated tumour cell adhesion to subendothelial matrix by pretreatment of platelets with antibodies to GPIb and/or GPIIb/IIIa suggesting that these platelet glycoproteins may participate in platelet facilitated metastasis.

4 Evidence for the Presence of Immunologically Related GPIb and GPIIb/IIIa on Nucleated Cells

Within the last several years evidence has emerged suggesting the presence of glycoproteins immunologically related to GPIIb/IIa on a variety of cell types. Tabilio et al. (1984) first demonstrated the presence of platelet glycoprotein Ib in a human leukemic cell line (HEL). Kieffer et al. (1986) recently purified the immunologically related glycoprotein Ib from a stable HEL subclone. Thiagarajan et al. (1987) purified a functionally active GPIIb/IIIa complex from HEl cells. Fitzgerald et al. (1985) have recently demonstrated the presence of GPIIIa on umbilical vein and bovine aortic endothelial cells. They have localized immunoligically related glycoproteins (i. e., vitronectin receptor) on the surface of human endothelial cells, smooth muscle cells and MG-63 fibrolasts (Charo et al. 1986). All of these proteins appear to be structurally and functionally related although not identical.

Therefore, glycoproteins immunologically related to platelet GPIIb/IIIa receptor complex appear to be present in a variety of cells. This family of adhesion receptors has been termed "integrins." Most of these glycoproteins are similar to GPIIb/IIIa in that they are complexes of two dissimilar subunits. All of the glycoproteins in this family appear to be structurally and functionally related. Most are two subunit molecules (an α subunit and a smaller β subunit). All mediate cellular interactions and most do so by binding to the RGD sequence on adhesive proteins such as fibronectin, fibrinogen, vitronectin, von Willebrand, etc. (Ruoslahti and Pierschbacher 1987).

5 Evidence for the Presence of IRGPIb and IRGPIIb/IIIa on Cells from Solid Tumours and Their Role in Tumour Cell-Platelet and Tumour Cell-Matrix Interactions

Platelet glycoprotein Ib and IIb/IIIa have well established roles as adhesion receptors in cell-cell and cell-matrix interactions. A fundamental characteristic of malignant neoplasms is their ability to metastasize. Because the metastatic process is a series of events requiring tumour cell-host cell interactions, it seems reasonable to investigate for the presence of these glycoproteins in cells from solid tumours. Glycoproteins immunologically related to platelet GPIb (i. e., IRGPIb) and platelet GPIIb/IIIa (i. e., IRPGIIb/IIIa) have been recently identified on cells from human (Grossi et al. 1988), rat (Chopra et al. 1988), and murine (unpublished observation) solid tumours. The evidence for the presence of these IRGP's has been obtained by (a) use of specific monoclonal and polyclonal antibodies which distinguish between the various integrin receptors (Grossi et al. 1988; Chopra et al. 1988) and (b) Northern blot analysis using cDNA probes for GPIIb and GPIIIa (Chorpra et al. 1988, Chang et al. 1988). Two tumour types used for study were a human colon carcinoma (Clone A) and a human cervical carcinoma (MS751). The carcinomas were chosen because in humans they are highly invasive, metastatic, and induce the aggregation of platelets.

All antibodies tested against IRGPIb on MS751 and Clone A cells demonstrated a positive reaction (Fig. 2. a, b). In general, a diffuse staining pattern was observed on both MS751 and Clone A cells for all antibodies against IRGPIb. In any sample of MS751 of Clone A cells we observed heterogeneity (i. e., high and low) in staining intensity. Similarly, MS751 and Clone A cells demonstrated positive staining for IRGPIIb/IIIa with all antibodies tested (Fig. 3a, b). In contrast to the staining pattern observed for IRGPIb, there were numerous islands

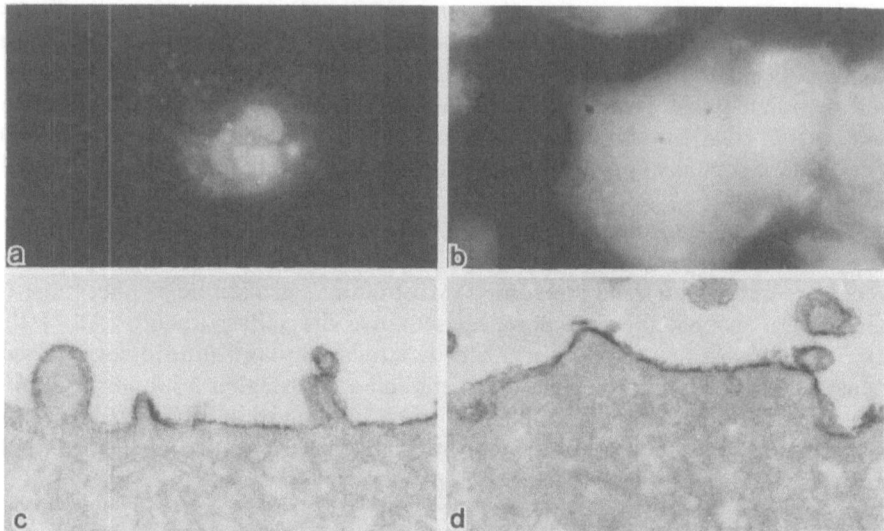

Fig. 2a–d. Immunological identification and localization of IRGPIb. Human cervical carcinoma (MS751; **a**) and human colon carcinoma (Clone A; **b**) cells demonstrate positive staining with all mAb and pAb tested. Immunocytochemical localization of IRGPIb demonstrate a diffuse distribution of reaction product on the surface MS751 **(c)** and Clone A **(d)** tumour cells

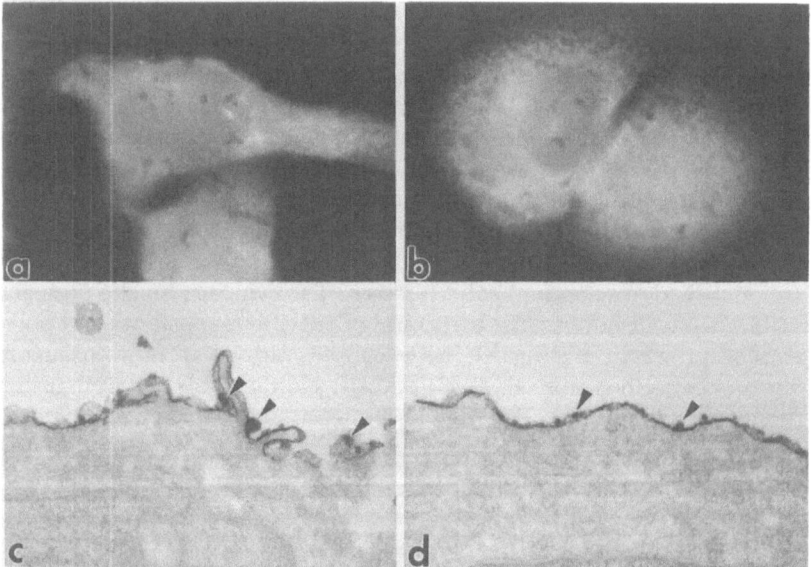

Fig. 3a–d. Immunological identification and localization of IRGPIIb/IIIa. Human cervical carcinoma (MS751; **a**) and human colon carcinoma (Clone A; **b**) cells demonstrate positive staining with all mAb and pAb tested. Immunocytochemical localization of IRGPIIb/IIIa demonstrate a punctate distribution of reaction product suggesting that IRGPIIb/IIIa forms aggregates *(arrows)* representing high receptor concentrations on the surface of MS751 **(c)** and Clone A **(d)** tumour cells

of high fluorescence intensity and regions of low fluorescence intensity for IRG-PIIb/IIIa on both cell types (Grossi et al. 1988). Similar to the staining patterns for IRGPIb, we observed heterogeneity in staining for IRGPIIb/IIIa among populations of MS751 and Clone A cells.

In order to localize the IRGP's on the surface of tumour cells, both Clone A and MS791 cells were processed for electron microscopic immunocytochemistry. In non-permeabilized platelets glycoproteins Ib and IIb/IIIa are localized to the platelet plasma membrane. Non-permeabilized Clone A and MS751 cells demonstrated a positive staining on their membrane surfaces when labelled with either monoclonal or polyclonal antibodies against platelet glycoprotein Ib or the IIb/IIIa complex (Figs. 2c,d and 3c,d). Antibody against platelet glycoprotein Ib demonstrated a homogenous labelling pattern on the membrane surface of both MS751 and Clone A cells (Fig. 2c,d). In contrast, cells labelled with antibodies against platelet GPIIb/IIIa were shown to have a punctate distribution of reaction product suggesting that IRGPIIb/IIIa forms aggregates of high receptor concentration on the tumour cell plasma membrane.

In other model systems we observed that pretreatment of tumour cells with a combination of antibodies against platelet glycoprotein Ib and antibody platelet glycoprotein IIb/IIIa resulted in a complete inhibition of TCIPA and inhibition of platelet adhesion to the tumour cell plasma membrane as determined ultrastructurally (Chopra et al. 1988 and date not shown). Although we do not have evidence for the identity of the linking molecules(s) (e.g., fibrinogen, etc.), it appears that IRGPIb and/or IRGPIIb/IIIa serve as receptors for platelet attachment to the tumour cell plasma membrane prior to TCIPA (Chopra

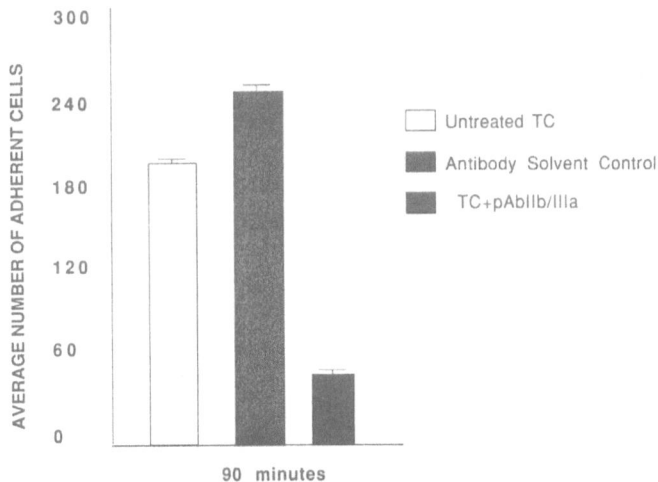

Fig. 4. Inhibition of Clone A tumour cell adhesion to fibronectin-coated plates by polyclonal antibody IIb/IIIa. Polyclonal antibody Ib and nonimmunized rabbit IgG were ineffective (data not shown)

et al. 1988). Since we have demonstrated the presence of IRGPIb and IRGPIIb/IIIa on the surface of both Clone A and MS751 cells, and localized them to the plasma membrane of both cell types, we propose that these glycoproteins may be modulators of tumour cell adhesion to subendothelial matrix through the same adhesion proteins which influence platelet function. Fibronectin is a component of the subendothelial matrix and is an adhesive protein which binds to platelet glycoprotein IIb/IIIa. Therefore, we tested for the ability of antibodies to GPIIb/IIIa to inhibit Clone A tumour cell attachment to a fibrinectin substratum. Clone A tumour cells pretreated with pAbIIb/IIIa demonstrated a marked reduction in their ability to adhere to fibronectin coated plates (Fig. 4; and Grossi et al. 1988). Similar results have been observed for tumour cell adhesion to subendothelial matrix (data not shown).

6 Role of 12-HETE in Tumour Cell Expression of IRGPIb and IRGPIIb/IIIa, Adhesion and Metastasis

Batchev et al. (1986) examined phorbol ester binding to two murine fibrosarcoma sublines. The two sublines are derived from the same tumour but they respond differentially to stimulation with phorbol esters. In one subline, phorbol esters stimulated a rapid attachment and spreading response concomitant with directional migration. The other line did not migrate in response to stimulation with phorbol esters and the attachment and spreading response was slower. The cell line which rapidly responded to phorbol ester was highly malignant when injected into syngeneic animals. In this highly malignant line, phorbol ester stimulated a rapid release of arachidonic acid and arachidonic acid conversion into cyclooxygenase and lipoxygenase products. These effects were not observed in the low malignant line (Batchev et al. 1986). These studies suggest a relationship between (a) metastatic ability and (b) ability of the cell to respond to external stimuli with rapid arachidonic acid metabolism. Therefore, we examined the effects of a phorbol ester (TPA) on adhesion of human and murine tumour cells to subendothelial matrix and matrix components (i.e., fibronectin). TPA (100 nM) increased adhesion of Clone A cells to fibronectin, an effect which was blocked by pAbGPIIb/IIIa (data not shown). The effect of TPA on tumour cell adhesion was also reduced by LOX inhibitors (data not shown). These results suggest to us that TPA by a LOX dependent mechanism may increase surface expression of IRGPIIb/IIIa receptors and/or "activate" existing IRGPIIb/IIIa receptors. To test this hypothesis, MS751 cells were stimulated with 12-HETE or TPA in the presence of a LOX inhibitor (i. e., quercetin) at concentrations which did not inhibit protein kinase C. Following stimulation the cells were labelled with a specific monoclonal antibody to GPIIb/IIIa, followed by a secondary antibody conjugated to fluorescein and quantitated by flow cytometric analysis. Both 12-HETE and TPA increased the surface expression of IRGPIIb/IIIa (Fig. 5). The TPA effect was reduced to near baseline levels of expression with 50 µM quercetin and reduced below baseline levels with 100 µM quercetin (Fig. 5).

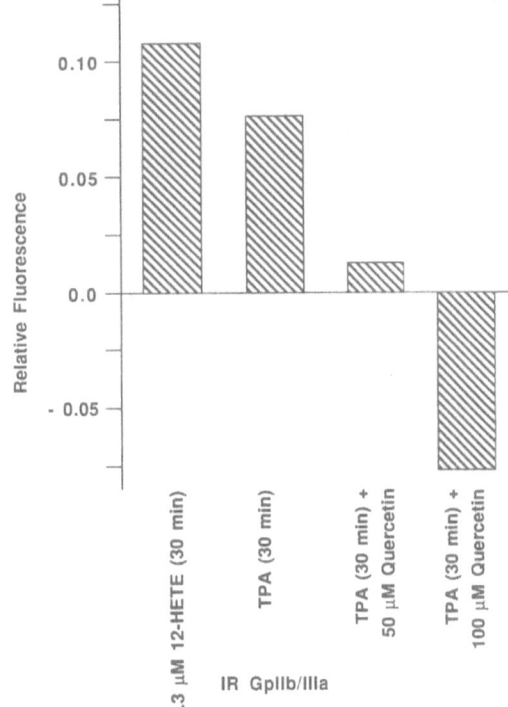

Fig. 5. Increased surface expression of IRGPIIb/IIIa induced by TPA or 12-HETE and inhibition of the TPA response by the LOX inhibitor, quercetin. Similar results were observed for expression of IRGPIb (data not shown)

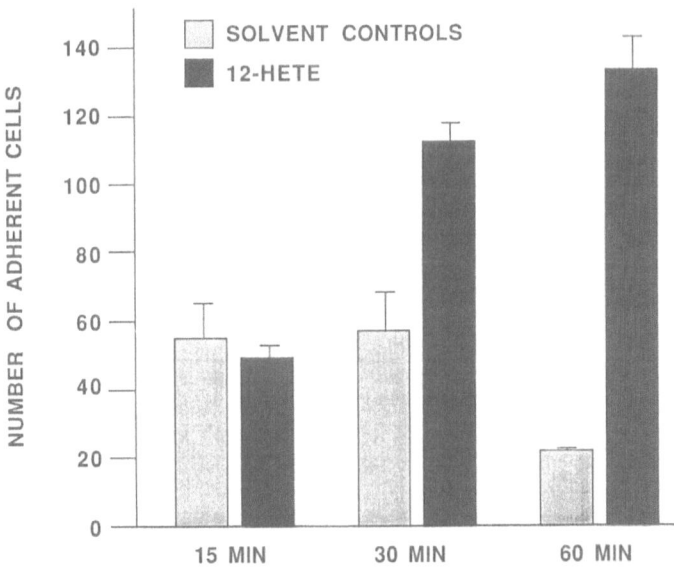

Fig. 6. Increased adhesion of MS751 cells to subendothelial matrix following stimulation with 12-HETE (2 μM)

In a companion experiment 12-HETE incresed adhesion of MS751 cells to subendothelial matrix (Fig. 6) an effect which was inhibited by pAbIIb/IIIa (data
not shown). The effects of TPA and 12-HETE on tumour cell expression of
IRGPIIb/IIIa and adhesion are dependent upon an intact tumour cell cytoskeleton (data not shown).

Onoda et al. (1988) reported that elutriated subpopulations of the B 16 amelanotic melanoma (B 16a) differ in their ability to form pulmonary tumour colonies
in the lung colony assay. The fraction designated B 16a-340 forms significantly
more colonies than the fraction designated B 16a-180 (Fig. 7). Therefore, we examined fractions B 16a-180 and B 16a-340 for the percentage of cells expressing
IRGPIb, IRGPIIb/IIIa and production of 12-HETE in response to 1 μM arachidonic acid. We previously reported that 12-HETE is the major LOX product in
B 16a tumour cells (Honn and Dunn 1982). The B 161-340 fraction exhibited a
greater percentage of cells positive for IRGPIb and IRGPIIb/IIIa and synthesized significantly greater amounts of 12-HETE in response to 1 μM arachidonic
acid (Fig. 7). In related studies, we demonstrated that exogenous 12-HETE increased surface expression of B 16a IRGPIIb/IIIa as well as their adhesion to
subendothelial matrix of fibronectin (data not shown).

7 Conclusion

Our laboratory has proposed that eicosanoid metabolism may be a causal factor
in tumour cell-platelet-endothelial cell/subendothelial matrix interactions during the process of hematogenous metastasis. We proposed a pivotal role for

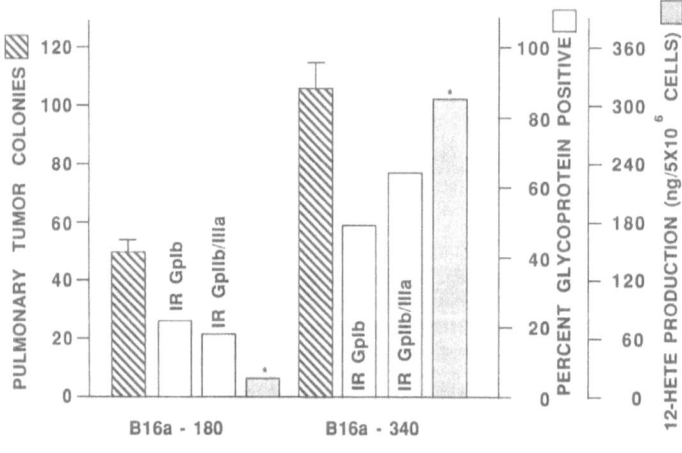

*1 μM ARACHIDONIC ACID STIMULATION OF 5X10^6 CELLS

Fig. 7. Correlation between increased synthesis of 12-HETE, expression of IRGPIb and IRG-
PIIb/IIIa and lung colony formation by B 16a cells in elutriated fractions derived from collagenase dispersed B 16a tumours

endothelial cell PGI_2 and LOX products as well as platelet TXA_2 (Honn et al. 1983). Exogenous PGI_2 inhibits lung colony formation (in vivo) and TCIPA and tumour cell adhesion to endothelial cells in vitro (Honn et al. 1981; Menter et al. 1984, 1987c,d and unpublished observations). While TXSI reduce metastasis in vivo (Honn 1983) these results may be due to their sensitization of platelets to endogenous PGI_2 (Honn et al. 1984; Onoda et al. 1984) as results with TXSI inhibition of TCIPA in vitro have been disapointing. However, the recent results of Steinert et al. (submitted) strongly suggest a role for both platelet LOX and COX pathways in TCIPA and may explain the previous lack of effect with COX or TXSI inhibitors alone. Recently, we have focused our attention on tumour cell eicosanoids and their role in metastasis. In this paper we presented evidence to support the hypothesis that tumour cell LOX metabolites (i.e., 12-HETE) play a major role in metastasis by regulating the surface expression of adhesive glycoproteins (e.g., IRGPIb, IRGPIIb/IIIa, etc.) and perhaps other cell surface molecules (e.g., proteinases) which are important in tumour cell metastasis (Fig. 8). The tumour cell cytoskeleton appears to be involved in the regulation of expression of some of these surface determinants (i.e., IRGPIIb/IIIa). These findings suggest new targets for the development of antimetastatic therapies.

Acknowledgement. Original results in this paper supported by grants from the National Institutes of Health (CA29997 and CA47115) and a grant from Harper/Grace Hospitals.

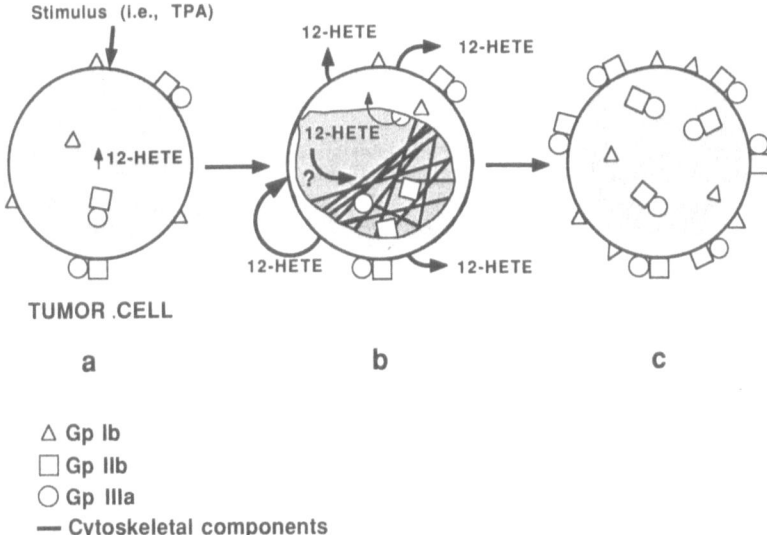

Fig. 8. Proposed hypothesis for 12-HETE regulation of surface expression of tumour cell adhesion receptors

References

Batchev AC, Riser DL, Hellner EG, Fliegiel SEG, Varani J (1986) Phorbol ester binding and phorbol ester-induced arachidonic acid metabolism in a highly responsive murine fibrosarcoma cell line and in a less-responsive variant. Clin Exptl Metastasis 4:51-61

Cavanaugh PG, Sloane BF, Bajkowski AS, Taylor JD, and Honn KV (1985) Purification and characterization of platelet aggregating activity from tumor cells: copurification with procoagulation activity. Thromb Res 37:309-326

Cavanaugh PG, Sloane BF, Honn KV (1988) Role of coagulation system in tumour cell induced platelet aggregation and metastasis. Hemostasis (in press)

Chang YS, Fitzgerald LA, Grossi IM, Sundram U, Murray JA, Honn KV (1988) Tumour cell expression of mRNAs coding for the integrin receptors. FASEB J 2:A1406

Charo IF, Fitzgerald LA, Steiner B, Raol Jr., SC, Bekeart LS, Phillips DR (1986) Platelet glycoproteins IIb and IIIa: evidence for a family of immunologically and structurally related glycoproteins in mammalian cells. Proc Natl Acad Sci USA 83:8351-8355

Chopra H, Hatfield JS, Chang YS, Grossi IM, Fitzgerald LA, O'Gara CY, Marnett LJ, Diglio CA, Taylor JD, Honn KV (1988) Role of tumour cell cytoskeleton and membrane glycoprotein IRGIIb/IIIa in platelet adhesion to tumour cell membrane and tumour cell induced platelet agregation. Cancer Res (in press)

Coller BS, Peerschke EI, Scudder LE, Sullivan CA (1983) A murine monoclonal antibody completely blocks the binding of fibrinogen to platelets and produces a thrombasthenic-like state in normal platelets and binds to glycoproteins IIb and/or IIIa. J Clin Invest 72:325-338

Drago JR, Al-Mondhiry HAB (1984) The effect of prostaglandin modulators on prostate tumour growth and metastasis. Anticancer Res 4:391-394

Estrada J, Nicolson GL (1984) Tumour cell-platelet aggregation does not correlate with metastatic potential of rat 13762 NF mammary adenocardinoma tumour cell clones. Int J Cancer 34:101-105

Fitzgerald LA, Charo IF, Phillips DR (1985) Human and bovine endothelial cells synthesize membrane proteins similar to human platelet glycoproteins IIb and IIIa. J Biol Chem 260:10893-10896

Fitzgerald LA, Phillips DR (1988) Structure and function of platelet membrane glycoproteins. In: Kunicki TJ, George J (eds) Platelet immunology. Lippincott, Philadelphia (in press)

Gasic GJ, Gasic TB, Stewart CC (1968) Antimetastatic effects associated with platelet reduction. Proc Natl Acad Sci USA 61:46-52

Gasic GJ, Boettinger D, Catalfamo JL, Gasic TB, Stewart GJ (1978) Aggregation of platelets and cell membrane vesiculation by rat cells transformed in vitro by Rous sarcoma virus. Cancer Res 38:2950-2955

George JN, Nurden AT, Phillips DR (1984) Molecular defects in interactions of platelets with the vessel wall. N Engl J Med 311:1084-1098

Grossi IM, Fitzgerald LA, Kendall A, Taylor JD, Sloane BF, Honn KV (1987a) Inhibition of human tumour cell induced platelet aggregation by antibodies to platelet glycoproteins Ib and IIb/IIIa. Proc Soc Exp Biol Med 186:378-383

Grossi IM, Honn KV, Sloane BF, Thompson J, Ohannesian D, Kendall A, Newcombe M (1987b) Role of platelet glycoproteins Ib and IIb/IIIa in tumour cell induced platelet aggregation and tumour cell adesion to extracellular matrix. Thromb Haemostas 58:507

Grossi IM, Hatfield JS, Fitzgerald LA, Newcombe M, Taylor JD, Honn KV (1988) Role of glycoproteins immunologically related to Ib and IIb/IIIa in tumour cell-platelet and tumour cell-matrix interactions. FASEB J (in press)

Hadjiagapiou C, Spector AA (1986) 12-Hydroxyeicosatetraenoic acid reduces prostacyclin production by endothelial cells. Prostaglandins 31:1135-1144

Honn KV (1983) Inhibition of tumour cell metastasis by modulation of the vascular prostacyclin/thromboxane A_2 system. Clin Exp Metastasis 1:103-114

Honn KV, Dunn JR (1982) Nafazatrom (Bay g 6575) inhibition of tumour cell lipoxygenase activity and cellular proliferation. FEBS Lett 139:65-68

Honn KV, Cicone B, Skoff A (1981) Prostaxyclin: a potent antimetastatic agent. Science 212:1270-1272

Honn KV, Meyer J, Neagos G, Henderson T, Westley C, Tatanatharathorn V (1982) Control of tumour growth and metastasis with prostacyclin and thromboxane synthase inhibitors: evidence for a new antitumour and antimetastatic agent (Bay g 6575). In: Jamieson G (ed) Interactions of platelets and tumour cells. Liss, New York, pp 295–331
Honn KV, Busse WD, Sloane BF (1983) Prostacyclin and thromboxanes. Implications for their role in tumour cell metastasis. Biochem Pharmacol 32:1–11
Honn KV, Menter DG, Onoda JM, Taylor JD, Sloane BF (1984) Role of prostacyclin as a natural deterrent to hematogenous tumour metastasis. In: Nicolson G, Milas L (eds) Cancer invasion metastasis. Raven, New York, pp 361–388
Honn KV, Onoda JM, Menter DG, Taylor JD, Sloane BF (1985) Prostacyclin in the control of tumour metastasis. In: Cohen MM (ed) Biological protection with prostaglandins, vol 1. CRC, Boca Raton, pp 91–110
Honn KV, Steinert BW, Onoda JM, Sloane BF (1987a) The role of platelets in metastatis. Biorheology 24:127–137
Honn KV, Steinert BW, Moin K, Onoda JM, Taylor JD, Sloane BF (1987b) The role of platelet cyclooxygenase and lipoxygenase in tumour cell induced platelet aggregation. Biochem Biopys Res Comm 145:384–389
Kieffer N, Debili N, Wicki A, Titeux M, Henri A, Michel Z, Gorius JB, Vainchanker W, Clemetson KJ (1986) Expression of platelet glycoproteins Ib-α in HEL cells. J Biol Chem 261:15854–15862
Maniglia CA, Loulakis PP, Sartorelli AC (1986) Interference with tumour cell-induced degradation of endothelial matrix on the antimetastatic action of nafazatrom. J Natl Cancer Inst 76:739–744
Marnett LJ, Siedlick PH, Ochs RC, Pagels WR, Daas M, Honn KV, Warnock RH, Tainer BE, Eling TE (1984) Mechanism of the stimulation of prostaglandin H synthase and prostacyclin synthase by the antithrombotic and antimetastatic agent, nafazatrom. Mol Pharmacol 26:328–335
Mehta P (1984) Evidence for altered arachidonic acid metabolism in tumour metastasis. In: Honn KV, Sloane BF (eds) Hemostatic Mechanisms of Metastasis. Nijhoff, Boston, pp 233–243
Mehta P, Springfield D, Ostrowski N (1983) Arterial proatacyclin generation is decreased in patients with malignant bone tumours. Cancer 52:1297–1300
Mehta P, Lawson D, Ward MB, Ambrose LL, Limura A (1986) Effects of thromboxane A₂ inhibition of osteogenic sarcoma cell-induced platelet aggregation. Cancer Res 46:5061–5063
Menter DG, Onoda JM, Taylor JD, Honn KV (1984) Effects of prostacyclin on tumour cell induced platelet aggregation. Cancer Res 44:450–456
Menter DG, Steinert BW, Sloane BF, Gundlach N, O'Gara CY, Narnett LJ, Diglio CA, Walz D, Taylor JD, Honn KV (1987a) Role of platelet membrane in enhancement of tumour cell adhesion to endothelial cell extracellular matrix. Cancer Res 47:6751–6762
Menter DG, Sloane BF, Steinert BW, Onoda J, Craig R, Harkins C, Taylor JD, Honn KV (1987b) Platelet enhancement of tumour cell adhesion to subendothelial matrix. Role of cytoskeleton and platelet membrane. J Natl Cancer Inst 79:1077–1090
Menter DG, Harkins C, Onoda J, Riorden W, Sloane BF, Taylor JD, Honn KV (1987c) Inhibition of tumour cell induced platelet aggregation by prostacyclin and carbacyclin: an ultrastructural study. Invasion Metastasis 7:109–128
Menter DG, Onoda JM, Moilanen D, Sloane BF, Taylor JD, Honn KV (1987d) Inhibition by prostacyclin of the tumour cell-induced platelet release reaction and platelet aggregation. J Natl. Cancer Inst 78:961969
Menter DG, Hatfield JS, Harkins C, Sloane BF, Taylor JD, Crissman JD, Honn KV (1987e) Tumour cell-platelet interactions in vitro and their relationship to in vivo arrest of hematogenously circulating tumour cells. Clin Expl Metastasis 5:65–78
Nurden AT, George JN, Phillips DR, 1986, Platelet membrane glycoproteins: their structure, function and modification in disease. In: Phillips DR, Schuman MS (eds) Biochemisty of platelets. Academic, Orlando pp 159–224
Onoda JM, Sloane BF, Honn KV (1984) Antithrombogenic effects of calcium channel blockers: synergism with prostacyclin and thromboxane synthase inhibitors. Thromb Res 34:367–378

Onoda JM, Nelson KK, Grossi IM, Umbarger LA, Taylor JD, Honn KV (1988) Separation of high and low metastatic subpoppulations from solid tumours by centrifugal elutriation. Proc Soc Exp Biol Med 187:250-255

Peterson DM, Stathopoulos NA, Georgio DD, Helums JD, Molke JL (1987) Shear-induced platelet aggregation requires von Willebrand factor and platelet membrane glycoproteins IIb and IIIa. Blood 69:625-628

Plow E, Ginsberg MH, Marguerie GA (1986) Expression and function of adhesive proteins on the platelet surface. In: Phillips DR, Schuman MS (eds) Biochemistry of platelets. Academic, Orlando, pp 225-256

Ruoslahti E, Pierschbacher MD (1987) New perspectives in cell adhesion: RGD and integrins. Science 238:491-497

Serneri GGN, Abbate R, Gensini GF, Panetta A, Coslo G, Costantini A, Carini M, Seloi C (1986)Altered 1-^{14}C arachidonic acid metabolism in arterial wall from patients with renal cell carcinoma. J Urol 135:1071-1074

Sloane BF, Honn KV (1985) Proteolytic enzymes and arachidonic acid metabolism. In: Lands WE (ed) Biochemistry of arachidonic acid metabolism. Nijhoff, Boston, pp 311-322

Sloane BF, Makim S, Dunn JR, Lacoste R, Theodorou M, Battista J, Alex R, Honn KV (1982) Lipoxygenase products as mediators of tumour cell lysosomal enzyme release: inhibition with nafazatrom. In: Powles TJ, Bockman RS, Honn KV, Ramwell P (eds) Prostaglandins and cancer. Liss, New York, pp 789-792

Steinert BW, Grossi IM, Umbarger LA, Fitzgerald LA, Honn KV Role of platelet eicosanoid metabolism and membrane glycoproteins in tumour cell induced platelet aggregation. (submitted)

Tabilio A, Rosa JP, Testa P, Kieffer U, Nurden N, DelCanizio AT, Breton-Gorius MT, Vainchenker W (1984) Expression of platelet membrane glycoproteins and α-granule proteins by a human erythroleukemia cell line (HEL). EMBO J 3:353-359

Tanaka NG, Tohgo A, and Ogawa H (1986) Platelet-aggregating activities of metastasizing tumour cells. V. In situ roles of platelets in hematogenous metastasis. Invasion Metastasis 6:209-224

Thiagarajan P, Shapiro S, Swererlitsch L, McCord S (1987) Human erythroleukemia cell line synthesizes a functionally active glycoprotein IIb/IIIa complex capable of binding fibrinogen. Biochem Biophys Acta 924:127-134

Tohgo A, Tanaka NG, Ogawa H (1986) Platelet aggregating activities of metastasizing tumour cells. IV. Effects of cell surface modification on thrombin generation, platelet aggregation ans dubsequent lung colonization. Invasion Metastasis 6:58-68

Alteration of Arachidonic Acid Metabolism in Rats after Inoculation of Tumour Cells and Their Subcellular Fractions: Role of Mononuclear Phagocytes as a Major Source of Enhanced Prostanoid Synthesis

S. K. Nigam and R. Averdunk

Prostaglandin Research, Department of Gynecological Endocrinology, Klinikum Steglitz, Free University Berlin, 1000 Berlin 45, FRG

1 Introduction

It is a well-established fact that eicosanoids participate in the initiation and promotion of carcinogenesis and that they have a close relationship to tumour promoters and growth factors. High concentrations of prostaglandins have been found in the blood or urine of tumour-bearing animals (Levine 1981). We have also shown in malignant tumours of the breast (Nigam et al. 1985) and the gastrointestinal tract (Nigam et al. 1987) that increased levels of eicosanoids remained unchanged even after surgical removal of the tumour. The mechanism for this elevation is still unknown. Clinical data, however, present evidence that about 50% of the patients with malignant breast cancer who had elevated levels of eicosanoids after surgery were later rehospitalized on account of recidivism (unpublished results).

Several factors may influence the growth potential of a tumour, e.g. host platelet–tumour cell interactions and tumour cell–vessel wall interactions. Eicosanoids such as thromboxane A_2 (TXA_2) and prostacyclin (PGI_2) produced by platelets, tumour cells and vessel walls were originally supported to be the key substances responsible for the regulation of tumour cell metastasis (Honn et al. 1983; Honn 1983). As TXA_2 is the principal cyclooxygenase metabolite of platelets and a potent inducer of platelet aggregation, thromboxane synthetase inhibitors have been suggested to be antimetastatic agents (Honn et al. 1984). However, conflicting reports on the effectiveness of such inhibitors have appeared in the literature (Honn et al. 1987).

In this paper we present evidence that not only tumour cells are capable of activating blood and tissue cells in rats thus leading to increased biosynthesis of eicosanoids, especially TXA_2, which is a prerequisite for the growth and metastasis of tumours. Our data also show that inoculation of subcellular tumour cell fractions causes augmentation of TXA_2 blood levels in rats. In order to determine whether the increased levels of eicosanoids found in the blood of rats originate from the activation of tumour cells or from the host-infiltrating cells, we used two different cell lines, a T cell line that does not produce eicosanoids in culture and a CAM cell line, an epithelial mammary cancer cell line of the rat, which produces high concentrations of eicosanoids in culture (Table 1).

Nigam et al. (Eds.), Eicosanoids,
Lipid Peroxidation and Cancer
© Springer-Verlag Berlin Heidelberg 1988

2 Methods

Cell cultures were grown in RPMI 1640 medium with 10% fetal calf serum. The lymphoma cell line has a generation time of 10–12 h and was grown in suspension cultures. The cells exhibited surface markers of an undifferentiated cell type. The cells grew as ascites or as a solid tumour in the rat. 5×10^9 to 2×10^{10} cells were harvested 2 weeks after inoculation of the rat with 1–5×10^7 tumour cells. The CAM cell line is grown as a monolayer culture with a generation time of approximately 20 h. It also grew as ascites or as a solid tumour in the rat after intraperitoneal inoculation of 2–5×10^6 tumour cells.

Lewis or Wistar rats kept on a normal diet were used for the experiments. The tumour cells or their subcellular fractions were given intraperitoneally. After cardiac puncture of anaesthetized animals, 5 ml blood was withdrawn into a polypropylene tube containing 0.2 ml 0.1 mM EDTA. Indomethacin (1 µg/ml blood) and meclofenamic acid (5 µg/ml blood) were immediately added to inhibit the in vitro activity of cyclooxygenase.

For inhibition studies the rats were treated with indomethacin, Iloprost (Schering, Berlin), or thromboxane receptor antagonists BM 13 177 (Boehringer, Mannheim) and SQ 29 548 (Squibb, USA) as indicated in Table 4 for 3 consecutive days.

Subcellular fractions were prepared by differential centrifugation after disrupting cells by osmotic swelling in 0.01 M Tris (HCl) buffer for 1 h at 4°C and subsequent homogenization at 10 strokes in a Potter homogenizator. The 150 000 g pellet was used as the membrane fraction. Membrane fractions were tested for 5-nucleotidase activity to indicate the purity of the membrane.

Prostaglandins were determined as described by Nigam et al. (1985) and Nigam (1987).

Statistical evaluations were made by Kruskal-Wallis non-parametric tests. Differences were considered significant when $p < 0.05$

3 Results

In culture lymphoma cells did not produce significant amount of prostanoids. In contrast, CAM cells produced high concentrations of prostanoids, especially 6-keto-PGF$_{1\alpha}$ and TXB$_2$ (Table 1). However, 10 days after intraperitoneal inoculation of 10^8 lymphoma cells into the rats a 20-fold increase in the plasma concentration of 6-keto-PGF$_{1\alpha}$, a 30-fold increase of TXB$_2$ and 42-fold increase of PGE$_2$ were found. Since CAM cells grow more slowly in the rat (detectable tumour 2 cm after 6 weeks), eicosanoid levels in plasma were measured 21 days after inoculation of 10^8 cells. The elevations in plasma concentration were of the same order as found after inoculation of lymphoma cells, e.g. 16-fold for 6-keto-PGF$_{1\alpha}$, 32-fold for TXB$_2$ and 45-fold for PGE$_2$ (Table 2).

Table 1. Concentration of eicosanoids (pg/ml) in supernatant of the tumour cell culture (mean values $\pm$ SEM; $n = 3$)

	6-keto-PGF$_{1\alpha}$	TXB$_2$	PGE$_2$
RPMI 1640 medium + FCS	8.5 $\pm$ 2.4	11.0 $\pm$ 1.9	3.6 $\pm$ 0.8
Lymphoma cells	6.5 $\pm$ 0.6	10.0 $\pm$ 1.1	8.7 $\pm$ 1.2
CAM cells	669 $\pm$ 48	99 $\pm$ 12	90 $\pm$ 16

After harvesting, 5×10^6 cells were washed four times with RPMI medium by centrifugation at 800 g for 10 min. The cells were then incubated with 10 μg/ml arachidonic acid for 2 h in the serum-free medium at 37°C. After pelleting cells, eicosanoids were determined in the supernatant.
FCS, fetal calf serum.

Table 2. Concentration of eicosanoids (pg/ml) in plasma of rats 10 days or 21 days after inoculation of 10^8 lymphoma cells or CAM cells respectively (mean values $\pm$ SEM; $n = 6$)

	6-keto-PGF$_{1\alpha}$	TXB$_2$	PGE$_2$
Control	2.0 $\pm$ 0.2	160 $\pm$ 22	47 $\pm$ 11
Lymphoma cells	50 $\pm$ 8.6[a]	5055 $\pm$ 121[b]	1969 $\pm$ 91[b]
CAM cells	32 $\pm$ 4.8[a]	5024 $\pm$ 188[b]	2117 $\pm$ 169[b]

Tests of significance (compared to control): [a] $p < 0.01$, [b] $p < 0.001$

Table 3. Time-dependent increase in concentration of prostanoids (pg/ml) in plasma of rats after inoculation of 10^8 lymphoma cells or CAM cells (mean values $\pm$ SEM; $n = 6$)

	Blood sampling time	6-keto-PGF$_{1\alpha}$	TXB$_2$	PGE$_2$
Control	–	2.5 $\pm$ 0.2	247 $\pm$ 41	50 $\pm$ 9
Lymphoma cells	2 h	40 $\pm$ 8.0[a]	3338 $\pm$ 211[b]	738 $\pm$ 63[b]
	8 h	33 $\pm$ 7.2[a]	2698 $\pm$ 143[b]	633 $\pm$ 108[b]
	1 d	31 $\pm$ 4.7[a]	2131 $\pm$ 141[b]	610 $\pm$ 89[b]
	3 d	26 $\pm$ 6.6[a]	1843 $\pm$ 128[b]	654 $\pm$ 56[b]
	6 d	90 $\pm$ 9.9[b]	4512 $\pm$ 264[b]	770 $\pm$ 87[b]
Control	–	2.3 $\pm$ 0.4	160 $\pm$ 33	47 $\pm$ 12
CAM cells	2 h	22.5 $\pm$ 3.6[a]	3000 $\pm$ 167[b]	190 $\pm$ 17[a]
	8 h	17.8 $\pm$ 2.4[a]	1020 $\pm$ 45[a]	320 $\pm$ 34[b]
	1 d	5.4 $\pm$ 0.9	382 $\pm$ 34	302 $\pm$ 28[b]
	3 d	27.2 $\pm$ 4.1[a]	4123 $\pm$ 202[b]	315 $\pm$ 41[b]
	6 d	67.5 $\pm$ 9.5[b]	4888 $\pm$ 312[b]	

Tests of significance (compared to controls): [a] $p < 0.01$, [b] $p < 0.001$

An increase in eicosanoid levels in the plasma was found as early as 2 h after the inoculation of tumour cells. It decreased slightly within the first 3 days for lymphoma cells and 1 day for CAM cells and then again increased markedly (Table 3). Plasma concentrations of eicosanoids correlated well with the number of cells inoculated and the duration of tumour growth.

From the above data it is evident that TXA_2 is the principal arachidonic acid metabolite in the circulation of rats and is possibly a potent inducer of tumour growth. We, therefore, used some agents (Table 4) which affect thromboxane synthesis and activity. Whereas indomethacin and Iloprost (gift from Schering, Berlin), had only partial success in controlling the tumour growth, the thromboxane receptor antagonists BM 13177 (gift from Boehringer, Mannheim) and SQ 29548 (gift from Squibb, USA) rendered no detectable tumours after 6 weeks.

In order to investigate whether whole tumour cells or only their cellular components are responsible for the in vivo augmentation of eicosanoid synthesis in the rat, we prepared subcellular fractions of the tumour cells (see Methods) and inoculated them intraperitoneally into the rat. Blood was withdrawn 2 h later for eicosanoid determination. The results are summarized in the Table 5. The plasma membrane-containing particulate fraction (150000 g pellet) shows the highest capacity to stimulate eicostanoid biosynthesis. Again TXA_2 is the predominant metabolite of the arachidonic acid cascade.

Table 4. Influence of agents affecting tumour growth ($n=6$). 10^8 lymphoma or CAM cells were inoculated i.p. into rats pretreated with the agents listed

	Tumour growth after 6 weeks (>2 cm)				TXB_2 (pg/ml)
	Lymphoma cells		CAM cells		
	Detected	Not detected	Detected	Not detected	
Indomethacin (0.2 mg/kg bwd)	6	–	5	1	157 ± 23
BM 13177 (50 mg/kg bwd)	–	6	–	6	2944 ± 131
SQ 29548 (10 mg/kg bwd)	–	6	–	6	3529 ± 189
Iloprost (250 µg/kg bwd)	2	4	2	4	1678 ± 212

TXB_2 values represent the mean $\pm$ SEM

Table 5. Concentration of eicosanoids (pg/ml) in plasma of rats 2 h after inoculation of subcellular fractions from 1×10^8 cells (mean values $\pm$ SEM; $n = 6$)

	Subcellular fraction	6-keto-PGF$_{1\alpha}$	TXB$_2$	PGE$_2$
Lymphoma cells	10000-g pellet	5.0 ± 0.8	407 ± 30	266 ± 27
	10000-g supernat.	6.1 ± 1.0	518 ± 56	190 ± 22
	150000-g supernat.	4.2 ± 0.4	394 ± 43	316 ± 54
	150000-g pellet	4.9 ± 1.2	2806 ± 167	436 ± 42
CAM cells	10000-g pellet	5.3 ± 1.2	728 ± 76	368 ± 44
	10000-g supernat.	4.2 ± 0.9	688 ± 81	349 ± 36
	150000-g supernat.	4.7 ± 0.9	408 ± 42	181 ± 32
	150000-g pellet	4.1 ± 0.4	3081 ± 217	1766 ± 158

4 Discussion

Several previous reports on the involvement of prostaglandin synthesis in experimental animal tumours as well as in cancer patients were either based on the measurement of prostaglandins in tumour biopsy material (Khan et al. 1982; Alam et al. 1982; Bockman 1983) or in the plasma of tumour-bearing rats (Kort et al. 1987). We also reported in a clinical study that prostaglandin levels were increased in patients with breast cancer (Nigam et al. 1985). A striking feature of the study was that prostaglandin levels remained unchanged after surgical removal of the tumour in patients with malignant tumours. We, therefore, believed that detached tumour cells might be interacting with other tissue and blood cells thus stimulating the prostaglandin systhesis.

This study demonstrates that arachidonic acid metabolites from the cyclooxygenase pathway were considerably increased in plasma when tumour cells were inoculated into the rat and no tumour was detectable in the early period of growth. As reported by other authors (Chiabrando et al. 1985; Chiabrando et al. 1987), we also found a preferential increase of a few eicosanoids, e.g. PGE$_2$ and TXB$_2$, during tumour growth, the abundant metabolite being TXB$_2$. Although a significant increase in the 6-keto-PGF$_{1\alpha}$ level was also observed, absolute values were relatively low. As a substantial increase in eicosanoid levels in rat plasma is reported for both lymphoma and CAM tumour cells and lymphoma cells do not produce eicosanoids (Table 1), a major contribution of tumour cells to the arachidonic acid metabolic pattern can thus be ruled out. This observation is in contradiction to a previous report (Chiabrando et al. 1987) in which in vitro cultured tumour cells were shown to produce very high amounts of TXB$_2$.

We believe that the increased overall metabolism of arachidonic acid observed in rats after inoculation of tumour cells could be attributed to an alteration in the host antitumour reaction. According to our experiments the major contribution thus apparently comes from mononuclear phagocytes capable of infiltrating tumours (Nigam et al., 1988). Mononuclear cells are present in both progressively or regressively growing tumours and have been shown to possess cytolytic activity for regressive tumours but not for progressive ones. Further evidence supporting this assumption comes from our investigations with the inoculation of subcellular fractions of tumour cells into the rats (Table 5). The $150\,000\,g$ membrane fraction stimulated eicosanoid synthesis in the rats 2 h after the inoculation, suggesting that host cells are activated in a manner similar to tumour cells. Thus, these findings rule out in vivo elevation of plasma eicosanoids due to increased overall metabolism of arachidonic acid in tumour cells.

The rapid preferential increase in concentration of eicosanoids in rat plasma within 2 h after inoculation of tumour cells (Table 3) or the membrane fraction (Table 5) might reflect in mononuclear cells (a) selective arachidonic acid cascade enzyme induction and/or (b) a shift in the endoperoxide metabolism towards the most active enzymatic pathway. In the case of growing tumours, preferential growth of selected cell subpopulations with different amounts of enzymes and fatty acid availabilities may be the major event.

Increased PGE_2 levels in rat plasma, often regarded as typically abundant in tumours, are indicative of lymphokine suppression which allows the growing tumour to escape from immunologic rejection (Goodwin and Ceuppens 1983; Berlinger 1984).

Thromboxane synthetase inhibitors have been reported to reduce tumour cell growth (Honn 1982), tumour size (Drago and Al-Mondhiry 1984) and metastases (Honn 1983). However, many of these antitumour agents were either non-specific, decreasing prostacyclin, an antimetastatic eicosanoid, or they diverted precursor endoperoxides to other prostaplandins such as PGE_1, PGD_2 and $PGF_{2\alpha}$ which may promote tumour growth. As cyclic endoperoxides have aggregating properties similar to TXA_2, the use of non-specific TXA_2 antagonists did not yield consistantly beneficial results.

In the present paper, we used four agents that inhibited TXA_2 in different ways. The non-specific thromboxane synthetase inhibitor indomethacin had practically no effect on the tumour growth under the experimental conditions. Iloprost, a prostacyclin analogue, proved to be partially successful as an antitumour agent (Table 4). One possible explanation for this partial success may be that the prostacyclin infusion initially triggers a reactive thromboxane synthesis in blood vessels (Nigam et al. 1988), causing platelet aggregation in the microenvironment which facilitates the attachment of the tumour cell to the vascular surface. Large amounts of TXB_2 have thus been shown to be generated locally in injured vessels (Mehta and Roberts 1983). In contrast to non-specific TXA_2 antagonists, thromboxane receptor antagonists, e.g. BM 13 177 and SQ 29 548, prevented tumour growth. These antagonists occupy the receptors which are shared both by TXA_2 and by precursor cyclic endoperoxide. The potential advantage of these agents lies in the fact that no diversion of endoperoxides to other prostaglandins or decrease in synthesis of prostacyclin is feasible. The tumour-promoting effect of TXA_2 and precursor endoperoxides is thus eliminated. High TXA_2 concentrations in rat plasma, however, remain unaffected.

In conclusion, this study demonstrates that not only tumour cells but also tumour cell-induced or cell membrane-induced activation of tissue and blood cells in the rat stimulated the synthesis of eicosanoids, especially TXA_2, which are responsible for the tumour growth and metastasis. The predominant cells involved in eicosanoid synthesis are the mononuclear phagocytes that infiltrate the tumour. The tumour growth caused by TXA_2 is, however, abolished by pretreatment of animals with TXA_2-receptor antagonists but not with non-specific thromboxane synthetase inhibitors. Our experiments suggest that platelet-tumour cell-vessel wall interactions may occur in the presence of uninhibited TXA_2 and cyclic endoperoxides. In addition, the activation of mononuclear phagocytes infiltrating the tumour must be prevented to inhibit tumour growth.

Besides the role of prostanoids in tumour growth and metastasis, little information exists about the function of lipoxygenase products in tumour cell–platelet–endothelial cell interactions. Leukotriene C_4 has been shown to be involved in the development of breast cancer (Nigam and Pickartz 1985). Recently, 12-hydroperoxyeicosatetraenoic acid (12-HETE) has been shown to play a major role in tumour cell metastasis by regulating the surface expression of adhesive glycoproteins (K.V. Honn, this volume). These findings suggest that pharmacological manipulations of arachidonic acid metabolism with the help of selective inhibitors may prove useful in future studies to evaluate the roles of eicosanoids and lipoxygenase products in tumour growth and metastasis.

Acknowledgements. This work was generously supported by the Deutsche Forschungsgemeinschaft, Bonn (Ni 242/2-1) and the Association for Internatioional Cancer Research, UK. The authors wish to thank Prof. T.F. Slater, UK, for helpful discussions. In addition, the technical assistance of Barbara Steiger, Heike Wyneken, Sabine Ziedrich and Marion Kruppa is thankfully acknowledged.

References

Alam M, Jogee M, MacGregor WG, Dowdell JW, Elder MG, Myatt L (1982) Peripheral plasma immunoreactive 6-oxo-prostaglandin $F_{1\alpha}$ and gynaecological tumours. Br J Cancer 45:384–389

Berlinger NT (1984) Deficient immunity in head and neck cancer due to excessive monocyte production of prostaglandins. Laryngoscope 94:1407–1410

Bockman RS (1983) Prostaglandins in cancer: a review. Cancer Invest 74:485–493

Chiabrando C, Broggini M, Castagnoli MN, Donelli MG, Noseda A, Vivintainer M, Garattini S, Fanelli R (1985) Prostaglandin and thromboxane synthesis by Lewis lung carcinoma during growth. Cancer Res 45:3605–3608

Chiabrando C, Broggini M, Castelli MG, Cozzi E, Castagnoli MN, Donelli MG, Garattini S, Giavazzi R, Fanelli R (1987) Prostaglandin and thromboxane snthesis by M 5076 ovarian reticulosarcoma during growth: effects of a thromboxane synthetase inhibitor. Cancer Res 47:988–991

Drago JR, Al-Mondhiry JHB (1984) The effect of prostaglandin modulators on prostate tumour growth and metastasis. Anticancer Res 4:391–394

Goodwin JS, Ceuppens J (1983) Regulation of the immune response by prostaglandins. J Clin Immunol 3:295–315

Honn KV (1982) Prostacyclin/thromboxane ratios in tumour growth and metastasis. In: Powles TJ, Bockman RS, Honn KV, Ramwell PW (eds) Prostaglandins and cancer. Alan R Liss, New York, pp 733–752

Honn KV (1983) Inhibition of tumour cell metastasis by modulation of the vascular prostacyclin thromboxane A_2 system. Clin Exp Metastasis 1:103–114

Honn KV, Busse WD, Sloane BF (1983) Prostacyclin and thromboxanes: implications for their
role in tumour cell metastasis. Biochem Pharmacol 32:1–11

Honn KV, Onoda JM, Menter DG, Taylor JD, Sloane BF (1984) Prostacyclin/thromboxanes
and tumour cell metastasis. In: Honn KV, Sloane BF (eds) Hemostatic mechanisms and metastasis. Nijhoff, Boston, pp 207–231

Honn KV, Steinert BW, Moin K, Onoda JM, Taylor JD, Sloane BF (1987) The role of platelet
cyclooxygenase and lipoxygenase pathways in tumour cell-induced platelet aggregation. Biochem Biophys Res Commun 145:384–389

Khan O, Hensby CN, Williams G (1982) Prostacyclin in prostatic cancer: a better marker than
bone scan or serum acid phosphatase. Br J Urol 54:26–31

Kort WJ, Weijma IM, Bijma AM, van Schalkwijk WP, Vijlstra SJ, Westdroek DL (1987) Growth
of an implanted fibro-sarcoma in rats is associated with high level of plasma prostaglandin E_2
and thromboxane B_2. Prostaglandins, Leukotrienes Med 28:25–34

Levine L (1981) Arachidonic acid transformation and tumour production. Adv Cancer Res
35:49–79

Mehta J, Roberts A (1983) Human vascular tissues produce thromboxane as well as prostacyclin. Am J Physiol 244:R839–R844

Nigam S (1987) Extraction of eicosanoids from biological samples. In: Benedetto C, Mc Do-
nald-Gibson RG, Nigam S, Slater TF, Prostaglandins and Related Substances: A Practical
Approach. IRL Press, Oxford, pp 45–52

Nigam S (1988) Reactive thromboxane generation in the rat is triggered by prostacyclin infusion
(submitted)

Nigam S, Averdunk R (1988) Functional capacity of tumour infiltrating cells for synthesis of
different eicosanoids is a major factor in the tumour growth. Proc. 2nd Int. Conf. on Leukotr.
and Prostanoids in Health and Disease. Zor U, Cohen F (Eds). Karger Verlag (in press)

Nigam S, Pickartz H (1985) The role of leukotriene C_4 in the development of breast cancer.
International conference on prostanoids and leukotrienes in health and disease, Tel Aviv,
pp 23

Nigam S, Becker R, Rosendahl U, Hammerstein J, Benedetto C, Barbero M, Slater TF (1985)
The concentrations of 6-keto-$PGF_{1\alpha}$ and TXB_2 in plasma samples from patients with benign
and malignant tumours of the breast. Prostaglandins 29:513–528

Nigam S, Rosendahl U, Benedetto C (1987) Involvement of arachidonic acid metabolites in
tumour growth and metastasis/ role of prostacyclin and thromboxane in gynecologial and
gastrointestinal cancer of humans. In: Muszbek L (Ed) Hemostasis and cancer. CRC Press,
Boca Raton, pp 231–242

Russel SW, Doe WF, McIntosh AT (1977) Functional characterization of a stable non-cytolytic
stage of macrophage activation in tumours. J Exp Med 146:1511–1520

*Platelet-Activating Factor;
Cellular Response to Arachidonate
Metabolism*

The Role of Platelet-Activating Factor and Structurally Related Alkyl Phospholipids in Immune and Cytotoxic Processes

D. Hosford, J. M. Mencia-Huerta and P. Braquet

Institut Henri Beaufort Research Laboratories, 17 avenue Descartes,
92350 Le Plessis-Robinson, France

1 Introduction

Alkyl phospholipids are a diverse chemical species characterized structurally by the presence of an ether bond at the sn-1 position of the glycerol moiety. This ether linkage confers unique pharmacological properties upon this group of compounds and distinguishes them from other phospholipids which usually possess an ester linkage at this position. Alkyl phospholipids are now known to be more abundant in various cell types than previously thought and there is considerable evidence, that apart from their function as membrane components, certain of these compounds play a crucial role in the regulation of various biological processes.

Pharmacological interest in these compounds developed in the 1960s; firstly when Munder et al. (1966) initiated studies on phospholipid metabolism in macrophages and secondly, when an interaction was noted between leukocytes and platelets of immunized rabbits, in which it was postulated that an unidentified soluble mediator was produced by leukocytes. This mediator was demonstrated to trigger the release of histamine from platelets (Henson 1970; Siraganian and Osler 1971; Benveniste et al. 1972). As we shall see, the studies of Munder and colleagues led to the discovery of the tumoricidal properties of certain synthetic alkyl phospholipids, while in 1972 and later, Benveniste and colleagues reported the methodology for preparing the principle released from the antigen-stimulated leukocytes (Benveniste et al. 1972; Benveniste 1974 a,b). This latter group named the mediator platelet-activating factor (PAF) and established that the substance was lipidic in nature and was released from basophils by an IgE-dependent mechanism. Semi-synthesis of the compound was achieved in the late 1970s (Benveniste et al. 1979; Blank et al. 1979; Demopoulos et al. 1979) and its structure subsequently established as 1-O-alkyl-2(R)-acetyl-glycero-3-phosphocholine (Hanahan et al. 1980), an alkyl phospholipid sensitive to phospholipase A_2 (Benveniste et al. 1982). Confirmation of the structure was obtained by the total synthesis of 1-O-alkyl-2(R)acetyl-glycero-phosphocholine, a compound exhibiting all the biological activities of natural PAF (Godfroid et al. 1980).

It is now known that a diverse range of inflammatory cells are capable of PAF production including neutrophils, eosinophils, monocytes, macrophages, platelets and endothelial cells (Camussi et al. 1981; Chap et al. 1981; Braquet et al. 1987) and it appears that many of the cells or tissues generating the mediator are

Nigam et al. (Eds.), Eicosanoids,
Lipid Peroxidation and Cancer
© Springer-Verlag Berlin Heidelberg 1988

themselves targets of PAF-induced bioactions (Handley et al. 1984). The mediator is not stored in the cells per se, but originates from a precursor, 1-O-alkyl-2(R)acyl-glycero-3-phosphocholine (alkyl-acyl-GPC). It is released in two steps following activation of (a) a phospholipase A_2, which converts alkyl-acyl-GPC to lyso-PAF and (b) acetyltransferase which acetylates the lyso-compound into PAF (Ninio et al. 1982; Albert and Snyder 1983).

PAF is a potent mediator of anaphylaxis and inflammation and is also implicated in shock, graft rejection, renal disease, ovoimplantation and certain disorders of the central nervous system (CNS) (reviewed in Snyder 1985; Braquet et al. 1987). Although its precise role in these processes awaits clarification, considerable advances have been made following the recent discovery of specific PAF-receptor antagonists (reviewed in Braquet 1987). the advent of these compounds has allowed a more detailed investigation of binding sites and an improved analysis of the circumstances in which PAF mediates the observed pathophysiological effects.

While considerable attention has been focussed on PAF as a mediator of inflammation, the studies of Munder and colleagues (reviewed in Munder et al. 1981; Berdel 1982) have led to the recognition that other closely related alkyl phospholipids may play an important role in various pathological conditions. In 1966, Munder et al. published the first of a series of reports on stimulus-induced phospholipid metabolism by macrophages. These authors noted that after phagocytosis by this cell type, phospholipase A_2 was activated which degraded cellular phosphatidylcholine and phosphatidylethanolamine to the corresponding lyso-compounds, 2-lysophosphatidylcholine (2-LPC) and 2-lysophosphatidylethanolamine (2-LPE) with the concomitant release of free fatty acids (Munder et al. 1969; 1976). These workers hypothesised that the accumulation of lyso-compounds in cell membranes might represent an endogenous principle in the host defence mechanisms. While naturally occurring 2-LPC is rapidly metabolised by a lysophospholipase (Hill and Lands 1970), analogues of 2-LPC have been synthesised which are not substrates for this enzyme (Weltzien and Westphal 1967). These synthetic alkyl lysophospholipids (ALP) possess an increased half-life in the organism (Arnold et al. 1978) and have been shown to modulate the cellular immune response (Berdel et al. 1980).

In this review we will consider the role of PAF and the structurally related ALP in modulating the complex system of host defenses, in particular the direct effects of PAF on cellular interactions in the immune response and the tumoricidal activity of ALP.

2 Direct Effects of PAF on the Cellular Immune Response

2.1 Lymphocyte Proliferation, Cytokine Production and Surface Antigen Expression

When PAF was added to human peripheral blood lymphocyte cultures (containing 5%–10% monocytes) stimulated with the mitogens phytohaemagglutinin (PHA) or concanavalin A (Con A), a concentration-dependent inhibition of lymphocyte proliferation was observed (Rola-Pleszczynski et al. 1987a). The IC_{50} ranged between 10^{-12} and 10^{-10} M and inhibition was observed when PAF was added within the first 3 h of a 72-h culture period. Addition of PAF at a later time had no significant effect on lymphocyte proliferation. The inhibition was not accompanied by cell toxicity as assessed by the trypan blue exclusion method and a similar suppression of lymphocyte proliferation was observed with 2-ethoxy-PAF, a non-hydrolysable PAF analogue (Pignol et al. 1987a). PAF-induced inhibition of lymphocyte proliferation was prevented by the use of the PAF receptor antagonist, BN 52021. Interestingly, suppression was also reversed by the cyclo-oxygenase inhibitor, indomethacin, suggesting that prostaglandins may have been involved as second messengers in the inhibition of lymphocyte proliferation modulated by PAF (Rola-Pleszczynski et al. 1987a). A similar effect of PAF and 2-ethoxy-PAF was observed when production of interleukin 2 (IL 2) by human lymphocytes was measured. In both cases suppression of IL 2 production was reversed by BN 52021 (Pignol et al. 1987a; Rola-Pleszczynski et al. 1987a). Recently, Dulioust et al. (1987) have reported that the mediator also suppressed $CD4^+$ T cell proliferation. However, this effect was apparent only at higher concentrations of PAF or when the mediator was added late during the culture cycle. In addition the PAF effect was not associated with an alteration in IL 2 production.

Since inhibition of lymphocyte proliferation and IL 2 production may be due to the activiation of T-suppressor cells, lymphocytes were preincubated with PAF for 3–18 h, washed, then added to fresh autologous lymphocytes and the cells stimulated with a mitogen. This co-culture assay demonstrated that PAF activated suppressor cells which subsequently inhibited lymphocyte proliferation in the absence of any further contract with the mediator (Rola-Pleszczynski et al. 1987b). This induction of suppressor cells was accompanied by an increase in the $CD8^+$ T / $CD4^+$ T cell ratio after 18–48 h. The PAF antagonists BN 52021 and WEB-2086 also generated some suppressor cell activity, although to a lesser extent than PAF (Rola-Pleszcznski et al. 1987b). Interestingly, PAF-preincubated $CD4^+$ T cells markedly enhanced lymphocyte proliferation and this effect was not blocked by BN 52021. This result is presently unexplained.

In a different system, PAF and two non-hydrolysable PAF analogues increased the proliferation of IL 2-stimulated human lymphoblasts, whereas some antagonists (CV-3988 and L-652,-731, but not WEB 2086 and BN 52021) inhibited it (Barrett et al. 1986; Ward et al. 1987). This effect of the antagonists was observed even when the drugs were added 48 h after the beginning of a 72-h culture period suggesting interference with a late event in T cell activation. Furthermore, it is possible that endogenously produced PAF may be involved in some step(s) of IL 2-induced proliferation of T lymphoblasts. This is indicated by inhibition of this process by the PAF synthesis inhibitor, L-648, 611 (Barrett et al. 1986; Ward et al. 1987). An important corroboration of some of the in vitro observations outlined above has been provided by the in vivo instillation of PAF into rats via subcutaneously implanted osmotic minipumps connected to the jugular vein (Pignol et al. 1988a). After seven days of instillation, PAF-treated rat splenocytes showed enhanced IL 2 production in response to Con A.

2.2 Interaction of PAF with Interleukin 1 Production

Since accessory cells are essential for many T cell functions, the effect of PAF on interleukin 1 (IL 1) production by monocytes/macrophages has been studied. A biphasic response was observed when increasing concentrations of PAF were added to lipopolysaccharide (LPS)-stimulated human monocytes (Pignol et al. 1987b). Low concentrations of PAF (10^{-12} to 10^{-10} M) significantly enhanced IL 1 production while high concentrations (10^{-8} to 10^{-6} M) were markedly inhibitory. These effects were more pronounced when platelets were added (10–30:1) to the monocyte cultures. All these effects of PAF on the IL 1 production were blocked by the PAF receptor antagonist BN 52021. Under certain experimental conditions, PAF or the PAF analogs, PR 1501 and PR 1502, but not (S) PAF (the optical isomer of PAF), were individually able to induce IL 1 production from monocytes although to a lesser extent than in combination with LPS (Barrett et al. 1987).

The PAF/IL-1 interaction may also play a role in the immune injury of the arterial wall observed in shock, acute respiratory distress syndrome or sepsis. Indeed, PAF is a potent amplifier of platelet and leukocyte responses. At very low concentrations (10^{-16} to 10^{-11} M), it dramatically potentiates not only the release of IL 1 by monocytes/macrophages but also the production of leukotrienes (Chilton et al. 1982) and free radicals ($O_2^{\cdot-}$, OH$\cdot$ (Vercelloti et al. 1986) from stimulated polymorphonuclear cells.

Cell-mediated graft rejection is another area in which the relationship between PAF and various cytokines may be important (Reviewed in Braquet et al. 1987). An initial step in rejection is the accumulation of lymphocytes within the graft, which occurs during the first 3–5 days after transplantation. After this period, the increase in lymphocytes in the graft results mainly from in situ proliferation. In cell-mediated rejection, the lymphocyte requires two signals from the macrophage in order to proliferate, foreign antigen presentation and IL 1 production. Once stimulated, the lymphocytes produce IL 2 which is needed for clonal proliferation of helper and cytotoxic T cells. As macrophages are stimulated by PAF to synthesize IL 1, the PAF antagonists may have an effect on lymphocyte proliferation via modulation of this process. Such direct or indirect effect of the mediator on IL 1 and IL 2 production may explain the protective action of BN 52021 on graft rejection (Foegh et al. 1986).

PAF is released from renal allografts during hyperacute rejection and as stated above, BN 52021 increases cardiac allograft survival in rats, acting synergistically with azathioprine and cyclosporin A (CSA) (Foegh et al. 1986). Recently, Pirotzky et al. (1988) demonstrated that BN 52021 prevented CSA-induced nephrotoxicity without altering the immunosuppressive effect of the drug. In addition, Pignol et al. (1988b) have shown that treatment with BN 52063 (a ginkgolide mixture of BN 52020, BN 52021 and BN 52022, molar ratio 2:2:1) in combination with CSA increased the immunosupression induced by the latter drug. Both of these results suggested that PAF is generated by CSA. Whether the in vivo effect of BN 52021 or BN 52063 is due to the antagonism of PAF-enhanced cytotoxic cell activity, the generation of suppressor cells or the action of the PAF antagonists on other cellular or fluid-phase factors involved in graft rejection remains to be defined.

2.3 Lymphocyte Cytotoxicity

The precise role of PAF in the generation of cytotoxic lymphocytes in vitro is still unclear, although, studies with BN 52021 are leading to an improved understanding of this process. This PAF antagonist potentiated alloantigen recognition in primary and secondary mixed lymphocyte cultures and enhances the generation of cytotoxic lymphocytes in vitro (Gebhard et al. 1988). The presence of BN 52021 throughout the duration of primary and secondary mixed lymphocyte cultures had the greatest enhancing effect on the proliferative capacity of the cells, as measured by [^{3}H]thymidine incorporation. Addition of BN 52021 24 h or more after culture initiation reduced its enhancing effect on lymphocyte proliferation. The removal of BN 52021 from cultures up to 48 h after initiation eliminated the potentiating effect produced by this antagonist.

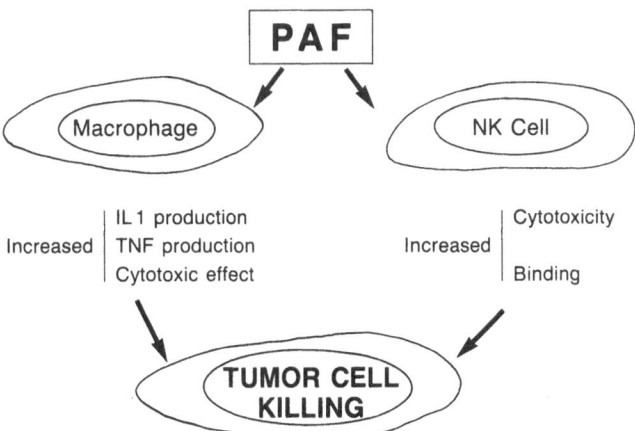

Fig. 1. Diagramatic representation of the effects of PAF on cell cytotoxicity

Similar effects were observed in the mixed cultures employed to generate cytotoxic T lymphocytes. Presence of BN 52021 during the entire 72-h culture period produced an enhanced level of cell mediated cytotoxicity, whereas the removal of the antagonist up to 24 h after culture initiation eliminated the effect.
Since the secondary mixed lymphocyte cultures and the bulk cultures used to
generate cytotoxic T lymphocytes already contained adequate IL 2 to support
the growth of the cells, the potentiating effect of BN 52021 on alloantigen recognition could not have been due to an enhanced production of this cytokine in
antagonist-treated cultures (Gebhard et al. 1988). The effects of PAF on lymphocyte cytotoxicity are summarized in Fig. 1.

2.4 Tumoricidal Processes

An important and pertinent area in which lymphocytes and various phagocytes
express their cytotoxicity is against tumour cells and thus recent studies have
examined the effect of PAF and PAF antagonist on cytotoxic effector cell functions. Natural killer (NK) cell-mediated lysis of the erythroleukemic target cell
line K562 was markedly enhanced by picomolar concentrations of PAF (Braquet
and Rola-Pleszczynski 1987). Preincubation of NK cells with PAF prior to culture with K562 target cells, or delayed addition of PAF also produced enhanced
NK activity, an effect that BN 52021 was able to block. In contrast, without PAF
addition, the NK-mediated cytotoxicity of rat and human lymphocytes against
YAC-1 or K562 target cells, respectively, was significantly diminished by BN
52021. Target cell lysis was reduced while the binding of target cells to effector
cells was unaffected. Furthermore, pretreatment of K562 cells with the PAF antagonist produces greater inhibition than pretreatment of the effector cells
(Mandi et al. 1988).
   Natural cytotoxic cells and macrophages mediate their cytotoxic activities, at
least in part, through synthesis and release of TNF. When preincubated, or co-
incubated with PAF, LPS-treated human monocytes and mouse peritoneal macrophages produced significantly higher quantities of TNF (M. Rola-Pleszczynski, personal commununication). In addition, large granular lymphocytes, which
comprise the effector NK cells responsible for lysis of K562 target cells, generate
PAF under certain conditions (Malavasi et al. 1986). Thus it appears that endogenous PAF may assist in the regulation of NK cell functions. These effects of
PAF on tumoricidal processes are summarized in Fig. 1.

3 ALP-Induced Tumour Necrosis

Like PAF, ALP show various biological activities including the capacity for
immunomodulation. They exhibit strong prophylactic and therapeutic effects on
the growth of animal tumours and thus may provide a new approach to cancer
therapy. In vitro these compounds have been shown to exert a cytostatic effect

on mouse methylcholanthrene- induced fibrosarcoma cells (Modollel et al. 1979), human leukemic cells (Andreesen et al. 1978; Tidwell et al. 1981) and on cells from various solid human malignomas such as hypernephronas (Berdel et al. 1981a) and urological (Berdel et al. 1981b) and gynaecological (Runge et al. 1980) tumours. Apart from direct cytostatic activity, ALP are also able to enhance the tumoricidal capacity of macrophages (Berdel et al. 1980). Unlike most antitumoral agents ALP do not appear to have a direct effect on DNA synthesis or function and are nonmutagenic.

3.1 Direct Cytotoxic Effects of ALP

A number of derivatives of ALP have been synthesised, such as 1-hexadecylthio-2-methoxymethyl-*rac*-glycero-3-phosphocholine (BM-14-440) (Herrmann 1986), SRI 62-834 and analogues (Houlihan 1986), compound 1-octadecyl-2-methyl-*rac*-glycero-3-phosphocholine (ET-18-OCH$_3$) (Storme 1986), and various *rac*-(2-alkoxyalkyl)phosphocholines (Bonjouklian et al. 1986). The presence of acetamine or methoxy substituents at the *sn*-2 position of ALP enhances the selective antitumour activity analogue of PAF, ET-18-OCH$_3$, one of the most effective ALP, has been shown to abolish [^{3}H]thymidine incorporation in various tumour cells and cause cell death. This compound, which is strongly associated with surface and intracellular membranes, also prevents uptake of essential nutrients (e.g. choline, palmitic acid) by HL-60 cells (Snyder et al. 1987), inhibits sialyl transferase (Bador et al. 1983) and the phospholipid-sensitive, Ca^{2+}-dependent protein kinase (Helfman et al. 1983).

A characteristic feature of the cytotoxic action of ALP on tumour cells is that of membrane disruption. Scanning electron microscopy has shown that neoplastically transformed cells possess numerous microvilli, membrane ruffles and blebs indicative of an actively moving membrane and transient surface alterations (Winslow et al. 1978; Spence and Coates 1981). When human brain tumour cells were incubated for 24 h with ET-18-OCH$_3$, these features were reduced in size and number and changes in the adherent properties of these cells were also detected (Berdel et al. 1984). A decrease in surface activity and membrane discontinuity was also observed in ET-18'OCH$_3$-incubated oesophagus carcinoma cells (Maistry et al. 1980), while cells of an acute myelomonocytic leukemia suffered membrane rupture when treated with this compound (Munder et al. 1979).

These observations imply direct cytotoxic surface activity by ALP, however, there is a striking difference in cytotoxicity between the potent lysophospholipids possessing an ether linkage at the *sn* 1 position of the glycerol and those almost rendered inactive by the presence of an ester linkage; both these chemical species exhibit identical surface activity (Berdel et al. 1980). Furthermore, ALP are bound to albumin and lipoproteins and are only exchanged slowly with cellular phospholipids, producing an extremely low concentration of ALP at the cell surface (Munder et al. 1981). Both of these facts tend to invalidate the theory of direct surface activity as the mechanism of ALP cytotoxicity.

The initial explanation of direct ALP action stemmed from the presence of the ether linkage. Neoplastic tissues do not contain the O-alkyl-cleavage enzyme required for ALP degradation, while this enzyme is present in normal cells, enabling ALP metabolism. This enzymatic difference was originally thought to account for the selectivity of their action (Snyder and Wood 1969; Soodsma et al. 1970) and allow ether linked lipids to accumulate in tumour cells blocking vital pathways of the phospholipid metabolism. In addition, the severe membrane damage observed was attributed to these compounds acting as antimetabolites of the synthesis of phosphatidylcholine (Modollel et al. 1979). This mechanism of ALP-mediated tumour cell destruction was supported not only by the fact that the ester linked analogue, 2-LPC showed no cytotoxicity but also by data demonstrating that pretreatment of ET-18-OH with an O-alkyl-cleavage enzyme preparation abolished its cytotoxic effect (Berdel et al. (1983).

However, it appears that the mechanism of direct ALP action is more complex than initially thought and cannot be solely attributed to the enzymatic difference between normal and malignant cells. While it is certain that neoplastic cells adsorb more ALP than normal cells, it has also been shown that normal lymphoblasts which possess high levels of the cleavage enzyme are susceptible to ALP cytotoxicity (Andreesen et al. 1979). Furthermore, due to substrate specificity of the O-alkyl-cleavage enzyme, some ALP substituted at sn-2 position are not subject to degradation (Lee et al. 1981). Thus, the disruption of phospholipid metabolism may result from metabolic alterations caused by ALP interaction with acetyl transferase (Bador et al. 1983), protein kinase (Helfman et al. 1983) and/or formation of diacyl-phosphatidylcholine and 1-O-alkyl-glycerol (Fleer et al. 1987) rather than intracellular accumulation of ALP per se.

3.2 Immunomodulated Cell Necrosis

The antitumour activity of ALP also appears to be mediated by the generation of highly tumoricidal immunocompetent cells from the monocyte/macrophage lineage (Munder 1986). Intraperitoneal injection of 2-LPC or ALP in mice induces temporary ascites which contains 60%–70% macrophages, the remainder being lymphocytes and a few granulocytes (Bausert 1978). These peritoneal cells significantly protect the animal from metastasis of syngeneic 3-LL cells when reinjected intravenously after the excision of the primary tumour (Berdel et al. 1980). If macrophages are removed from the peritoneal cell suspension the remaining lymphocytes and granulocytes are shown to be ineffective. Furthermore, pure syngeneic bone marrow macrophages treated in vitro with ALP afford significantly higher protection against metastasis in mice than untreated cells (Berdel et al. 1980). Macrophages can also be activated by incubation with ALP to induce in vitro necrosis of 3-LL syngeneic tumour cells (Berdel et al. 1979), methylcholanthrene-induced sarcomas (Munder et al. 1976) and cells from human hypernephromas (Berdel et al. 1981a) and lung carcinomas (Berdel et al. 1980). It is interesting to note that only alkyl-analogues are capable of generating macrophage cytotoxicity. The efficacy of ALP in this process is positively correlated with the number of carbon atoms in the alkyl chain at the sn-1 position, while removal of the substituant at the sn-2 position renders the compound virtually ineffective (Munder et al. 1981). The process of macrophage activation is still unclear (Munder 1986), but a stimulation of IL 1 production by ALP is possible (Pignol et al. 1987a). Differences in the enzymatic composition of normal and tumour cells as described above may also be important.

Finally, it is interesting to note that the uptake of PAF or its methoxy analogue by HL-60 cells is not inhibited by the PAF antagonists, BN 52021 and kadsurenone (Record et al. 1986) suggesting that the uptake of ether lipids is not mediated by PAF receptors. Indeed, PAF and related agonists are poor antitumour agents and an inverse relationship appears to exist between alkyl phospholipids exhibiting PAF activity (i.e. those being hypotensive, inflammatory and involved in allergic reactions) and those possessing selective antitumour properties. The hypothetical actions of ALP on tumour necrosis are presented in Fig. 2.

4 Conclusion

The involvement of PAF and related alkyl phospholipids in different pathological conditions is now apparent, however, their precise role in such processes awaits clarification. It is evident from the data reviewed here that PAF exerts various effects either directly or indirectly on many organ and tissue targets. In addition to directly modulating the immune response, PAF can also stimulate production of other mediators which, in turn, are able to regulate lymphocyte functions. These include cationic proteins and other suppressive factors released from activated eosinophils, and leukotrienes, prostaglandins and neuropeptides. Since the initial studies linking PAF production with allergic reactions, the number of processes in which the mediator may be operative has broadened to include shock, graft rejection, renal pathology and CNS disorders as well as the

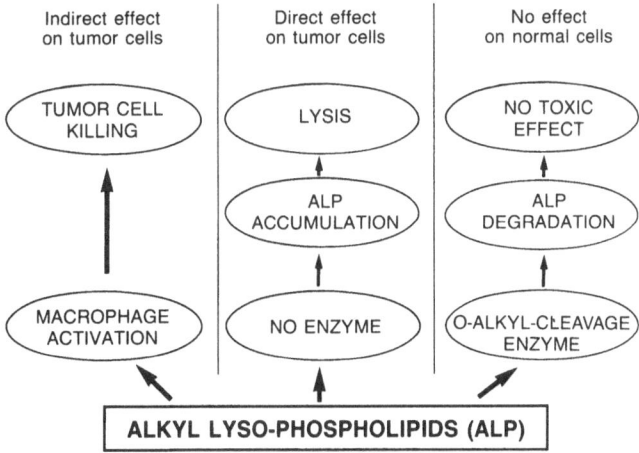

Fig. 2. Diagramatic representation of the effects of ALP on tumour necrosis

cellular immune response described here. The advent of specific PAF agonists, antagonists and synthesis-inhibitors has enhanced the prospects of defining the role of PAF in these phenomena. Structurally related PAF analogues, the ALP seem promising anticancer agents, emphasizing the potentially diverse biological activities of alkyl phospholipids. Whether other individual classes of phospholipids play specific regulatory roles other than their function as membrane components remains an intriguing question. Certainly, unravelling the biological properties of these compounds offers a unique and exciting challenge to pharmacological research.

References

Albert DH, Snyder F (1983) Biosynthesis of 1-alkyl-2-acetyl-*sn*-glycero-3-phosphocholine (platelet-activating factor) from 1-alkyl-2-acyl-*sn*-glycero-3-phosphocholine by rat alveolar macrophages. Phospholipase A_2 and acetyltransferase activitgies during phagocytosis and ionophore stimulation. J Biol Chem 25:97–102

Andreesen R, Modolell M, Weltzien HU, Eibl H, Common HH, Löhr GW, Munder PG (1978) Selective destruction of human leukemic cells by alkyl-lysophospholipids. Cancer Res 38:3894–3899

Andreesen R, Modolell M, Weltzien HU, Munder PG (1979) Alkyl-lysophospholipid induced suppression of human lymphocyte response to mitogens and selective killing of lymphoblasts. Immunobiology 156:498–508

Arnold B, Reuther R, Weltzien HU (1978) Distribution and metabolism of synthetic alkyl analogs of lysophosphatidylcholine in mice. Biochim Biophys Acta 530:47–55

Bador H, Morelis R, Louisot P (1983) Biochemical evidence for the role of alkyl-lysophospholipids on liver sialyltransferase. Int J Biochem 15:1137–1142

Barrett ML, Lewis DP, Ward S, Westwick J (1986) Platelet-activating factor modulates interleukin-2-induced proliferation of human T-lymphoblasts. Br J Pharmacol 89:505P

Barrett ML, Lewis GP, Ward S, Westwick J (1987) Platelet-activating factor induces Interleukin 1 production from human adherent macrophages. Br J Pharmacol 90:113P

Bausert W (1978) Der Einfluß von synthetischen Lysolecithin Analoga auf das Wachstum experimenteller Tumoren der Maus. PhD Thesis, Universita of Freiburg

Benveniste J (1974a) Characteristics and semi purification of platelet-activating factor from human and rabbit leukocytes. Fed Proc 33:797

Benveniste J (1974b) Platelet-activating factor, a new mediator of anaphylaxis and immune complex deposition from rabbit and human basophils. Nature 249:581–582

Benveniste J, Henson PM, Cochrane CG (1972) Leukocyte-dependent histamine release from rabbit platelets: the role of IgE, basophils and a platelet-activating factor. J Exp Med 136:1356–1377

Benveniste J, Tencé M, Varenne P, Bidault J, Boullet C, Polonsky J (1979) Semi-synthesis and proposed structure of platelet-activating factor (PAF): PAF-acether and alkyl ether analog of lysophosphatidylcholine. C R Acad Sci [D] 289:1037–1040

Benveniste J, Chignard M, Le Couedic JP, Vargaftig BB (1982) Biosynthesis of platelet-activating factor (PAF-acether). Il Involvement of phospholipase A_2 in the formation of PAF-acether and lyso-PAF-acether from rabbit platelets. Thromb Res 25:375–385

Berdel WE (1982) Antineoplastic activity of synthetic lysophospholipid analogs. Blut 44:71–78

Berdel WE, Fink U, Munder PG (1979) Synthetic lysophospholipids render macrophages cytotoxic. 5 Meeting of the International Society of Haematology, European and African Division

Berdel WE, Bausert WR, Weltzien HU, Modolell ML, Widmann KH, Munder PG (1980) The influence of alkyl-lysophospholipids and lysophospholipid-activated macrophages on the development of metastasis of Lewis lung carcinoma. Eur J Cancer 16:1199–1204

Berdel WE, Funk U, Egger B, Reichert A, Munder PG, Rastetter J (1981a) Alkyl-lysophospholipids inhibit the growth of hypernephroid carconomas in vitro. J Cancer Res Clin Oncol 101:325–330

Berdel WE, Funk U, Egger B, Reichert A, Munder PG, Rastetter J (1981b) Inhibition by alkyllysophospholipids of tritiated thymidine uptake in cells of human malignant urologic tumors. INCI 66:813–817

Berdel WE, Greiner E, Fink U, Stavrou D, Reichert A, Rastetter J, Hoffman DR, Snyder (1983) Cytotoxicity of alkyl-lysophospholipid derivates and low-alkyl-cleavage enzyme activities in rat brain tumor cells. Cancer Res 43:541–545

Berdel WE, Greiner E, Fink U, Zänker KS, Stavrou D, Trappe A, Fahlbusch R, et al. (1984) Cytotoxic effects of alkyl-lysophospholipids in human brain tumor cells. Oncology 41:140–145

Blank ML, Snyder F, Byers LW, Brooks B, Muirhead EE (1979) Antihypertensive activity of an alkyl ether analog of phosphatidylcholine. Biochem Biophys Res Commun 90:1194–1200

Bonjouklian R, Phillips ML, Kuhler KM, Grindey GB, Poore GA, Schultz RM, Altom MG (1986) Studies of the antitumor activity of (2-alkoxyalkyl)- and (2-alkoxyalkeny)-phosphocholines. J Med Chem 29:2472–2477

Braquet P (1987) The Ginkgolides: potent platelet-activating factor antagonists insolated from *Ginkgo biloba* L: Chemistry, pharmacology and clinical applications. Drugs Future 12:643–699

Braquet P, Rola-Pleszczynski M (1987) Platelet-activating factor and cellular immune responses. Immunol Today 8:345–352

Braquet P, Touqui L, Shen TY, Vargaftig BB (1987) Perspectives in platelet-activating factor research. Pharmacol Rev 39:97–145

Camussi G, Aglietta M, Coda R, Bussolino F, Piacibello W, Tetta C (1981) Release of platelet-activating factor (PAF) and histamine II. The cellular origin of human PAF: monocytes, polymorphonuclear neutrophils and basophils. Immunology 42:191

Chap H, Mauco G, Simon MF, Benveniste J, Douste-Blazy L (1981) Biosynthetic labelling of platelet-activating factor from radioactive acetate by stimulated platelets. Nature 289:312–314

Chilton FH, O'Flaherty JT, Walsh CE, Thomas MJ, Wykle RL, DeChatelet LR, Waite BM (1982) Platelet-activating factor stimulation of the lipoxygenase pathway in polymorphonuclear leukocytes by 1-O-alkyl-2-O-acetyl-*sn*-glycero-3-phosphocholine. J Biol Chem 257:5402–5407

Demopoulos CA, Pinckard RN, Hanahan DJ (1979) Platelet-activating factor. Evidence for 1-O-alkyl-2-acetyl-*sn*-glycerol-3-phosphorylcholine as the active component (a new class of lipid chemical mediators). J Biol Chem 254:9355–9358

Dulioust A, Vivier E, Salem P, Derynckz S, Benveniste J, Thomas Y (1987) Inhibition of human T4$^+$ cell proliferation by PAF-acether (platelet-activating factor). Fed Proc 46:2731

Fleer EAM, Unger C, Kim DJ, Eibl H (1987) Metabolism of etherlipids: a cause of cell death in leukemic cells? Proceedings of the 1st International Symposium on Ether-Lipids in Oncology, Gottingen

Foegh ML, Khirabadi BS, Rowles JR, Braquet P, Ramwell PW (1986) Prolongation of cardiac allograft survival with BN 52021, a specific antagonist of platelet-activating factor. Transplantation 42:86–88

Gebhardt PB, Bazan HEP, Braquet P, Bazan NG (1988) Platelet-activating factor suppresses cell mediated immune reactions in vivo. In: Braquet P (ed) New trends in lipid mediator research, vol 1. Karger, Basel (in press)

Godfroid JJ, Heymans F, Michel E, Redeuilh C, Steiner E, Benveniste J (1980) Platelet-activating factor (PAF-acether): Total synthesis of 1-O-octadecyl-2-O-acethyl-*sn*-glycero-3-phosphocholine. FEBS Lett 116:161–164

Hanahan DJ, Demopoulos CA, Liehr J, Pnckard RN (1980) Identification of platelet-activating factor isolated from rabbit. J Biol Chem 255:5514–5516

Handley DA, Arbeeny CM, Lee ML, van Valen RG, Saunders RN (1984) Efect of platelet-activating factor on endothelial permeability to plasma macromolecules. Immunopharmacology 8:137–142

Helfman D, Barnes K, Rinkade J, Vogler W, Shojl M, Kuo J (1983) Phospholipid-sensitive Ca^{2+}-dependent protein phosphorylation system in various types of leukemic cells from human patients and in human leukemic cell lines HL60 and K562, and its inhibition by alkyl-lysophospholipids. Cancer Res 43:2955–2961

Henson PM (1970) Release of vasoactive amines from rabbit platelets induced by sensitized mononuclear leukocytes and antigen. J Exp Med 131:287–306

Herrmann DBJ (1986) Preclinical studies with the thio ether phospholipid analogues BM 41440. 2nd International Conference on Platelet-Activating Factor and Structurally Related Alkyl Ether Lipids, Gatlinburg

Hill EE, Lands WEM (1970) Phospholipid metabolism. In: Wakil SJ (ed) Lipid metabolism. Academic, New York, p 185

Hoffman DR, Stanley JD, Berchtold R, Snyder F (1984) Cytotoxicity of ether-linked phytanyl phospholipid analogs and related derivatives in human HL-60 leukemia cells and polymorphonuclear neutrophils. Res Commun Chem Pathol Pharmacol 44:293

Houlihan WJ (1986) Antitumour activity of SRI 62-834, a cyclic ether analog of ET 18-OCH. Proceedings of the 1st International Symposium on Ether Lipids in Oncology, Gottingen

Lee RC, Blank ML, Fitzgerald V, Snyder F (1981) Substrate specificity in the biocleavage of the O-alkyl bond: 1-alkyl-2-acetyl-sn-glycero-3-phosphocholine (a hypotensive and platelet-activating lipid) and its metabolites. Arch Biochem Biophys 208:353–357

Maistry L, Robinson KM, Evers P, Munder PG, Andreesen R (1980) Morphologic effects of an antitumor agent on human esophagel carcinoma cells in vitro. Scan Electron Microsc 3:109–114

Malavasi F, Tetta C, Funaro A, Bellone G, Ferrero E, Franzone AC, Dellabona P, et al. (1986) Fc receptor triggering induces expression of surface activation antigens and release of platelet-activating factor in large granular lymphocytes. Proc Natl Acad Sci USA 83:2443–2447

Mandi Y, Farkas G, Koltai M, Braquet P, Beladi L (1988) The effect of BN 52021, a PAF-acether antagonist, on natural killer activity. In: Braquet P (ed) New trends in lipid mediator research, vol 1. Karger, Basel (in press)

Modollel M, Andreesen R, Pahlke W, Brugger U, Munder PG (1979) Disturbance of phospholipid metabolism during the selective destruction of tumor cells induced by alkyl-lysophospholipids. Cancer Res 39:4681–4686

Munder PG (1986) Studies on the antitumoral activity of alkyl-lysophospholipids (Abstr). 2nd International Conference on Platelet-Activating Factor and Structurally Related Alkyl Ether Lipids, Batlinburg, Tennessee

Munder PG, Modolell M, Ferber E, Fischer H (1966) Phospholipide in quarzgeschädigten Makrophagen. Biochemistry 344:310

Munder PG, Ferber E, Modolell M, Fischer H (1969) The influence of various adjuvants on the metabolism of phospholipids in macrophages. Int Arch Allergy 36:117

Munder PG, Weltzien HU, Modolell M (1976) Lysolecithin analogs: a new class of immunopotentiators. In: Miescher PA (ed) 7th International symposium on immunopathology. Schwabe, Basel, pp 411–424

Munder PG, Modolell M, Andreesen R, Weltzien HU, Westphal O (1979) Lysophosphatidylcholine (lysolecithin) and its synthetic analogues, immune modulating and other biologic effects. Springer Semin Immunopathol 2:187–203

Munder PG, Modolell M, Bausert W, Oettgen HF, Westphal O (1981) Alkyl-lysophospholipids in cancer therapy. In: Hersh EM (ed) Augmenting agents in cancer therapy. Raven, New York, pp 411–458

Ninio E, Mencia-Huerta JM, Heymans F, Benveniste J (1982) Biosynthesis of platelet-activating factor. I. Evidence for an acetyltransferase activity in murine macrophages. Biochim Biophys Acta 710:23–31

Pignol B, Henane S, Mencia-Huerta JM, Braquet P, Rola-Pleszczynski M (1987a) Platelet-activating factor (PAF-acether) inhibits interleukin 2 (IL 2) production and proliferation of human lymphocytes (Abstr). International Congress of Pharmacology, Sydney

Pignol B, Henane S, Mencia-Huerta JM, Rola-Pleszczynski M, Braquet P (1987b) Efect of PAF-acether (platelet-activating factor) and its specific antagonist, BN 52021, on interleukin 1 (IL 1) synthesis and release by rat monocytes. Prostaglandins 33:931–939

Pignol B, Henane S, Sorlin B (1988a) Effect of long-term in vivo treatment with platelet-activating factor on interleukin 1 and interleukin 2 production by rat splenocytes. In: Braquet P (ed) New trends in lipid mediator research, vol 1. Karger, Basel, (in press)

Pignol B, Henane S, Pirotzky E, Mencia-Huerta JM, Braquet P (1988b) Potentiation of immunosuppressive action of cyclosporine A by platelet-activating factor antagonists: an approach of the mechanism of action of these drugs in graft rejection. Transplant Proc, (in press)

Pirotzky E, Colliez P, Guilmard C, Schaeverbeke J, Braquet P (1988) Cyclosporine-induced nephrotoxicity: preventive effect of a PAF-acether antagonist, BN 52063. Transplant Proc, (in press)

Record M, Wagner M, Snyder F (1986) A kinetic study of the uptake and subcellular distribution of an antitumor PAF-analog ([³H]alkyl-2-methoxy-GPC) in HL-60 cells (Abstr). 2nd International Conference on Platelet-Activating Factor and Structurally Related Akyl Ether Lipids, Gatlinburg

Rola-Pleszczynski M, Pignol B, Pouliot C, Braquet P (1987a) Inhibition of human lymphocyte proliferation and interleukin 2 production by platelet-activating factor (PAF-acether): reversal by a specific antagonist, BN 52021. Biochem Biophys Res Commun 142:754–760

Rola-Pleszczynski M, Pouliot C, Pignol B, Braquet P (1987b) Platelet-activating factor induces human suppressor cell activity. Fed Proc 46:743

Runge MH, Andreesen R, Pfleiderer A, Munder PG (1980) Destruction of human solid tumors by alkyl-lysophospholipids. JNCI 64:1301–1306

Sirastetganian RP, Osler AG (1971) Destruction of rabbit platelets in the allergic response of sensitized leukocytes. J Immunol 106:1244–1251

Snyder F (1985) Chemical and biochemical aspects of platelet-activating factor: a novel class of acethyl ether-linked choline-phospholipids. Med Res Rev 5:107–140

Snyder F, Wood R (1969) Alkyl and alkyl-l-enyl ethers of glycerol in lipids from normal and neoplastic human tissues. Cancer Res 29:251–257

Snyder F, Record M, Smith Z, Blank ML, Hoffman DR (1987) Selective cytotoxic action of ether-lipid analogs of PAF: mechanistic studies related to their metabolism. Subcellular localization and effects on cellular transport systems. Aktuel Onkol, (in press)

Soodsma JF, Plantadosi C, Snyder F (1970) The biocleavage of alkyl glyceryl ethers in Morris hepatomas and other transplantable neoplasms. Cancer Res 30:309–311

Spence AM, Coates PW (1981) Scanning and transmission electron microscopy of cloned rat astrocytoma cells treated with dibutyryl cyclic AMP in vitro. J Cancer Res Clin Oncol 100:51–58

Storme G (1986) ET-18-OCH₃, an alkyl-lysophospholipid (ALP), and other ether lipid derivates have various anti-invasive effects in different cell lines. Proceedings of the 1st International Symposium on Ether Lipids in Oncology, Gottingen

Tidwell T, Guzman G, Volgler WR (1981) The effects of alkyl-lysophospholipids on leukemic cell lines. I. Differential action on two human leukemic cell lines, HL 60 and K 562. Blood 57:794–797

Vercelloti GM, Huh PW, Yin HQ, Nelson RD, Jacob HS (1986) Enhancement of PMN-mediated endothelial damage by platelet-activating factor (PAF): PAF primes PMN responses to activating stimuli. 28th Annual Meeting of the American Society of Hematology

Ward SG, Lewis GP, Westwick J (1987) A role of platelet-activating factor (PAF) in human T lymphocyte proliferation. In: Paubert-Braquet M, Braquet P, Demling R, Fletcher R, Foegh M (eds) Lipid mediators in immunology of burn and sepsis

Weltzien HU, Westphal O (1967) Synthesen von Cholinphosphatiden. IV. O-methylierte und O-acetylierte Lysolecithne. Justus Liebigs Ann Chem 709:240–243

Winslow DP, Roscoe JP, Rowles PM (1978) Changes in surface morphology associated with ethylnitrosurea-induced malignant transformation of cultured rat brain cells studies by scanning electron microscopy. Br J Exp Pathol 59:530–539

Platelet-Activating Factor in Renal Pathophysiology: The Effects of PAF Antagonists

G. E. Plante and P. Sirois

Departments of Physiology, Pharmacology, and Medicine, University of Sherbrooke, Sherbrooke, Québec J1H 5N4, Canada

1 Introduction

Platelet-activating factor (PAF) is one of the most recently described potential mediators of acute or chronic renal disorders (for review Plante and Hebert 1988; Braquet et al. 1987). This finding may be of critical importance, not only for understanding new pathophysiologic pathways in the development of renal diseases, but also, and perhaps more importantly, for the eventual use of PAF antagonists in the prevention and treatment of such diseases (Braquet 1987).

In a variety of normal experimental animals, PAF exerts potent systemic and renal physiological actions, in particular on renal hemodynamics (blood flow and glomerular filtration) and transport of fluid and electrolytes (Hebert et al. 1987; Plante et al. 1986). Most of these actions, and certainly the hemodynamic adaptation that follows PAF injection or infusion, appear to be mediated by a cascade of autacoids, including prostaglandins and leukotrienes (Plante et al. 1986; Hebert et al. 1988).

2 PAF in the Regulation of Body Fluid Volumes

PAF is one of the most potent endogenous compounds capable of shifting plasma from the intravascular to the interstitial compartments (Hwang et al. 1985). Therefore, its potential role in the development of abnormal distribution of body fluids, such as peripheral or localized oedema, as well as arterial hypertension, a peculiar form of fluid and sodium retention (DeWardener and Clarkson 1985), has to be carefully examined.

Nigam et al. (Eds.), Eicosanoids,
Lipid Peroxidation and Cancer
© Springer-Verlag Berlin Heidelberg 1988

2.1 Extra-renal Movements of Fluid and Solutes

Because of its well known effect on vascular permeability (Hwang et al. 1985), PAF may be involved in the control of fluid and electrolyte movements across capillaries, from intravascular towards interstitial space, and vice-versa. It appears, however, that the PAF-induced extravasation of plasma proteins is different from one microcirculation network to the other, suggesting specific organ responses to the glycerophospholipid (Sirois et al. 1988). Using Evans Blue diffusion into a variety of organs, in control and PAF-treated rats (intravenous bolus injections of 0.8 μg/kg), Sirois and co-workers were able to demonstrate an heterogeneous distribution of this marker in different tissues. Evans Blue concentration (μg/g dry weight tissue) increased 3- and 8-fold, respectively, in the duodenum and the pancreas of PAF-injected animals, as compared to control rats. In most other internal organs (heart, lungs, liver, spleen and kidneys), PAF failed to alter significantly the extravasation of Evans Blue. In a different model (Plante et al. 1988a), we documented an effect of PAF (administered in the abdominal cavity) on the clearance of fluid and solutes by the peritoneal membrane: fluid, urea and phosphate removal in 1.5% dextrose peritoneal solutions increased by 13%, 33%, and 70%, respectively. These results indicate that PAF may be involved in controlling the movement of fluid and solutes from the intravascular to other body compartments, in a relatively specific manner.

2.2 Renal Movements of Fluid and Solutes

We demonstrated that intravenous bolus injections of PAF are associated with dose-dependent reduction of urine flow and sodium excretion in the anesthetized dog (Hebert et al. 1987) and rat (Plante et al. 1988a). Previous investigators have attributed these changes to the profound reduction of renal blood flow and glomerular filtration which occur following PAF (Besin et al. 1983; Scherf et al. 1986). In our studies, however, it was possible to dissociate the renal hemodynamic from the tubular effects of PAF. It is likely, therefore, that PAF exerts direct tubular actions, resulting in enhanced movement of fluid and solutes from the tubular lumen towards the peritubular capillary. Since the kidney is capable of producing PAF (Pirotzky et al. 1984a, b), these results indicate that the glycerophospholipid could participate in the regulation of fluid and solutes transport across the tubular epithelium in the normal or diseased kidney.

2.3 PAF in Arterial Hypertension

A number of investigators (for review see DeWardener and Clarkson 1985) consider that essential or renal-associated arterial hypertension could be due to abnormal fluid balance or distribution between body compartments. Because of the potential roles of PAF in regulating fluid movements across several microcirculation networks, as well as fluid transfer in the gastrointestinal tract and the kidney, it is of interest to look at the eventual role of this endogenous compound on blood pressure. We examined the influence of BN 52021, a PAF-receptor antagonist (Braquet et al. 1985), on blood pressure and renal parameters of "uremic" rats (Plante et al. 1988b). Using the Bricker model to induce a chronic reduction of renal mass (20% of normal glomerular filtration), we demonstrated that BN 52021 administered daily for 4 weeks by gastric tube (3 mg/kg) significantly attenuated (196 ± 5 mmHg) the rise of systolic pressure expected to occur under such conditions in untreated animals (236 ± 15 mmHg). Although several other dysfunctions, due to a variety of other potential mediators, are expected in this complex model, the fact that PAF antagonism resulted in significant blood pressure differences is interesting, with respect to the potential role of this compound in the pathophysiology of kidney-mediated arterial hypertension.

3 PAF in the Development of Renal Disorders

In addition to the fact that isolated kidneys (Pirotzky et al. 1984a), and renal cells (Pirotzky et al. 1984b) are capable of PAF synthesis and degradation, the recent demonstration of specific effects of this endogenous coumpound on the contraction of cultured mesangial cells (Schlondorff et al. 1984), as well as the in vivo actions of PAF on renal function parameters (Hebert et al. 1987; Plante et al. 1986), both suggest that the glycerophospholipid could also play a role in renal pathophysiology, and therefore, justifies appropriate exploration.

3.1 Glomerular Immune Injury

Several pieces of experimental evidence support the critical role of PAF in mediating renal immune injury (Camussi et al. 1982, 1984). This effect can either be direct, through stimulation of filtration barrier (Farquhar 1975), or indirect, through activation of circulating blood cells (leukocytes, platelets). Increased permeability to plasma proteins secondary to loss of glomerular anionic charges (Camussi et al. 1984) has been documented following PAF injection.

3.2 Hyperfiltration and Progressive Renal Failure

The role of hyperfiltration of the remnant "intact" nephrons in initiating the development of progressive loss of renal function has been extensively studied over the past years (for review see Hostetter et al. 1981). Increased glomerular permeability to plasma proteins appears to be one of the initial critical alteration which leads to glomerular sclerosis. Since PAF has been shown to alter glomerular permeability, we found of interest to examine the influence of PAF antagonism on functional adaptation of the remnant nephrons, using again the Bricker model. BN 52021 was administered daily (3 mg/kg during four weeks) by gastric tube to conscious rats previously subjected to ⅘ amputation of the renal mass (20% of normal glomerular filtration). As compared to similarly treated control animals receiving only the vehicle, BN 52021-treated rats exhibited less absolute (0.5 ± 0.1 versus 1.2 μEq/min) and fractional (0.9 ± 0.1 versus 2.2%) urinary sodium excretion. This finding points to an eventual role of PAF in the adaptation of the remnant kidney vis-à-vis the maintenance of sodium balance, a mostly important function, especially in the diseased kidney (Bricker et al. 1978).

3.3 Ureteral Obstruction

Acute or chronic, unilateral or bilateral ureteral obstruction is associated with important physiological disturbances, in part mediated by profound changes in the local production of vasoactive mediators (Harris and Gill 1981). The role of PAF in initiating the renal production of prostaglandins has been recently proposed in a unilateral ureteral obstruction model (Weisman et al. 1985). It is believed that the local production and release of vasodilatory prostaglandins is involved in the adaptation of glomerular ultrafiltration determinants following elevation of intratubular pressure (Harris and Gill 1981). In the same manner, it is likely that renal prostaglandins are also involved in the diuretic and natriuretic response, which also characterizes the release of ureteral obstruction. These observations represent additional evidence supporting the critical role of PAF which appears to trigger the renal humoral cascade involved in several other pathophysiological conditions (Hebert et al. 1987; Plante et al. 1986; Brezis et al. 1984).

3.4 Renal Vein Obstruction

Renal vein obstruction represents a relatively frequent complication of the nephrotic syndrome (Trew et al. 1978), presumably because of the hypercoagulation state that characterizes this condition (Thomson et al. 1974). The renal dysfunctions seen in renal vein thrombosis include a mild decrease of filtration rate, associated with more profound reduction of urine volume and sodium excretion (Jobin et al. 1978). These abnormalities can be reproduced almost identically by intrarenal infusion or bolus injections of PAF (Hebert et al. 1987). Although indirect, these observations suggest that PAF may also be involved in the pathophysiology of renal vein obstruction.

3.5 Transplant Rejection

Recent studies indicate that PAF is released during acute allograft rejection (Ito et al. 1984) Treatment of animal cardiac allograft recipients with BN 52021 significantly reduces the incidence and severity of rejection, even when the PAF antagonist is administered as the sole immunosuppressive agent (Foegh et al. 1986). This finding is of considerable interest, not only for understanding the pathophysiology of graft rejection, but also, for the eventual use of PAF antagonists in the management of organ transplantation.

3.6 Renal Ischemia

The pathophysiology of acute renal failure has not yet been entirely established, in particular, the role played by humoral substances possibly responsible for the disorganized recovery pattern that follows renal ischemia (Brezis et al. 1984). The potential role of PAF in the pathophysiology of acute renal failure due to ischemic injury, and the eventual protective effect of PAF antagonist have not yet been examined. We recently studied in a "two-kidney-one-renal-artery-occlusion" anesthetized rat model, the effect of BN 52021 on the recovery pattern of renal function following 30 minutes of complete occlusion of the left renal artery (Plante et al. 1988c). Eighty minutes after the release of such an occlusion, left kidney glomerular filtration only recovers to 38% of control values in normal untreated animals, whereas in BN 52021-treated rats, this renal parameter recovers to 69% of pre-ischemic calues. In addition, contralateral renal adaptation to left kidney ischemia differs in control and BN 52021-treated rats. In the latter animals, right kidney glomerular filtration increases by 30% above control values, as opposed to control rats, where this parameter even decreases slightly. These results suggest that PAF is involved in the pathophysiology of renal failure in this experimental model. Furthermore, the protective effect of BN 52021 demonstrated in this study, may prove to be of clinical interest in the prevention and treatment of acute renal failure, as well as in the preservation of organs for transplantation.

4 Conclusion

The results available in the current litterature related to PAF, suggest, unanimously, that the glycerophospholipid first identified by Benveniste et al. (1972) plays a major role in regulating the vascular permeability of several critical organs, including the gastrointestinal tract, the lungs, the kidneys. It becomes therefore of importance to include PAF in future studies dealing with body fluid volume disturbances (extracellular and intracellular oedema), arterial hypertension, pathophysiology of immune renal injury, transplant rejection, and finally, renal ischemic aggression and organ transplantation.

References

Benveniste J, Menson PM, Cochrane CG (1972) Leukocyte-dependent histamine release from rabbit platelets. The role of IgE, basophils and a platelet activating factor. J Exp Med 136:1356

Bessin P, Bonnet J, Apffel D, Soulard C, Desgroux L, Pelas I, Benveniste J (1983) Acute circulatory collapse caused by platelet-activating factor (PAF-Acether) in dogs. Eur J Pharmacol 68:403

Braquet P (1987) The Gingkolides: potent platelet-activating factor antagonists isolated from Gingko biloba L.: Chemistry, pharmacology and clinical applications. Drugs Future 12:643

Braquet P, Spinnewyn B, Braquet M, Bourgain RH, Taylor JE, Etienne A, Drieu K (1985) BN 52021 and related compounds: a new series of highly specific PAF-acether receptor antagonists isolated from Gingko biloba. Blood Vessels 16:559

Braquet P, Touqui L. Shen TY, Vargaftig BB (1987) Perspectives in platelet-activating factor research. Pharmacol Rev 39:97

Brezis M, Rosen S, Silva P, Epstein FH (1984) Renal ischemia: a new perspective. Kidney Int 26:375

Bricker NS, Fine LG, Kaplan M, Epstein M, Bourgoignie JJ, Light A (1978) "Magnification phenomenon" in chronic renal disease. N Engl J Med 299:1287

Camussi G, Tetta C, Derigitus C, Bussolino F, Segolomi G, Vercellone A (1982) Platelet-activating factor (PAF) in experimentally-induced rabbit acute serum sickness: role of basophil-derived PAF in immune complex deposition. J Immunol 128:86

Camussi G, Tetta C, Coda R, Segoloni GP, Vercellone A (1984) Platelet-activating factor-induced loss of glomerular anionic charges. Kidney Int 25:73

DeWardener HE, Clarkson EM (1985) Concept of natriuretic hormone. Physiol Rev 65:658

Farquhar MG (1975) The primary glomerular filtration barrier: basement membrane or epithelial slits. Kidney Int 8:197

Foegh ML, Khirabadi BS, Rowles JR, Braquet P, Ramwell P (1986) Prolongation of cardiac allograft survival with BN 52021, a specific antagonist of platelet-activating factor. Transplantation 42:86

Harris RH, Gill JM (1981) Changes in glomerular filtration rate during complete ureteral obstruction in rats. Kidney Int 19:603

Hebert RL, Sirois P, Braquet P, Plante GE (1987) Hemodynamic effects of PAF-acether on the dog kidney. Prostaglandins Leukotrienes Med 26:189

Hebert RL, Sirois P. Plante GE (1988) Inhibition of PAF-induced renal hemodanymic and tubular dysfunctions with L 655240, a new thromboxane/prostaglandin endoperoxide antagonist. Prostaglandins, (in press)

Hostetter TH, Olson JL, Rennke HG, Venkatachalam MA, Brenner BM (1981) Hyperfiltration in remnant nephrons: a potentially adverse response to renal ablation. Am J Physiol 241:F85

Hwang SB, Li CL, Lam MH, Shen TY (1985) Characterization of cutaneous vascular permeability induced by platelet-activating factor in guinea pigs and rats, and its inhibition by a platelet-activating factor receptor antagonist. Lab Invest 52:617

Ito S, Camussi G, Tetta C, Milgrom F, Andres G (1984) Hyperacute renal allograft rejection in the rabbit: the role of platelet-activating factor and of cationic proteins derived from polymorphonuclear leukocytes and from platelets. Lab Invest 51:148

Jobin J, Hemmings R, Plante GE (1978) Effect of renal vein pressure on urinary sodium and hydrogen excretion. Can J Physiol Pharmacol 56:30

Pirotzky E, Bidault J, Burtin C, Gubler MC, Benveniste J (1984a) Release of platelet-activating factor, slow reacting substance, and vasoactive species from isolated rat kidneys. Kidney Int 25:404

Pirotzky E, Ninio E, Bidault J, Pfister P, Benveniste J (1984b) Biosynthesis of platelet-activating factor. VI. Precursor of platelet-activating factor and acetyl transferase activity in isolated rat kidney cells. Lab Invest 51:567

Plante GE, Hebert RL (1988) Platelet-activating factor in renal physiology and pathophysiology. Interest of the Gingkolides. In: Biology and Clinical Perspectives Braquet P, Gingkolides (ed). Telesymposia Proceedings, (in press)

Plante GE, Hebert RL, Lamoureux C, Braquet P (1986) Hemodynamic effects of PAF-acether. Pharmacol Res Commun Suppl 18:173

Plante GE, Prevost C, Braquet P, Sirois P (1988a) Increased peritoneal permeability with PAF-acether. Proceedings of the 8th National Conference CAPD, (in press)

Plante GE, Lussier YA, Chainey A, Boisclair L, Sirois P, Braquet P (1988b) BN 52021 reduces the development of hypertension and prevents hypernatriuresis in "uremic rats" Prostaglandins Leukotrienes Med (in press)

Plante Ge, Sirois P, Braquet P (1988c) Platelet-activating factor antagonism with BN 52021 protects the kidney against acute ischemic injury. Prostaglandins leukotrienes Med (in press)

Scherf H, Nies AS, Schwertschlag U, Hughes M, Gerber JG (1986) Hemodynamic effects of platelet-activating factor on the dog kidney in vivo. Hypertension 9:82

Schlondorff D, Satriano JA, Hagege J, Perez J Baud L (1984) Effect of platelet-activating factor and serum treated zymosan on prostaglandin E2 synthesis, arachidonic acid release and contraction of cultured rat mesangial cells. J Clin Invest 73:1227

Sirois P, Sirois M, Jancar S, Braquet P. PLante GE (1988) PAF increases vascular permeability in selected tissues: effect of BN 52021 and L-655240. In: There is a case for PAF-acether antagonists. Paris

Thomson C, Forbes CD, Prentice CRM, Kennedy AC (1974) Changes in blood coagulation and fibrinolysis in the nephrotic syndrome. Q J Med 43:399

Trew PA, Biava CG, Jacobs RP, Hopper J (1978) Renal vein thrombosis in membranous glomerulopathy. Incidence and association. Medicine (Baltimore) 57:69

Weisman SM, Felsen D, Vaughan ED (1985) Platelet-activating factor is a potent stimulus for renal prostaglandin synthesis: possible significance in unilateral ureteral ligation. J Pharmacol Exp Ther 235:10

Interaction of Macrophages and Mast Cells in the Production of Prostaglandins and Leukotrienes

B. A. Jakschik, Y. Wei and L. F. Owens

Department of Pharmacology, Washington University School of Medicine, 660 South Euclid Avenue, St. Louis, Missouri 63110, USA

1 Introduction

It is well established that macrophages are very important in host defence, including the elimination of tumour cells. Macrophages contribute to host defence by phagocytosis, cytotoxic activity and immune regulation (Allison et al. 1978; Nathan et al. 1980). Mast cells are best characterized for their role in immediate hypersensitivity reactions. However, they also have other functions in host defence. They have been shown to phagocytose, though less efficiently than macrophages (Otani et al. 1982), and kill tumour cells (Tharp et al. 1986). With certain types of tumours, the mast cell number is greatly increased (Burtin 1986). The functional significance of mast cells in tumour control has been confirmed by the observation that the tumour incidence is increased in mast cell deficient mice (Burtin 1986). Mast cells also seem to be important in the initiation of inflammation, and certain tumours are associated with intense inflammatory reactions. We have observed that the influx of polymorphonuclear leukocytes in inflammation is markedly reduced in mast cell deficient mice (Qureshi and Jakschik 1987). Mast cells as well as macrophages are positioned in the tissue where noxious material may enter. They have the capacity to initiate processes which could eliminate unwanted material or cells.

Both mast cells and macrophages avidly metabolize arachidonic acid to prostaglandins (PG), thromboxane (TX) and leukotrienes (LT). Certain prostaglandins are thought to modulate cell multiplication and differentiation as well as the immune response (Goodwin 1981). Thromboxane, a potent vasoconstrictor, will decrease blood flow and, therefore, may hinder metastasis of tumours. The leukotrienes also have potent biological actions which may affect tumour growth and spreading. LTB_4 especially seems to modulate immune functions at very low concentrations. LTB_4 is not only a potent chemotactic factor (Ford-Hutchinson et al. 1980), but has also been shown to augment the killer function of lymphocytes (Rola-Pleszczynski 1985; Rola-Pleszczynski and Gagnon 1986), enhance the production of interleukins 1 and 2 and interferon, as well as to induce suppressor lymphocytes (Farrar and Humes 1985; Rola-Pleszczynski 1985).

In this study we have investigated the modulation of eicosanoid release as macrophages and mast cells interact with each other in IgE-antigen mediated reactions and in phagocytosis of zymosan.

Nigam et al. (Eds.), Eicosanoids,
Lipid Peroxidation and Cancer
© Springer-Verlag Berlin Heidelberg 1988

2 Interaction of Mast Cells and Macrophages in Immediate Hypersensitivity Reactions

Mast cells activity in IgE-antigen mediated reaction has been studied extensively. A great variety of eicosonoids are produced during this process. None of the types of mast cells (connective tissue or mucosal) studied has the capacity to generate all the different eicosanoids released in anaphylaxis. These findings suggest that not only mast cells but also other cell types may participate in this reaction. Macrophages can be found in tissues in close proximity to mast cells (Baggiolini et al. 1982; Jakschik et al. 1987a). Therefore, we investigated a possible interaction of mast cells and macrophages in immediate hypersensitivity reactions. Mouse bone marrow-derived mast cells were sensitized with monoclonal anti-dinitrophenyl (DNP) IgE and challenged with DNP-BSA alone or in the presence of macrophages. When stimulated alone mast cells synthesized LTC_4, LTB_4, TXB_2 and PGD_2 (Jakschik et al. 1987a; Wei et al. 1986). Stimulation in the presence of resident peritoneal macrophages also produced PGE_2

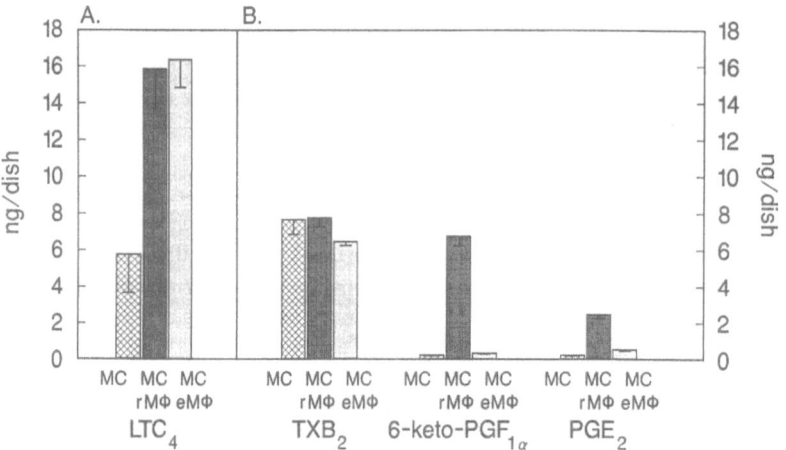

Fig. 1A, B. Mast cell-macrophage interaction in eicosanoid production in immediate hypersensitivity. Mouse (BALB/c) bone marrow derived mast cells *(MC)* were cultured as described (Wei et al 1986), passively sensitized with monoclonal anti-DNP IgE and then stimulated (0.5×10^6 cells/well) with DNP-BSA, 10 mg/ml, in the presence or absence of mouse resident *(r)* or elicited *(e)* peritoneal macrophages *(Mϕ)*. The macrophages (0.5×10^6/well) were purified by allowing them to attach for 2 h to plastic wells and then incubated overnight. The supernatant was analyzed by specific radioimmunoassay for leukotrienes, thromboxane and prostaglandins. The data represent mean ± SEM, $n=3$. **A** 2 min after challenge. **B** 1 h after challenge. This was repeated four times with similar results

and 6-keto-PGF$_{1\alpha}$ (Fig. 1). Insignificant amounts of these additional prostaglandins were observed when elicited macrophages were used instead of the resident ones (Jakschik et al. 1987b). It has been reported that eicosanoid production by elicited peritoneal macrophages is markedly reduced as compared to resident macrophages (Humes et al. 1980). These data suggest that the resident macrophages contributed PGE$_2$ and 6-keto-PGF$_{1\alpha}$. Macrophages also enhanced leukotriene production early in the time course (2 min) (Fig. 1). This was probably due to to an enhancement of the 5-lipoxygenase activity in the mast cells and not due to production of leukotrienes by macrophages. This conclusion was drawn because resident and elicited macrophages had a similar effect (Jakschik et al. 1987b; Wei et al. 1986). The granule release reaction, measured by hexosaminidase release, was not altered by the presence of resident or elicited macrophages.

The contribution of macrophages to eicosanoid production was further tested by prelabelling either mast cells or macrophages with [^{3}H]arachidonic acid. When mast cells were prelabelled, the labelled eicosanoids released after stimulation in the presence or absence of macrophages included leukotrienes, TXB$_2$ and PGD$_2$. Therefore, the same arachidonic acid metabolites contained label independent of the presence or absence of macrophages. When unlabelled mast cells were stimulated in the presence of labelled macrophages, label from arachidonic acid was found in PGE$_2$, 6-keto-PGF$_{1\alpha}$ and TXB$_2$, but not in leukotrienes or PGD$_2$. Therefore, the labelling experiments confirm that macrophages are stimulated by the mast cells or their products to release arachidonic acid and convert it to prostaglandins but not leukotrienes. Very little if any arachidonic acid was shuttled from mast cells to macrophages or vice versa (Jakschik et al. 1987a).

In order to investigate the mechanism by which macrophages are stimulated during immediate hypersensitivity reactions, three questions were addressed: (a) Does a soluble factor released by mast cells stimulate macrophages? (b) Does a direct mast cell-macrophage interaction involving the IgE-antigen complex on the mast cells take place? (c) Do the granules released from mast cells stimulate macrophages? In order to address the first question supernatants from antigen challenged mast cells were transferred to macrophages. These supernatants did not stimulate macrophages to produce prostaglandins or leukotrienes. Similar data were obtained when mast cells were added to Millicells which were placed in the wells containing macrophages (mast cells and macrophages separated by a 0.45 μm filter) and then stimulated (Jakschik et al. 1987a). Therefore, a soluble mast cell factor did not seem to be responsible for the stimulation of macrophages.

A direct mast cell-macrophage interaction via IgE-antigen bound to the mast cells was tested by using RBL-1 cells instead of mast cells. These cells bind IgE and antigen but do not release mediators upon antigen challenge. Addition of antigen to sensitized RBL-1 cells incubated together with macrophages did not cause any prostaglandin synthesis (Jakschik et al. 1987a) Therefore, a direct mast cell macrophage interaction was not the mechanism by which macrophages were stimulated.

It has been reported earlier that macrophages can phagocytose mast cell granules (Baggiolini et al. 1982), and phagocytosis stimulates macrophages to release eicosanoids (Humes et al. 1980; Rouzer et al. 1980; Scott et al. 1980). Therefore, the effect of mast cell granules on arachidonic acid metabolism by macrophages was investigated. Since granules of bone marrow mast cells are irregular in shape they are difficult to purify but granules from rat peritoneal mast cells can be readily isolated (Kruger et al. 1980). Therefore, the effect of granules from peritoneal mast cells was tested and compared to the effect of the 10 000 × g pellet of bone marrow mast cells. This pellet is enriched in granules and membranes. Resident macrophages were stimulated by granules from peritoneal mast cells to synthesize PGE_2 and 6-keto-$PGF_{1\alpha}$ in a dose dependent manner, but not leukotrienes. Similar results were obtained with the 10 000 × g pellet of bone marrow mast cells, but not when membrane fractions of either type of mast cells were used (Jakschick et al. 1987a). The finding that mast cell granules did not cause leukotriene release from macrophages was surprising. It has previously been thought that particulate stiumuli, such as zymosan, cause macrophages to produce large amounts of leukotrienes (Rouzer et al. 1980; Scott et al. 1980). Similar results were obtained with zymosan in our laboratory (Jakschik et al. 1987a; Wei et al. 1986). However soluble factors such as phorbol myristate acetate and lipopolysaccharide stimulate only prostaglandin release (Humes et al. 1982). Therefore, mast cell granules seem to act on macrophages in a different manner from zymosan, but similar to soluble factors. The arachidonic acid metabolites synthesized by macrophages upon addition of mast cell granules were the same as those contributed by macrophages during immediate hypersensitivity reactions. Therefore, it appears that macrophages take up mast cell granules and are stimulated by them to produce prostaglandins but not leukotrienes.

3 Mast Cells and Macrophages and Phagocytosis

Phagocytosis is an important aspect of host defence. Macrophages are very efficient in this process and synthesize large amounts of leukotrienes and cyclooxygenase products (Humes et al. 1980; Rouzer et al. 1980; Scott et al. 1980). Any contribution of eicosanoids by mast cells to this reaction has not been investigated until now. Therefore, macrophages or mast cells alone or together were exposed to zymosan, and eicosanoid production was measured by radioimmunoassay. With regard to cyclo-oxygenase products (Fig. 2), both macrophages and mast cells produced TXB_2. When the two cell types were stimulated together, the amount was additive. However, the amount of thromboxane generated by mast cells was less than after antigen challenge. Macrophages also released PGE_2 and 6-keto-$PGF_{1\alpha}$ and mast cells PGD_2 (Figs. 2, 3). In some experiments either mast cells or macrophages were prelabelled with [^{3}H]arachidonic acid and the cells were stimulated alone or together. Analysis by thin layer chromatography showed that arachidonic acid utilized in prostaglandin or thromboxane production was not shuttled from one cell type to the other when macrophages and mast cells were stimulated together (Fig. 3).

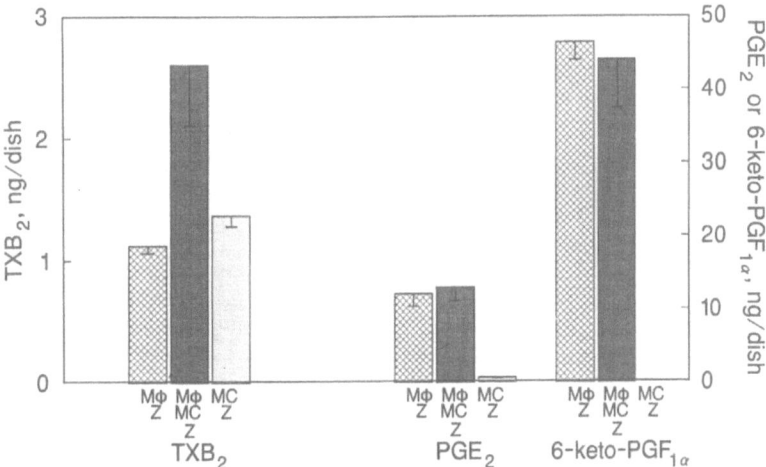

Fig. 2. Cyclo-oxygenase products synthesized by macrophages and mast cells during phagocytosis. Mouse resident peritoneal macrophages ($M\phi$, 0.7×10^6/well), purified by attachment (see legend of Fig. 1) and mouse bone marrow derived mast cells *(MC)* were stimulated with 100 µg/ml zymosan. The two cell types were exposed to zymosan alone and together for 2 h. The supernatants were examined by specific radioimmunoassay. The data represent mean ±SEM, $n=3$. This experiment was repeated three times with similar results

Zymosan challenge of macrophages alone or in the presence of mast cells also caused the release of LTC_4 and LTB_4. In some experiments (two out of four) the amount of LTC_4 and LTB_4 released when both cell types were stimulated together was greater than when macrophages were stimulated alone. However, when mast cells alone were exposed to zymosan no leukotriene production was observed (Fig. 4). Supernatants from incubations utilizing either prelabelled macrophages or mast cells were also examined by HPLC. [^{3}H]arachidonic acid was incorporated into LTC_4 either when prelabelled macrophages were stimulated alone or in the presence of unlabelled mast cells or when unlabelled macrophages were stimulated in the presence of labelled mast cells (Fig. 5). No radioactive peak co-migrated with LTC_4 when zymosan was added to mast cells alone. Apparently insufficient amounts of LTB_4 were formed to detect a distinct peak either by UV absorbance or radioactivity.

The finding that no radioactivy co-migrated with LTC_4 in mast cell supernatants after zymosan stimulation agrees well with the data obtained by radioimmunoassay. This confirms that mast cells do not synthesize leukotrienes under these conditions, but do release prostaglandins. It was rather surprising to find label originating from mast cells in LTC_4 generated when macrophages and mast cells together were incubated with zymosan. The mechanism of the contribution of [^{3}H]arachidonic acid under these conditions is not known at the present time. It is possible that macrophages or their products released during phagocytosis stimulate mast cells to synthesize LTC_4, or arachidonic acid from mast cells is shuttled to macrophages and converted by them to LTC_4. The latter, however, seems unlikely since little if any shuttling was observed in the analysis by thin

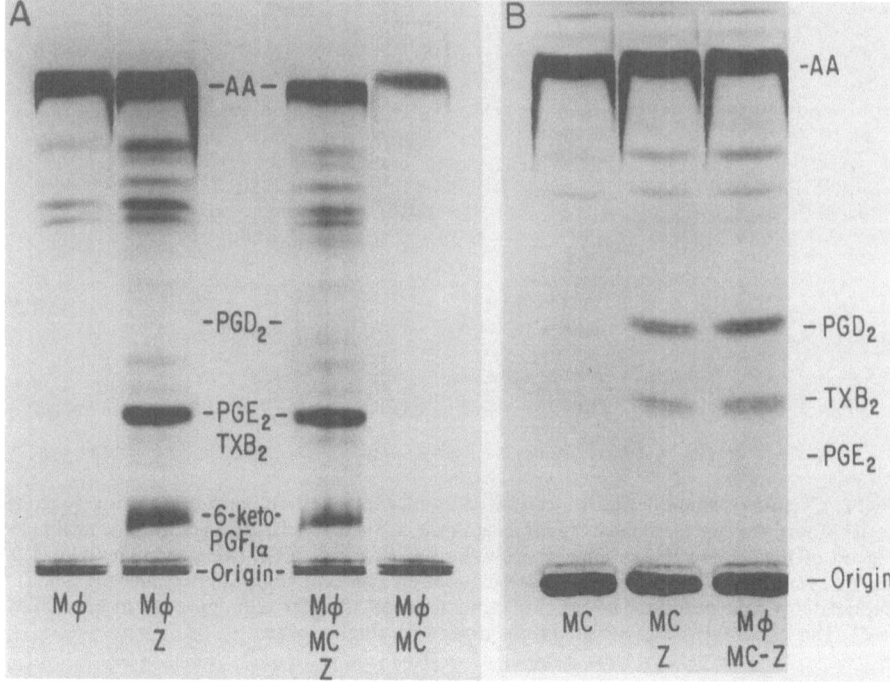

Fig. 3. Determination of cell source of released arachidonic acid. Mouse resident peritoneal macrophages *(Mφ)* and mast cells *(MC)* were prepared as decribed in the legend of Fig. 1. Mast cells (4 h) and macrophages (overnight) were incubated with 10 μCi/ml [³H]arachidonic acid (specific activity 240 Ci/mmol). Any arachidonic acid that was not taken up by the cells was washed away. Prelabelled macrophages (0.7×10^6/well) alone or in the presence of unlabelled mast cells (0.7×10^6/well) and prelabelled mast cells alone or in the presence of unlabelled macrophages were exposed to zymosan for 2 h. The supernatants were extracted with acetone: chloroform (pH 3.4). Thin-layer chromatography was performed in two solvent systems: benzene:dioxane:acetic acid (60:30:3) (BDA) which separates TXB_2 from other prostaglandins; and in the organic phase of ethyl acetate:2,2,4-trimethyl pentane:acetic acid:water (110:50:20:100) (A9) which separates 6-keto-$PGF_{1\alpha}$ from other prostaglandins. Products were identified by co-migration with standards. **A** Prelabelled macrophages, solvent system A9. **B** Prelabelled mast cells, solvent system BDA. This experiment was repeated with similar results

layer chromatography (Fig. 3). The contribution of LTC_4 by mast cells would agree with the observation that radioimmunoassayable LTC_4 was significantly increased in some experiments when macrophages and mast cells were stimulated together by zymosan as compared to macrophages alone. The variability may be due to metabolism of LTC_4. Experiments are in progress to further investigate the mechanism of LTC_4 formation under these circumstances.

The amount of LTC_4 synthesized by macrophages due to zymosan was comparable to that released by mast cells after antigen challenge. However, mast cells produced approximately 30 times as much LTB_4 in IgE-antigen mediated reactions as macrophages during phagocytosis of zymosan (Fig. 4). Therefore, mast cells may be of great importance in calling in polymorphonuclear leukocytes during inflammatory reaction and in other host defence functions modula-

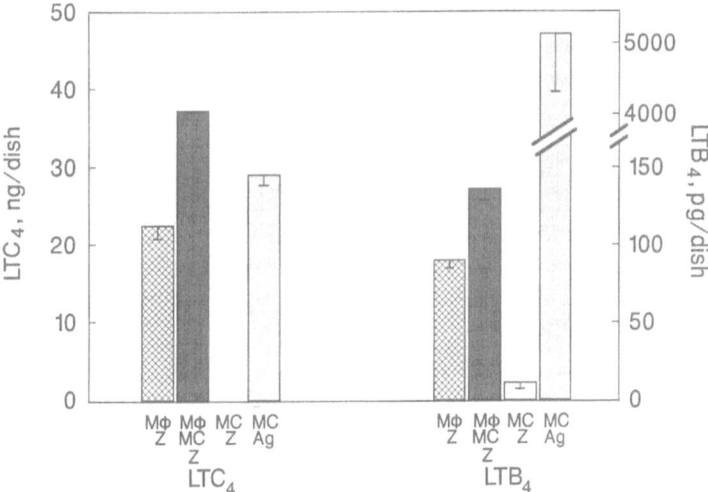

Fig. 4. Leukotriene synthesis during phagocytosis of zymosan. Macrophages *(M\phi)* and mast cells *(MC)* were prepared and stimulated as described in the legend of Fig. 2. The supernatants were examined by specific radioimmunoassay for leukotrienes. The data represent mean ± SEM, *n* = 3. This experiment was repeated three times. In two of these four experiments there was no enhancement of leukotriene formation when macrophages and mast cells were stimulated together as compared to macrophages alone

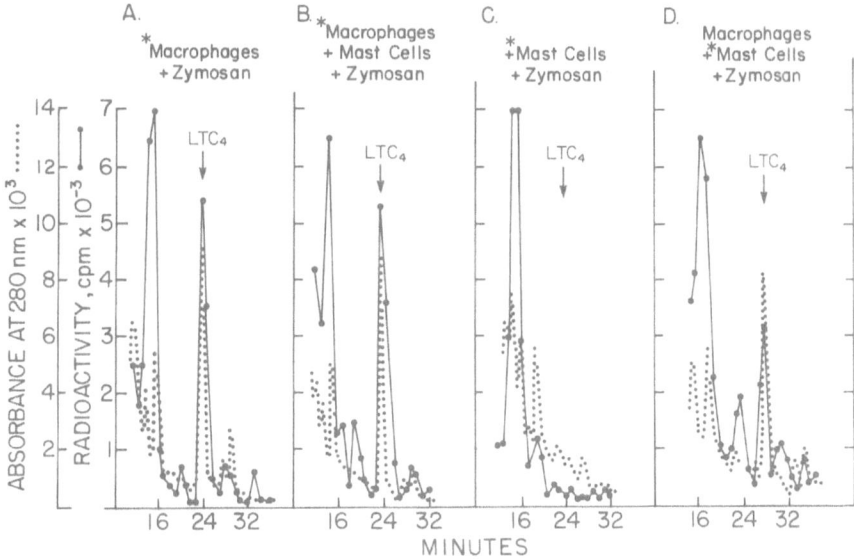

Fig. 5A–D. Incorporation of radioactivity from prelabelled macrophages or mast cells into LTC$_4$. Macrophages and mast cells were prepared, prelabelled and stimulated as described in the legend of Fig. 3. The supernatants from 12 wells were pooled and concentrated on ODS-C18 extraction columns and analyzed by reverse phase HPLC (Wei et al. 1986). Leukotrienes were identified by co-migration with standards. This experiment was repeated twice with similar results. **A, B** macrophages were prelabelled. **D, E** Mast cells were prelabelled

ted by LTB_4. We have shown that the influx of polymorphonuclear leukocytes is markedly reduced in mast cell deficient mice (Qureshi and Jakschik 1987). Mast cells are strategically located in tissue at possible entry points of noxious substances. They are found in the skin, the gut, and in the vicinity of blood vessels at the predominant site for vascular permeability changes for macromolecules and immigration of leukocytes from the blood vessel into the tissue. These cells may initiate an inflammatory response which would be beneficial by localizing and/ or removing the noxious agent. This process would be aided by the accumulation of leukocytes which are called in by LTB_4. This leukotriene would also be available to enhance the production of interleukins and interferon and the cytoxicity of lymphocytes.

4 Conclusion

The experiments discussed above demonstrate that mast cells and macrophages interact in immediate hypersensitivity reactions. Macrophages appear to take up the granules released by mast cells and to be stimulated by them to synthesize prostaglandins but not leukotrienes. This accounts, at least in part, for the observation that eicosanoids other than those synthesized by mast cells have been detected in anaphylactic reactions. During phagocytosis of zymosan eicosanoids are not only produced by macrophages but also mast cells. However, the latter release only TXB_2 and PGD_2 and not leukotrienes. The incorporation of $[^3H]$arachidonic acid from mast cells into LTC_4 when macrophages together with mast cells were exposed to zymosan, suggest an interaction between those two cell types also in this process. The nature of this interaction is not understood at the present time. The experiments clearly demonstrate that under certain biological conditions differential stimulation of eicosanoid production can occur. During phagocytosis of zymosan macrophages synthesize both prostaglandins and leukotrienes, but during the uptake of mast cell granules only prostaglandins. Similarly mast cells release leukotrienes and prostaglandins during IgE-antigen mediated reactions, but only prostaglandins when stimulated by zymosan.

Acknowledgement. This work was supported by NIH grant HL 31922.

References

Allison AC, Ferluga J, Prydz H, Schorlemmer HU (1978) The role of macrophage activation in chronic inflammation. Agents Action 8:27–35

Baggiolini M, Horisberger U, Ulrich M (1982) Phagocytosis of mast cell granules by mononuclear phagocytes, neutrophils and eosinophils during anaphylaxis. Int Arch Allergy Appl Immunol 67:219–226

Burtin C (1986) Mast cells and tumor growth. Ann Inst Pasteur Immunol. 137 D:289–294

Farrar WL, Humes JL (1985) The role of arachidonic acid metabolism in the activities of interleukin 1 and 2. J Immunol 135:1153–1159

Ford-Hutchinson AW, Bray M, Doig MV, Shipley ME, Smith MJ (1980) Leukotriene B, a potent chemokinetic and aggregating substance released from polymorphonuclear leukocytes. Nature 286:264–265

Goodwin JS (1981) Prostaglandins and host defence in cancer. Med Clin North Am 65:829–843

Humes JL, Burger S, Galavage M, Kuehl FA, Wrightman PD, Davies P, Bonney RJ (1980) The diminished production of arachidonic acid oxygenation products by elicited mouse peritoneal macrophages: possible mechanisms. J Immunol 124:2110–2116

Humes JL, Sadowski S, Galavage M, Goldberg M, Subers E, Bonney RJ, Kuehl FA Jr (1982) Evidence for two sources of arachidonic acid for oxidative metabolism by mouse peritoneal macrophages. J Biol Chem 257:1591–1594

Jakschik BA, Rengers TA, Pinski JR (1987a) Nature of the mast cell-macrophage interaction in immediate hypersensitivity. Adv Prostaglandin Thromboxane Leukotriene Res 17:180–185

Jakschik BA, WEi YF, Rengers TA (1987b) Modulation of eicosanoid production due to mast cell-macrophage interaction. Adv Prostaglandin Thromboxane Leukotriene Res 16:141–152

Kruger PG, Lagunoff D, Wan H (1980) Isolation of intact mast cell granules with intact membranes. Exp Cell Res 129:83–93

Nathan CF, Murray HW, Cohn ZA (1980) The macrophage as an effector cell. N Engl J Med 303:622–626

Otani I, Conrad HD, Carlo JR, Segal DM, Ruddy S (1982) Phagocytosis by rat peritoneal mast cells: independence of IgG Fc-mediated and C3-mediated signals. J Immunol 129:2109–2112

Qureshi R, Jakschik BA (1987) The role of mast cells in inflammation. Fed Proc 46:985

Rola-Pleszczynski M (1985) Immunoregulation by leukotrienes and other lipoxygenase metabolites. Immunol Today. 6:302–307

Rola-Pleszczynski M, Gagnon L (1986) Natural killer function modulated by leukotriene B_4: Mechanism of action. Transplant Proc [Suppl 4] 18:44–48

Rouzer AZ, Scott WA, Hamill AL, Cohn ZA (1980) Dynamics of leukotriene C_4 production by macrophages. J Exp Med 152:1236–1247

Scott WA, Zrike JM, Kempe J, Cohn ZA (1980) Regulation of arachidonic acid metabolites in macrophages. J Exp Med 152:324–335

Tharp M, Thiele D, Charley M, Sullivan T (1986) Studies on the mechanism of mast cell killing of tumor cells. Fed Proc 45:850

Wei Y, Heghinian K, Bell RL (1986) Contribution of macrophages to immediate hypersensitivity reaction. J Immunol 137:1993–2000

Prostaglandin E$_2$ and Thromboxane A$_2$ Production by Human Peritoneal Macrophages: Effect of Mitomycin C, Nitrogen Mustard and Cyclosporin A

M. L. Foegh[1] and J. Winchester[2]

1 Introduction

Eicosanoids are extensively implicated in experimental tumour growth and tumour metastasis (Powles et al. 1982). Macrophages are present in solid tumours and eicosanoids may exert local effects on cell growth and differentiation. These cells have the capacity for sustained release of eicosanoids (Feuerstein et al. 1981) which is probably due to the abundance of arachidonate in macrophage phospholipid. The possibility arises that metastatic compounds may exert some of their effects through an action on eicosanoid synthesis. Therefore, we studied the effect of mitocycin C and a nitrogen mustard (mechlorethamine) on the release of two important cyclo-oxygenase products, namely prostaglandin E$_2$ (PGE$_2$) and thromboxane A$_2$ (TXA$_2$) from human macrophages which were obtained from the peritoneal dialysate of patients with kidney failure (Maddox et al. 1984). We compared the effect of these two chemotherapeutic agents with that of the immunosuppresant drug cyclosporin A which is an inhibitor of lymphocyte proliferation. Both PGE$_2$ and TXA$_2$ were selected due to their opposing effects on the immune response, wherein PGE$_2$ inhibits and TXA$_2$ increases immunoreactivity (Foegh et al. 1984).

2 Methods

Human peritoneal macrophages were harvested from the waste dialysis bags of patients undergoing continuous ambulatory peritoneal dialysis, by the procedure previously described (Maddox et al. 1984; Foegh et al. 1983). The mononuclear cells were purified by discontinuous-density gradient contrifugation on 50% Percol. Cells obtained in this manner were >90% macrophages and <10% lymphocytes. The purified cells (10^6 cells/ml) were incubated in RPMI 1640 (Flow Laboratories, McLean, VA, USA) with or without the calcium ionophore A23187

[1] Department of Surgery, Division of Transplantation
[2] Department of Medicine, Division of Nephrology,
  Georgetown University Medical Center, Washington, DC 20007, USA

Nigam et al. (Eds.), Eicosanoids,
Lipid Peroxidation and Cancer
© Springer-Verlag Berlin Heidelberg 1988

(50 ng/ml; Calbiochem Boehringer, San Diego, CA, USA) at 37 °C for one hour. After incubation for 1 h the cyclooxygenase inhibitor indomethacin ($5.6 \times 10^-$ M)., was added to prevent further formation of PGE_2 and TXA_2. Three different concentrations of mitocycin C, mechlorethamine and cyclosporin A were added to the incubation media prior to stimulation of the cells with the calcium ionophore, A23187. All incubations were performed in duplicate. The RIA values were not different. The cell suspension was centrifuged at 40 °C for 10 min and the supernatant stored at -20 °C until determination by RIA of PGE_2 and TXB_2, which is the stable breakdown product of TXA_2. Cell viability was determined at the end of each experiment by trypan blue staining.

3 Results

Cyclosporin A increased thromboxane synthesis by human peritoneal macrophages in a concentration-dependent manner. The maximal concentration of cyclosporin used in these experiments increased TXB_2 three-fold, which is the amount elicited by A23187 alone. When A23187 and cyclosporin A were combined, a further increased release of TXB_2 was elicited which was six-fold the amount released in their absence Table 1. Thus, cyclosporin A clearly increases thromboxane synthesis from both unstimulated and activated peritoneal macrophages. However, cyclosporin A did not significantly modify PGE_2 synthesis. This may be due to the fact that synthesis was already elevated under the incubation conditions.

Mitomycin C, in contrast, clearly stimulated PGE_2 synthesis in a dose-dependent manner. This effect of mitocycin C on PGE_2 synthesis was also demonstrated when the macrophages were activated with A23187. However, the situation with respect to TXB_2 was quite different. In both the absence and presence of A23187 the synthesis of TXB_2 was inhibited in a concentration-dependent manner (Table 2).

Mechlorethamine at concentrations ≥ 20 µg/ml inhibited both PGE_2 and TXB_2 production by peritoneal macrophages whether the cells were stimulated

Table 1. Basal and calcium ionophore (A23187)-stimulated release of PGE_2 and TXB_2 from human peritoneal macrophages incubated with cyclosporin A (mean ± SEM)

Cyclosporin A (µg/ml)	PGE_2 − (pg/10^6 cells^{-1} h^{-1})	+ (pg/10^6 cells^{-1} h^{-1})	TXB_2 − (pg/10^6 cells^{-1} h^{-1})	+ (pg/10^6 cells^{-1} h^{-1})
0	938 ± 183	2224 ± 458	466 ± 168	2081 ± 370
1.2	984 ± 218	2529 ± 77	553 ± 205	2180 ± 233
12 10⁻	1258 ± 314	2767 ± 650	670 ± 262	2115 ± 399
120 10⁻	1305 ± 319	2895 ± 409	1404 ± 391	2940 ± 329

−, Without calcium ionophore A23187.
+, With calcium ionophore A23187.

Table 2. Basal and calcium ionophore (A23187)-stimulated release of PGE$_2$ and TXB$_2$ from human peritoneal macrophages incubated with mitomycin C (data indicates mean $\pm$ SEM)

Mitomycin C (µg/ml)	PGE$_2$ − (pg/10^6 cells^{-1}h^{-1})	+ (pg/10^6 cells^{-1}h^{-1})	TXB$_2$ − (pg/10^6 cells^{-1}h^{-1})	+ (pg/10^6 cells^{-1}h^{-1})
0	303 ± 97	890 ± 269	212 ± 20	1164 ± 132
20	392 ± 157	1321 ± 287	211 ± 5	538 ± 144
100	613 ± 170	1645 ± 579	118 ± 31	226 ± 60
500	1573 ± 422	1996 ± 430	28 ± 15	54 ± 30

−, Without calcium ionophore A23187.
+, With calcium ionophore A23187.

with A23187 or not (data not shown). The treated cells exhibited different degrees of staining with trypan blue, indicating the toxicity of mechlorethamine to the macrophages.

4 Discussion

Human peritoneal macrophages harvested from the waste dialysis bags of end-stage renal failure patients were used to evaluate the effect of three drugs commonly employed to inhibit proliferation. The three drugs were found to have markedly different effects. In terms of transplantation, the most clinically relevant data are clearly those which indicate that cyclosporin A increases TXA$_2$ synthesis, even when the macrophages are almost fully activated. The high basal values for PGE$_2$ precluded detection of a stimulatory effect of cyclosporin A. These high basal values for PGE$_2$ may reflect the status of the peritoneal dialysis patient from whom the macrophages were harvested for these experiments. However, cyclosporin A has been reported to increase PGE synthesis by human monocytes (Whisler et al. 1984), whereas Fan and Lewis (1985) found cyclosporin A to inhibit prostacyclin synthesis by macrophages. The high TXB$_2$ values speak to cyclosporin A nephrotoxicity which may be due to the effect of TXA$_2$ on the glomerulus. Thromboxane synthase inhibitors and receptor antagonists attenuate the fall in glomerular filtration rate elicited by high doses of this immunosuppressant drug in rats (Coffman et al. 1985). Platelet-activating factor (PAF) antagonists also protect against cyclosporin A toxicity (Pirotzky et al. 1987). PAF is known to release TXB$_2$ and the PAF antagonists reduce TXB$_2$ release in guinea pigs (Sirois et al. 1987). The manner whereby PAF antagonists attenuate cyclosporin A-induced nephrotoxicity is still unknown but may by preventing PAF-induced release of arachidonate.

    The situation, with respect to mitocycin C is quite different, for here the drug has the opposite effect and appears to act like a thromboxane synthase inhibitor. The analogy is even closer when the PGE$_2$ data are inspected since this is in accord with the notion that endoperoxide is shunted from the thromboxane pathway to PGE$_2$. Thromboxane synthase is a P450 enzyme, and the inhibitory effect of mitomycin C on thromboxane may be related to the redox effects of mitomycin C (Trush et al. 1982).

Mitomycin C increased the synthesis of PGE$_2$ but may inhibit prostacyclin as well as TXA$_2$ because prostacyclin synthase, like thromboxane synthase, is a P450 enzyme (Ullrich et al. 1981) and therefore may be susceptible to interference by mitomycin C. Further studies are indicated since the functionality of the gastric mucosa's microcirculation is dependent on prostacyclin. If the synthesis of prostacyclin and the perfusion of the mucosa is compromised by mitomycin then it is possible that the clinical effectiveness of mitocycin C in treating gastric carcinoma may be increased by co-administration of a prostacyclin analogue. This possibility has received support from Ichihashi et al. (1983) who found administration of PGE$_1$ in combination with mitomycin significantly suppressed the growth of Yoshida sacroma cells transplanted subcutaneously when compared with use of mitomycin only.

Mechlorethamine was unexpectedly toxic to these human peritoneal macrophages as shown by trypan blue staining.

In conclusion, cyclosporin A increased thromboxane synthesis and mitomycin increased PGE$_2$ synthesis by human peritoneal macrophages. The action of cyclosporin A may be to promote arachidonate release and if so one can anticipate increased synthesis of all the eicosanoids. The situation is different with respect to mitomycin since the possibility exists that the drug is affecting P450 enzyme mechanisms associated with thromboxane and prostacyclin. These findings indicate the need for further study since they suggest means of modifying the clinical effectiveness of mitomycin.

References

Coffman TM, Yarger WE, Klotman PE (1985) Functional role of thromboxane production by acutely rejecting renal allografts in rats. J Clin Invest 75:1242–1248

Fan T-PD, Lewis GP (1985) Mechanism of cyclosporin A induced inhibition of prostacyclin synthesis by macrophages. Prostaglandins 30:735–747

Feuerstein N, Foegh M, Ramwell PW (1981) Leukotrienes C$_4$ and D$_4$ induce prostaglandin and thromboxane release from rat peritoneal macrophages. Br J Pharmacol 72:389–391

Foegh ML, Maddox YT, Winchester J, Rakowski F, Schreiner G, Ramwell PW (1983) Prostaglandin and thromboxane release from human peritoneal macrophages. Adv Prostaglandin, Thromboxane Leukotriene Res 12:45–49

Foegh ML, Alijani MR, Helfrich GB, Khirabadi BS, Lim K, Ramwell PW (1986) Lipid mediators in organ transplantation. Transpl Proc 5 [Suppl 4]:20–24

Ichihashi H, Kuzuya H, Kondo T (1983) Enhancement of the cytocidal effect of mitomycin C by means of simultaneous administration of prostaglandin E$_1$. Gann 74:169–175

Maddox YT, Goegh ML, Zeligs B, Zmudka M, Bellanti J, Ramwell PW (1984) A routine source of human peritoneal macrophages. Scand J Immunol 19:23–29

Pirotzky E, Collier P, Guilmard C, Schaeuerbeke J, Mencia-Huerta J, Braquet P (1987) Protection of cyclosporin-induced nephrotoxicity by BN 52021 in the rat. In: Second International Congress on Cyclosporine, p. 40

Powles TJ, Bockman RS, Houn KV, Ramwell PW (eds) (1982) Prostaglandins and Cancer. Liss, New York

Sirois P, Harczy M, Braquet P, Borgeat P, Maclouf J, Pradelles P (1987) Inhibition of the release of prostaglandins, thromboxane and leukotrienes from anaphylactic lungs by a platelet activating factor (PAF) antagonist and its stimulation by PAF. Adv Prostaglandin Thromboxane Leukotriene Res 17:1033–1032

Trush MA, Mimnang LEG, Ginsburg E, Gram TE (1982) Studies on the in vitro interaction of mitomycin C, mitofurantoin and paraquat with pulmonary microsomes. Biochem Pharmacol 31:805–814

Whisler RL, Lindsey JA, Proctor KVW, Morisaki N, Cornwell D (1984) Characteristics of cyclosporine induction of increased prostaglandin levels from human peripheral blood monocytes. Transplantation 38:377–381

Ullrich V, Castle L, Weber P (1981) Spectral evidence in the cytochrome P450 nature of prostacyclin synthetase. Biochem Pharmacol 30:2033–2036

Uterine Lipoxygenase Pathways

A. Morgan[1], S. Flatman[1], R. Hammond[2], R. G. McDonald-Gibson[1], J. Osypiw[1], S. Tay[3] and T. F. Slater[1]

1 Introduction

Prostaglandins (PGs), produced via the cyclo-oxygenase pathway of arachidonic acid metabolism, have been implicated in reproductive physiology since they were first discovered in semen in the mid 1930s and it is now widely accepted that they play important roles in processes such as ovulation, luteinization, implantation and parturition (Poyser 1981). In recent years, however, the presence of active lipoxygenase pathways for polyunsaturated fatty acids such as arachidonic acid (AA) has also been demonstrated in both human and animal reproductive tissues (Morgan et al. 1987). The uterus of various species was shown to respond to lipoxygenase products (Weichman and Tucker 1982) and as with PGs, their formation appears to be hormone-dependent (Thaler-Dao et al. 1985; Morgan et al. 1986) and varies during the ovarian cycle (Morgan et al. 1984; Rees et al. 1987). Such data stimulate interest in the possible involvement of products of this pathway in uterine physiology/pathophysiology.

Human cervical tissue is active in metabolising arachidonic acid (Flatman et al. 1986) and an area of particular interest is the possible link between disturbances in eicosanoid formation and development of cervical carcinoma. Such an association was originally suggested by Slater following observations of significant quantitative differences between normal and abnormal cervical tissues in the presence of a prostaglandin-associated organic peroxyl radical (Slater and Cook 1970; Benedetto et al. 1981; Tomasi et al. 1984). Prostaglandins have long been thought to be involved in processes such as carcinogenesis, cell replication and differentiation and host-tumour interactions (Bockman 1983; Honn et al. 1983) whilst a link between lipoxygenase products and cancer has been the matter of recent investigations (Werner et al. 1985).

The purpose of this article is to outline some of our work on the 12-lipoxygenase pathway of human cervix and rat uterus.

[1] Department of Biology and Biochemistry, Brunel University, Uxbridge, Middlesex UB8 3PH, UK
[2] Department of Gynaecology, Hillingdon Hospital, Uxbridge, Middlesex UB8 3NN, UK
[3] Department of Pathology, Whittington Hospital, London N19 5NF, UK

Nigam et al. (Eds.), Eicosanoids,
Lipid Peroxidation and Cancer
© Springer-Verlag Berlin Heidelberg 1988

2 Lipoxygenase Activity of Human Uterine Cervix

2.1 Products Synthesised

Lipoxygenase activity in the human uterus was first reported by Saeed and Mitchell (1982a) who demonstrated in cervical tissue the synthesis of products cochromatographing with 12- and 5-HETE. The presence of a very active 12-lipoxygenase was also shown by our group following radio-TLC and RP-HPLC analysis of metabolites of arachidonic acid produced by homogenates of cervix tissue (Morgan et al. 1984; Flatman et al. 1988). Synthesis of 12-HETE was subsequently confirmed using GC-MS, in collaboration with Dr. S. Nigam of the Klinikum Steglitz, Free University, Berlin. The cervical lipoxygenase activity appears to be primarily localised in the squamous epithelium and uppermost stroma of the ectocervix, with comparatively low levels in the endometrium and myometrium (Morgan et al. 1984) where the major and minor products were reported to be 12- and 5-HETE respectively (Demers et al. 1984).

2.2 Cellular Origin of Cervical Lipoxygenase

Recent studies by our group have been directed at determining the cellular origin of the cervical lipoxygenase. Cells present in cervical scrapes were incubated with [^{14}C]AA and the products extracted and analysed by TLC and RP-HPLC using procedures previously reported (Flatman et al. 1986). There was a great variation between samples in the percentage of [^{14}C]AA conversion into products. In all cases, however, the formation of the major metabolite was inhibited by nordihydroquaiaretic acid and it co-migrated with 12-HETE in a number of different TLC solvent systems (Fig. 1a). The synthesis of 12-HETE was also demonstrated by radio-RP-HPLC (Fig. 1b); the 12-lipoxygenase activity is consistent with the presence of cervical epithelial cells in the cell suspensions studied.

2.3 Properties of Isolated Lipoxygenase

Human cervical lipoxygenase has been isolated to a high degree of purity (Flatman et al. 1986) and found to have optimum activity at pH 6.5, a K_m of 15 µM for AA, and to be stimulated by ATP and Ca^{2+}. Analysis of subcellular fractions showed a preponderance of activity in membrane fractions. Following purification of the enzyme there was a loss in the peroxidase activity responsible for the reduction of 12-HPETE to 12-HETE. Analysis of products resulting from the incubation of AA with the purified lipoxygenase revealed metabolites chromatographing with di-HETEs and trihydroxyeicosatrienoic acid isomers as well as low levels of 12-HPETE. The addition of GSH and GSH peroxidase brought a decrease in the amounts of more polar products and a proportional increase in

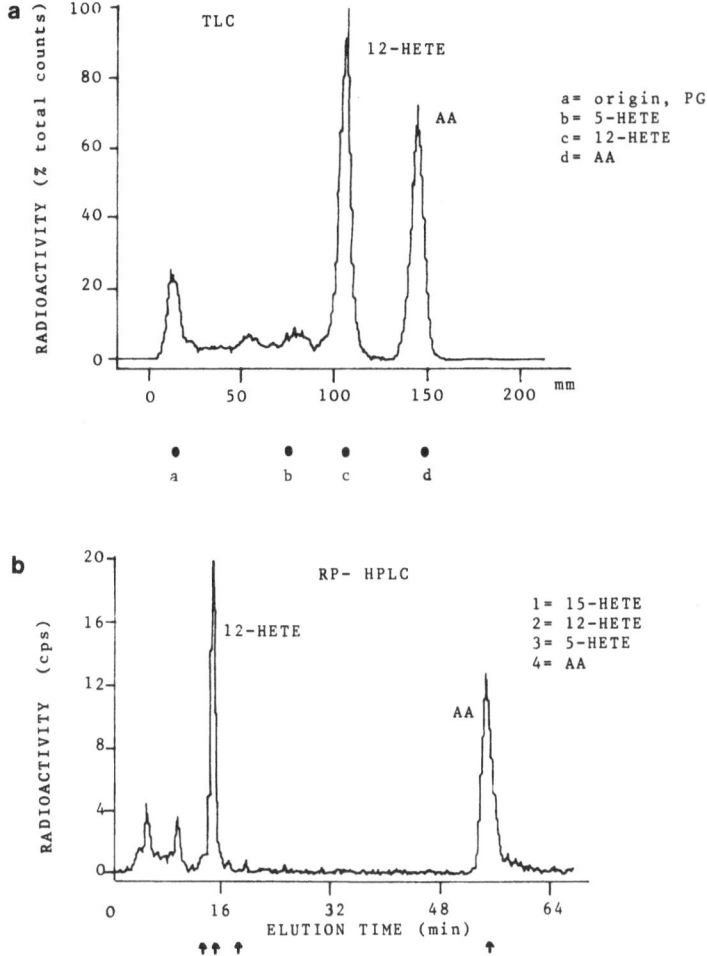

Fig. 1. [¹⁴C]Arachidonic acid metabolism by epithelial cells present in cervical scrapes. Epithelial cells suspended in PBS (1–5 µg DNA) were incubated with 10 µM [¹⁴C]AA (0.1 µCi) for 45 min at 37°C. Extracts were analysed by TLC (**a**) and RP-HPLC (**b**). Reference standards chromatographed at the positions indicated

12-HETE. The results with the pure enzyme suggest that in the absence of peroxidase activity breakdown of the hydroperoxide to more polar products was taking place.

2.4 Lipoxygenase Activity During the Menstrual Cycle

Measurement of [¹⁴C]AA metabolism by homogenised cervix specimens obtained from patients undergoing hysterectomy for non-malignant causes showed

an enormous variation between samples, both in total conversion to lipoxygenase products and in conversion to 12-HETE (Flatman et al. 1988; Table 1). Uterine cyclo-oxygenase activity is known to be higher in the secretory rather than
the proliferative phase of the cycle and this is an agreement with our observations. Regarding cervical lipoxygenase activity, however, there were no significant differences between the two groups classified as proliferative and secretory
on the basis of the histological appearance of the endometrium. Radio-TLC,
however, is not a good method for quantitative evaluation of absolute metabolite
levels (Hurst et al. 1987) and it is possible that measurement of 12-HETE by
HPLC would have revealed a different pattern and this is under study. It is also
of interest to note the lower lipoxygenase activity in samples taken from post-
menopausal patients. Even though the number of patients used in the study is
only 3 the results are consistent with data obtained using cervical scrapes from
post-menopausal women and suggest the possibility that the enzyme level or
expression is hormone-dependent. Recent studies by Rees and co-workers on
human endometrium have in fact shown a variation in leukotriene release during
the menstrual cycle (Rees et al. 1987).

3 Rat Uterine Lipoxygenase Activity

3.1 Variations During Pregnancy

In contrast to human cervical lipoxygenase activity, metabolism of [^{14}C]AA into
12-HETE by rat uterine homogenates was found to vary during the oestrous
cycle and was highest at dioestrous (Morgan et al. 1984). Subsequent evaluation
of uterine lipoxygenase activity during rat pregnancy revealed a particularly interesting pattern (Fig. 2). [^{14}C]12-HETE synthesis was very high at pre-implanta-
tion stages (pregnancy days 3 and 4) but this was followed by a dramatic de-

Table 1. [^{14}C]Arachidonic acid metabolism by incubated human cervical and uterine homogenates

Menstrual cycle phase	Number of samples	% Conversion [^{14}C]arachidonic acid		
		Cervix		Uterus PGs
		Total	12-HETE	
Proliferative	17	43 ± 23	24 ± 15	3 ± 2
Secretory	17	36 ± 29	18 ± 16	13 ± 7
Post-menopausal	3	∼0.7	∼0.7	0

Values are means ± SD
Ratio of 12-HETE:total ∼0.5–0.56

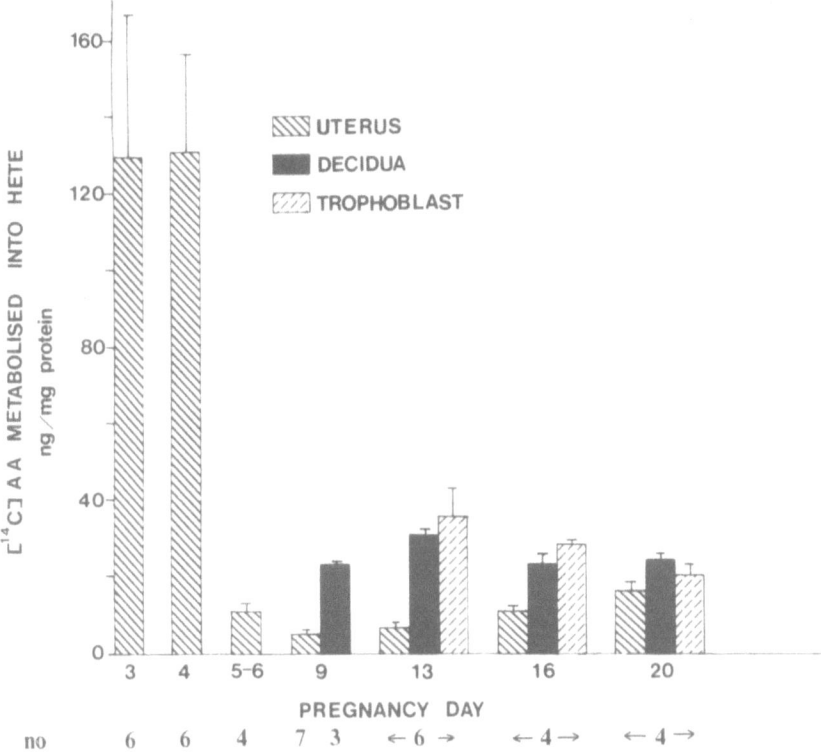

Fig. 2. [^{14}C]Arachidonic acid (AA) metabolism into 12-HETE by homogenised rat uterus, decidua and trophoblast during the course of pregnancy (mean $\pm$ s.e.m.). Homogenates were incubated with 15 μM [^{14}C]AA (0.15 μCi) for 30 min at 37°C. Following extraction metabolites were analysed by TLC

crease around the time of implantation (day 5) and levels remained low for the remainder of gestation. After implantation relatively high lipoxygenase activity was found in homogenates of decidua and trophoblast both of which are tissues showing bursts of active cell proliferation and differentiation. RP-HPLC analysis of arachidonate metabolites synthesised by homogenate pools largley confirmed the radio-TLC data. A noticeable observation however was the substantially higher amounts of 12-HETE as compared to [^{14}C]12-HETE suggesting synthesis from endogenous substrate. Measurement of endogenous non-esterified arachidonic acid by gas chromatography (McDonald-Gibson 1987) showed the presence of relatively high levels before incubation, with further increases during incubation.

Regarding identification of lipoxygenase products, RP-HPLC analysis demonstrated that 12-HETE was the major product in extracts of incubated rat uterine, decidual and trophoblast homogenates (Fig. 3), an exception being synthesis by decidua of day 13 of pregnancy, of a product co-eluting nearer to 15- rather than 12-HETE. Mono-HETEs, including 12-HETE have also been shown to be synthesised to varying degrees by intrauterine tissues of other species, such

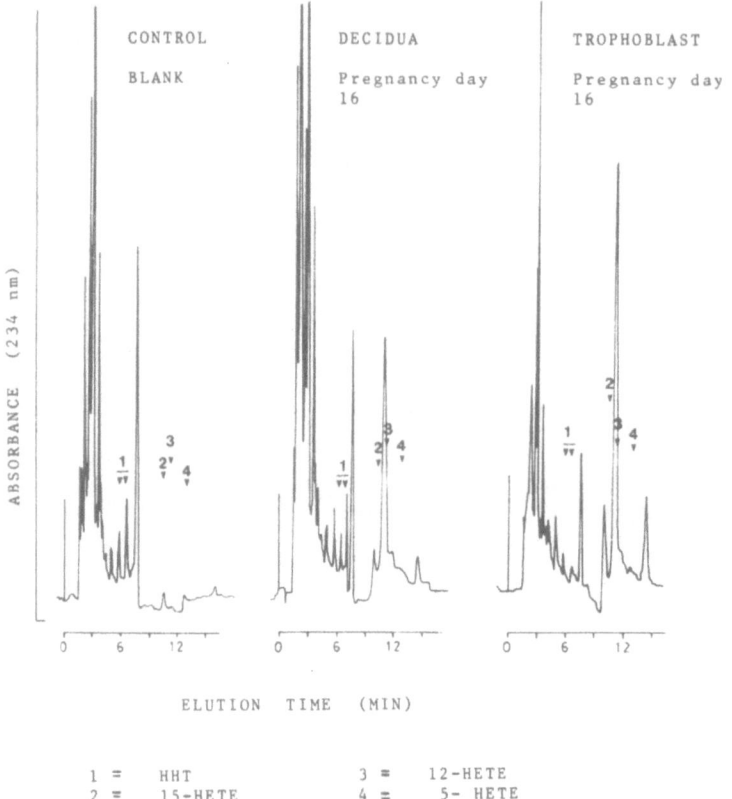

Fig. 3. Lipoxygenase products synthesised by incubated rat decidua and trophoblast homogenates. Extracts. were analysed by RP-HPLC

as human foetal membranes, decidua vera and placenta (Saeed and Mitchell 1982b; Rose et al. 1984; Mitchell and Grzyboski 1987) and sheep maternal cotyledon (Mitchel et al. 1987). Other lipoxygenase products reported in intrauterine tissues include 15-HETE and LTB$_4$ (Mitchell et al. 1987, Mitchell and Grzyboski 1987).

Speculations as to the possible roles of lipoxygenase products during pregnancy include an involvement in the control of haemodynamics, muscle tone and immune reactions within the utero-placental complex (Mitchell and Grzynboski 1987; Morgan et al. 1987).

3.2 Lipoxygenase Versus Cyclo-oxygenase Pathway

Lipoxygenase products can act in opposition or synergy with cyclo-oxygenase products (Williams 1983) and biosynthetic interactions are known to take place between enzymes and products of the two pathways of arachidonic acid metabolism (Kuehl et al. 1984). For example 12-HPETE is thought to be a regulator for the production of prostacyclin (PGI_2) (Turk et al. 1980). In rat uterus PGI_2 is the major cyclo-oxygenase product (Poyser and Scott 1980). In the non-pregnant state we found synthesis of 6-keto-PGF_1 (6KF), the stable metabolite of PGI_2, to be quantitatively less significant than synthesis of 12-HETE at all stages of the oestrous cycle (Morgan 1985), in particular at dioestrous ($>6:1$). During pregnancy however there is a reversal of this ratio following implantation. On days 3 and 4 we find the ratio of 12-HETE:6KF to be around $10:1$ whilst on day 6 it is decreased to $1:1$ and by day 20 there is a further reversal in favour of 6KF ($1:7$) due to the increased PGI_2 production towards the end of pregnancy. A similar pattern was reported by Vesin and Harbon (1982) for rat uterus while synthesis of a mixture of HETEs by rabbit foetal membranes was higher in middle rather than late pregnancy (Elliot et al. 1984) which again is the opposite to cyclo-oxygenase product formation.

4 Lipoxygenase Activity in Relation to Cervical Cancer

As mentioned earlier previous studies by Slater demonstrated the presence of a strong free radical signal in normal cervix tissue which was absent in invasive cancer (Slater and Cook 1970; Benedetto et al. 1981). Subsequent studies indicated an association betwwen the free radical signal and reactions of the arachidonate cascade (Tomasi et al. 1984). The discovery of a very active lipoxygenase pathway in the ectocervix has therefore aroused interest in the possible link between disturbance in AA metabolism and the development of cervical intraepithelial neoplasia (CIN) which if untreated can progress to invasive cancer.

In recent studies we investigated [^{14}C]AA metabolism by cervical tissue from normal patients and from patients with histological and colposcopic evidence of CIN and cervical cancer (Table 2). Homogenates used in the study were very dilute (1:16 tissue weight:buffer volume) hence endogenous substrate would be expected to be very low compared to the exogenously added [^{14}C]AA (15 µM).

Table 2. Lipoxygenase activity of human cervix homogenates incubated with [^{14}C]arachidonic acid

	Number of samples	% Conversion [^{14}C]arachidonic acid		
		12-HETE	Total	Ratio 12-HETE:total
Normal	8	4.2 ± 1.0	16.3 ± 3.8	0.26 ± 0.02
CIN 1/2	8	3.9 ± 1.2	24.8 ± 8.6	0.17 ± 0.01[a]
CIN 3	13	4.2 ± 1.4	16.9 ± 4.6	0.26 ± 0.04
Invasive cancer	5	2.7 ± 0.7	11.7 ± 1.8	0.23 ± 0.04

Values are means $\pm$ SEM
CIN, Cervical intraepithelial neoplasia
[a] $p < 0.01$

There were no significant differences between the various groups either in the total conversion into lipoxygenase products or in the conversion into 12-HETE. There was however a significant decrease in CIN 1/2 samples, in the ratio of 12-HETE to total lipoxygenase product formation, due to a higher level of polar products. This suggests either the presence of insuficient GSH or the loss of peroxidase activity within the CIN 1/2 tissue. Further studies are necessary to establish whether this finding is of pathophysiological importance.

References

Benedetto C, Bocci A, Dianzani MU, Chiringhello B, Slater TF, Tomasi A, Vannini V (1981) Electron-spin resonance studies on normal human-uterus and on benign and malignant uterine tumors. Cancer Res 41:2936-2942

Bockman RS (1983) Prostaglandins in cancer – a review. Cancer Invest 1:485-493

Demers LM, Rees MCP, Turnbull AC (1984) Arachidonic-acid metabolism by the non-pregnant human uterus. Prostaglandins Leukotrienes Med 14:175-180

Elliot WJ, McLaughlin LL, Bloch MH, Needleman P (1984) Arachidonic acid metabolism by rabbit fetal membranes of various gestational ages. Prostaglandins d27:27-36

Flatman S, Hurst JS, McDonald-Gibson RG, Jonas GEG, Slater TF (1986) Biochemical studies on a 12-lipoxygenase in human uterine cervix. Biochim Biophys Acta 883:7-14

Flatman S, Morgan A, McDonald-Gibson RG, Davey A, Jonas GEG, Slater TF (1988) 12-Lipoxygenase activity in human uterine cervix. Prostaglandins Leukotrienes and Essential Fatty Acids 32:87-94

Honn KV, Busse WD, Sloane BF (1983) Prostacyclin and thromboxanes. Implications for their role in tumor-cell metastasis. Biochem Pharmacol 32:1-11

Hurst JS, Flatman S, McDonald-Gibson RG (1987) Thin-layer chromatogrphy (including radio thin-layer chromatography and autoradiography) of prostaglandins and related compounds. In: Benedetto C, McDonald-Gibson RG, Nigam S, Slater TF (eds) Prostaglandins and related substances, IRL, pp 53-73

Kuehl FA, Dougherty HW, Ham EA (1984) Interactions between prostaglandins and leukotrienes. Biochem Pharmacol 33:1-5

McDonald-Gibson RG (1987) Quantitative measurement of arachidonic acid in tissues or fluids. In: Benedetto C, McDonald-Gibson RG, Nigam S, Slater TF (eds) Prostaglandins and related substances. IRL, pp 259-265

Mitchell MD, Grzyboski CF (1987) Arachidonic acid metabolism by lipoxygenase pathways in uterine and intrauterine tissues of pregnant sheep. Prostaglandins Leukotrienes Med 28:303-312

Mitchell MD, Grzyboski CF, Dedhar CM, Hunter JA (1987) Prostaglandins Leukotrienes Med 27:197-207

Morgan A (1985) Aspects of uterine arachidonic acid metabolism. PhD Thesis, Brunel University

Morgan A, Flatman S, McDonald-Gibson RG, Slater TF (1984) Uterine arachidonic acid lipoxygenase activity. In: McBrien DCH, Slater TF (eds) Cancer of the uterine cervix. Academic, New York, pp 255-263

Morgan A, McDonald-Gibson RG, Flatman S, Slater TF (1986) Uterine lipoxygenase activity:steroid hormone effects and pregnancy. In: Katori M, Yamamoto S, Hayaishi O (eds) Challenging frontiers for prostaglandin research. Gendai Iryosha, Tokyo, p 175

Morgan A, Flatman S, McDonald-Gibson RG (1987) Uterine lipoxygenase activity and function. Med Sci Res 15:467-470

Poyser NL (1981) Prostaglandins in reproduction. Research Studies, Chichester

Poyser NL, Scott FM (1980) Prostaglandin and thromboxane production by the rat uterus and ovary in vitro during the estrous cycle. J Reprod Fertil 60:33-40

Rees MCP, Marzo VP, Tippins JR, Morris HR, Turnbull AC (1987) Leukotriene release by non pregnant human endometrium and myometrium. Adv Prostaglandin Thromboxane Leukotriene Res 17A:1125–1128

Rose MP, Elder MG, Myatt L (1984) Lipoxygenase and cyclo-oxygenase products of arachidonic acid in human intra-uterine tissues. Clin Exp Hypertens [B]3:243

Saeed SA, Mitchell MD (1982a) New aspects of arachidonic acid metabolism in human uterine cervix. Eur J Pharmacol 81:515–516

Saeed SA, Mitchell MD (1982b) Formation of arachidonate lipoxygenase metabolites by human foetal membranes, uterine decidua vera and placenta. Prostaglandins Leukotrienes Med 8:635–640

Slater TF, Cook JWR (1970) Electron spin resonance studies on normal and malignant human tissues and in normal and damaged rat liver. In: Evans DMD (ed) Cytology automation. Livingstone, Edinburgh, pp 108–120

Thaler-Dao H, Jouanen A, Benmehdi F, Crastes de Paulet A (1985) Regulation by oestradiol of the lipoxygenase pathway in the rat uterus. Prostaglandins Leukotrienes Med 18:59–64

Tomasi A, Benedetto C, Slater TF (1984) Identification of a strong free radical signal that is detectable in human uterine cervix. In: McBrien DCH, Slater TF (eds) Cancer of the uterine cervix. Academic, London, pp 265–275

Turk J, Wyche A, Needleman P (1980) Inactivation of vascular prostacyclin synthase by platelet lipoxygenase products. Biochem Biophys Res Commun 95:1628–1632

Vesin HF, Harbon S (1982) An inbalance of the lipoxygenase and cyclo-oxygenase pathways in pregnant rat. 5th International Conference on Prostaglandins, May 18–21. Lorenzini, Florence, p 404 (Abstract book)

Weichman BM, Tucker SS (1982) Contraction of guinea pig uterus by synthetic leukotrienes. Prostaglandins 24:245–253

Werner EJ, Walenga RW, Dubowy RL, Boone S, Stuart MJ (1985) Inhibition of human malignant neuroblastoma cell DNA synthesis by lipoxygenase metabolites of arachidonic acid. Cancer Res 45:561–563

Williams TJ (1983) Interactions between prostaglandins leukotrienes and other mediators of inflammation. Br Med Bull 39:239–242

Modulation of Ocular Eicosanoid Biosynthesis by Oxidants and Anti-oxidants

J. S. Hurst, C. A. Paterson, C. S. Short and M. D. Hughes

Kentucky Lions Eye Research Institute, Department of Ophthalmology,
University of Louisville School of Medicine, Louisville, Kentucky, USA

1 Introduction

The association betwen oxidative damage and inflammation is well documented
(Bragt et al. 1980), and arachidonate metabolites are important modulators of
the inflammatory response in animal and human tissues including the eye (Eakins et al. 1972; Masuda et al. 1973). Ocular tissues generate eicosanoids by both
cyclo-oxygenase and lipoxygenase pathways (Bhattacherjee et al. 1979; Kass
and Holmberg 1979; Kulkarni et al. 1984). Since it has been suggested that arachidonate metabolism is regulated by the endogenous peroxide concentration
metabolism is regulated by the endogenous peroxide concentration (Hemler and
Lands 1980) we proposed to investigate the effect of changing the oxidative environment in vitro upon arachidonate metabolism in the iris-ciliary body and
cornea; for comparison lung, spleen and platelet were also examined. The oxidant studied was hydrogen peroxide and the anti-oxidants employed were ascorbate, glutathione and ebselen, a seleno-organic compound with established glutathione peroxidase-like activity (Wendel et al. 1984). Much of the data to be
given here has been previously presented (Hurst et al. 1987b) and has been submitted for publication in detail elsewhere.

2 Materials and Methods

Unlabelled arachidonic acid and chemicals of the highest available purity were
obtained from Sigma (St. Louis, Missouri). Organic solvents (HPLC grade) and
silica gel TLC plates were purchased from EM Science (Cherry Hill, New Jersey). Prostaglandin standards were obtained from Seragen (Boston, Massachusetts). [^{14}C]arachidonate (specific activity 58.3 mCi/mM) was obtained from
Amersham (Arlington Heights, Illinois).

Male albino rabbits (1.5–2.5 kg) were sacrificed with an intravenous injection
of T-61 (American Hoechst, Somerville, New Jersey). The anterior segments of

Nigam et al. (Eds.), Eicosanoids,
Lipid Peroxidation and Cancer
© Springer-Verlag Berlin Heidelberg 1988

the eyes were removed and the iris-ciliary bodies and corneas rapidly dissected away, rinsed in cold 0.15 M saline, blotted and weighed. Tissues were pooled and homogenised at 0–4°C (60 mg of tissue/ml 50 mM phosphate buffer, pH 7.5, corresponding to approximately 2 mg protein; Lowry et al. 1951) by a Polytron homogeniser (Brinkman, Westbury, New York). Spleens and lungs were rapidly removed and homogenised (1:8, w/v) in cold phosphate buffer as above. Platelets were prepared as previously reported (Hurst et al. 1987a).

Tissue homogenates (0.4 ml) were immediately incubated aerobically with sodium [^{14}C]arachidonate (0.1 μCi) in a total volume of 0.5 ml at 37°C for 30 min. The reactions were stopped by acidification with HCl to be within the pH range 3–4. The eicosanoids were extracted with six volumes ethyl acetate. The solvent was evaporated under nitrogen and the extract reconstituted with 30 μl ethyl acetate and applied equally to two TLC plates. Plates were developed in two solvent systems. The first combination was solvent system (A): ethyl acetate/ trimethyl pentane/water/acetic acid (55:25:50:10 by vol) and solvent system (P): diethyl ether/petroleum ether/acetic acid (60:40:1, by vol). The second combination was solvent system (A) and (B): benzene, dioxan/acetic acid (30:20:1 by vol). The chromatograms were scanned and analyzed by an ISOM-ESS IM-3016 radio-TLC analyser (IN/US, Fairfield, New Jersey). Radioactive peaks that co-chromatographed with unlabelled authentic standard compounds were visualised by exposure to iodine vapour.

3 Results

Radio-TLC scans of extracts from control ocular tissue homogenates incubated with [^{14}C]arachidonate in the absence of added co-factors show the typical pattern of [^{14}C]eicosanoid biosynthesis by the iris-ciliary body and cornea respectively (Figs. 1, 2). The iris-ciliary body produced material that co-migrated with 6-keto-PGF$_{1\alpha}$ (a stable metabolite of prostacyclin), PGF$_{2\alpha}$, PGE$_2$, PGD$_2$ and HHT. The principal cyclo-oxygenase metabolite of the cornea was PGD$_2$ and this tissue, unlike the iris-ciliary body, generated appreciable amounts of material co-chromatographing with 12 and 15-HETE. This compound, however, had the same reverse-phase HPLC retention time as 12-HETE standard. HPLC analysis was performed as previously reported (Flatman et al. 1986). The profile of arachidonate metabolism by the ocular tissues observed is qualitatively similar to that previously demonstrated (Bhattacherjee et al. 1979; Kass and Holmberg 1979; Williams et al. 1985).

3.1 Hydrogen Peroxide

Arachidonate metabolism by the iris-ciliary body was not affected by hydrogen peroxide at concentrations less than 1 mM. The overall metabolism of arachidonate was, however, decreased when the concentration of hydrogen peroxide

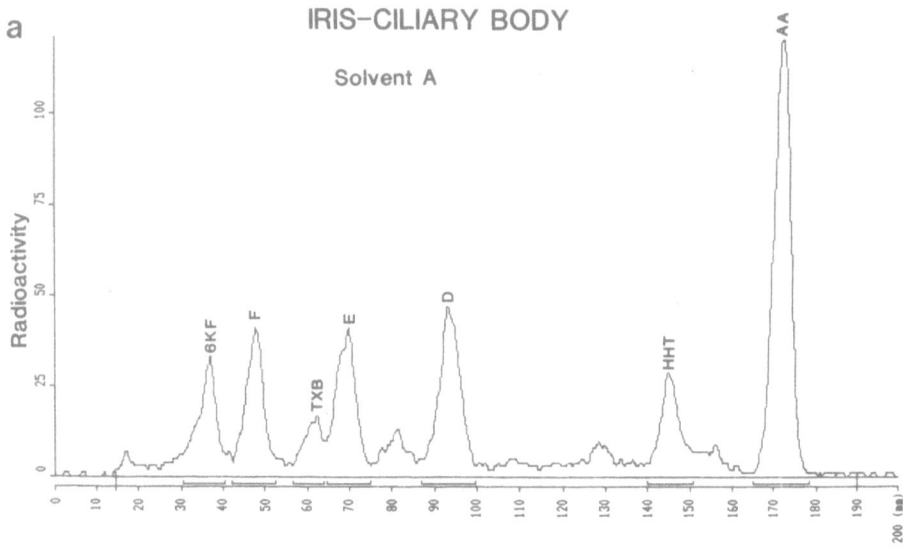

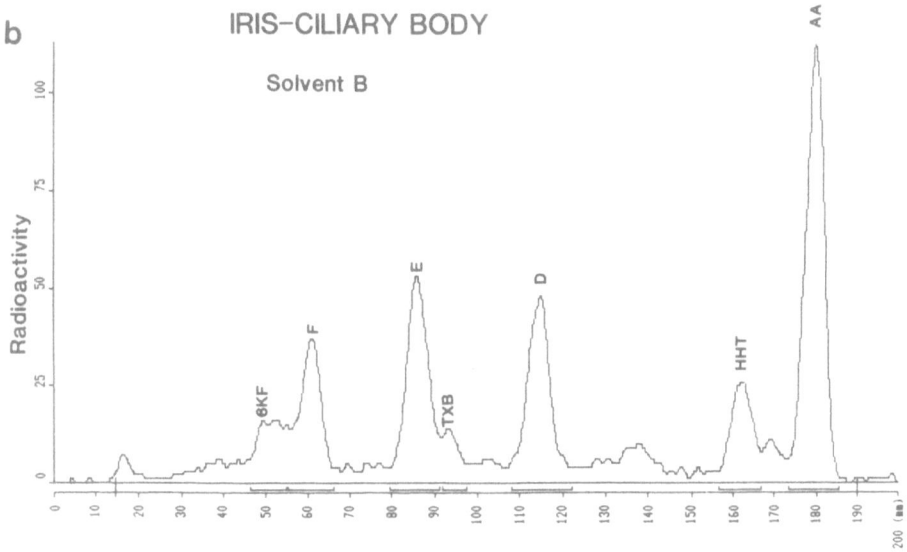

Fig. 1. Scans of radio-TLC extracts from rabbit iris-ciliary body homogenates incubated with [^{14}C]arachidonate (0.1 µCi, 3.43 µM) developed in TLC solvents A **(a)** and B **(b)**

was increased to 1 mM and this correlated with reduced formation of prosta-glandins, with the exception of 6-keto-PGF$_{1\alpha}$ and TXB$_2$ which were not signifi-cantly inhibited (Fig. 3). Hydrogen peroxide at 100 µM significantly inhibited arachidonate metabolism by the cornea (Fig. 4). Hydrogen peroxide at 1 mM

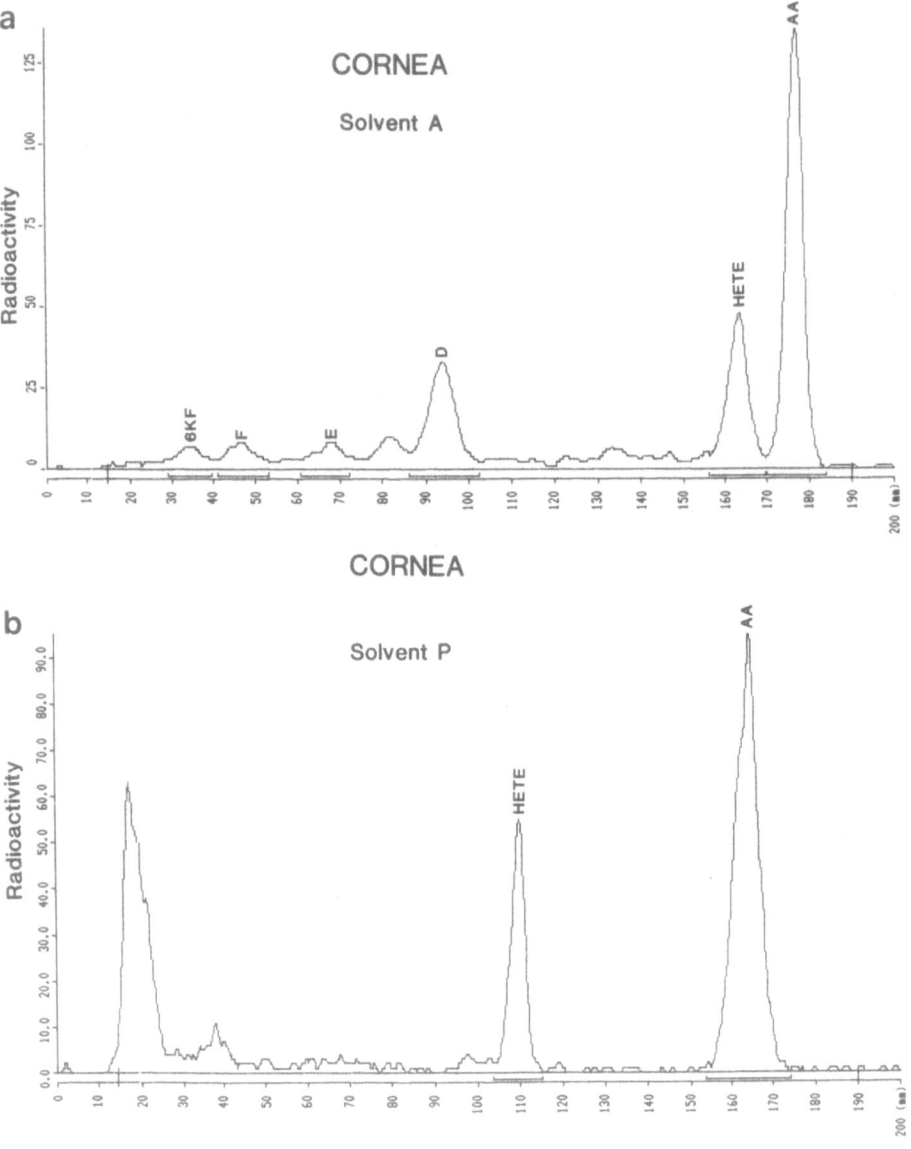

Fig. 2. Scans of radio-TLC extracts from rabbit cornea homogenate incubated with [^{14}C]arachidonate (0.1 μCi, 3.43 μM) developed in TLC solvents A **(a)** and P **(b)**

strongly suppressed arachidonate metabolic activity by the cornea and biosynthesis of both cyclo-oxygenase products and 12-HETE was potently inhibited (Figs. 4, 5). The generation of eicosanoids by the non-ocular tissues was not suppressed by hydrogen peroxide to the same extent as the ocular tissues.

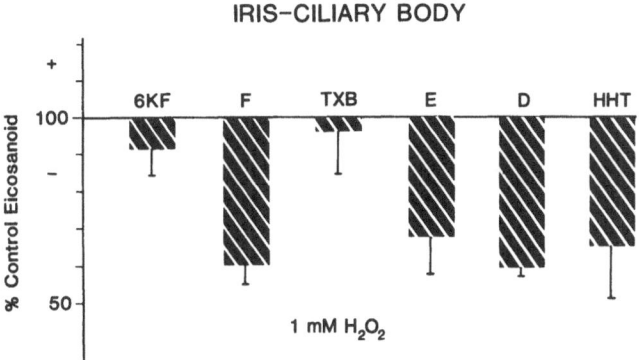

Fig. 3. Effect of hydrogen peroxide (1 mM) on the metabolism of [^{14}C]arachidonate by iris-ciliary body. Data from radio-TLC scans of extracts from incubators of iris-ciliary body homogenates. *Error bars* indicate SD ($n = 4$)

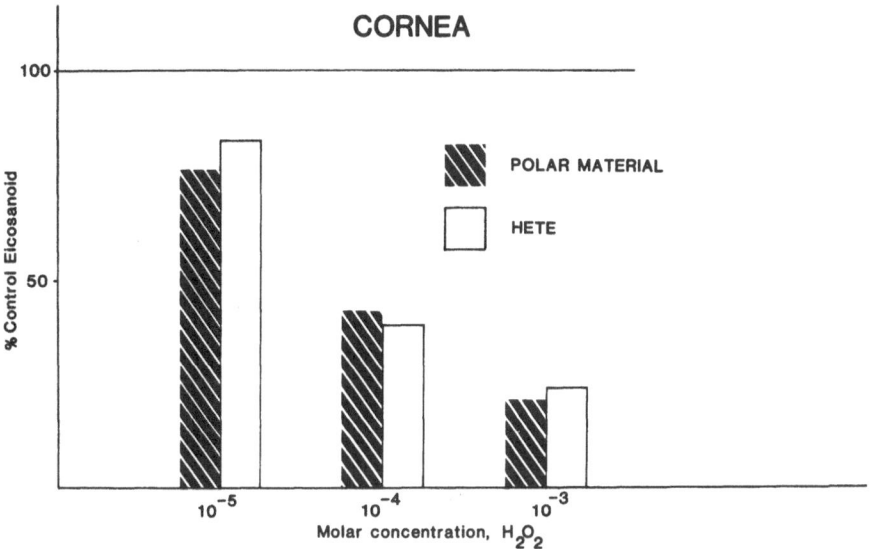

Fig. 4. Effect of hydrogen peroxide (10^{-5}, 10^{-4}, 10^{-3}M) on the metabolism of [^{14}C]arachidonate by rabbit cornea. Data from radio-TLC scans of extracts from cornea homogenate, developed in solvent P

3.2 Ascorbate

Ascorbate (1 mM) stimulated the formation of both 6-keto-PGF$_{1\alpha}$ and HHT by the iris-ciliary body but apparently did not affect the biosynthesis of other eicosanoids (Fig. 6). The biosynthesis of prostaglandins by the cornea was generally enhanced in the presence of ascorbate (1 mM) whereas the generation of 12-HETE was inhibited (Fig. 7). Spleen and platelet metabolism of arachidonate

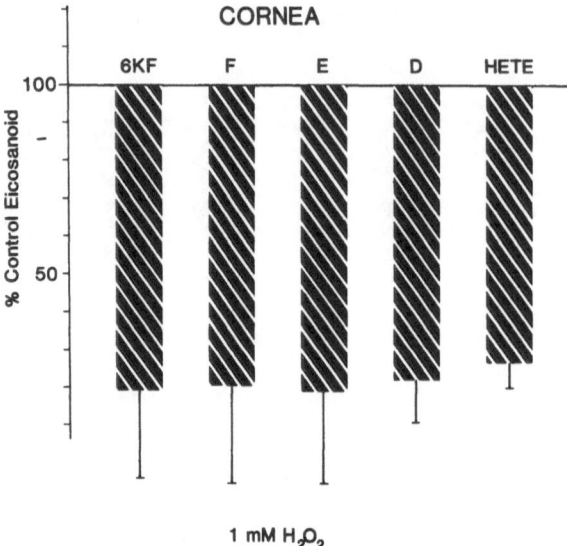

Fig. 5. Effect of hydrogen peroxide (1 mM) on the metabolism of [^{14}C]arachidonate by rabbit cornea. Data from radio-TLC scans of extracts from incubations of cornea homogenate. *Error bars* indicate SD ($n=4$)

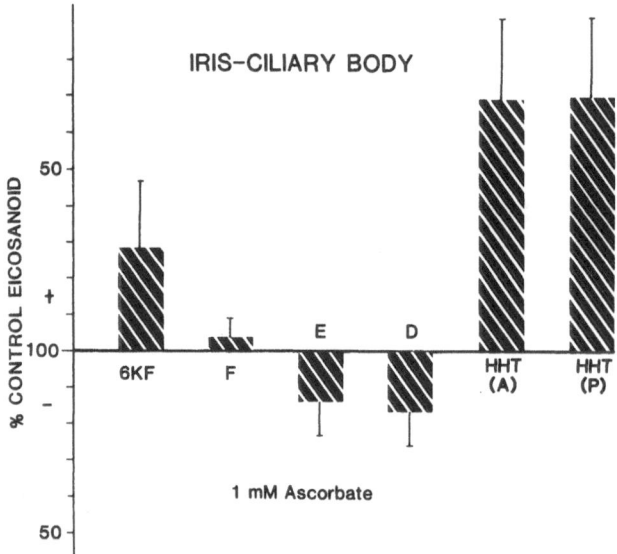

Fig. 6. Effect of ascorbate (1 mM) on the metabolism of [^{14}C]arachidonate by rabbit iris ciliary body. Data from radio-TLC scans of extracts from iris-ciliary body homogenate developed in solvents A and P. *Error bars* indicate SD ($n=4$)

was not apparently influenced by ascorbate. There was, however, a small increase in the formation of both 6-keto-PGF$_{1\alpha}$ and HHT by the lung in the presence of 1 mM ascorbate.

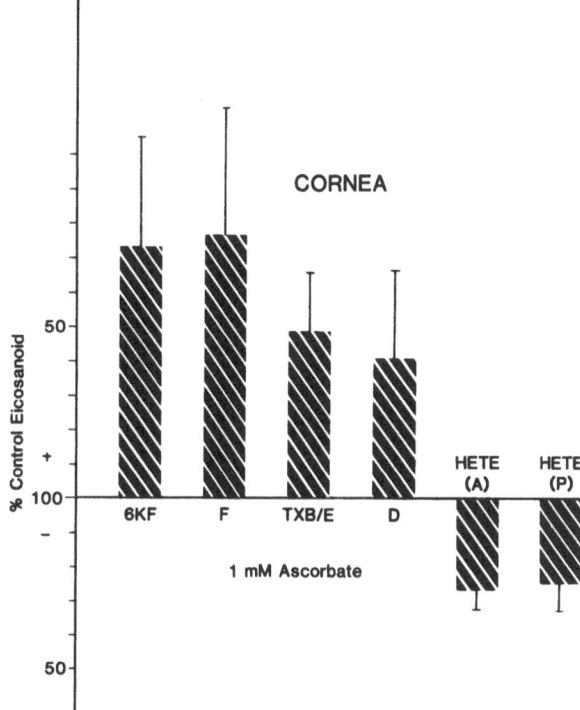

Fig. 7. Effect of ascorbate (1 mM) on the metabolism of [^{14}C]arachidonate by rabbit cornea. Data from radio-TLC scans of extracts from incubation of cornea homogenate developed in solvents A and P. *Error bars* indicate SD (*n* = 4)

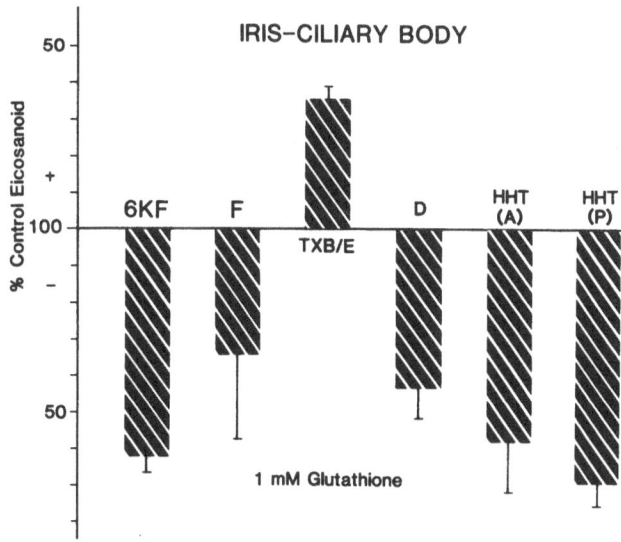

Fig. 8. Effect of reduced glutathione (1 mM) on the metabolism of [^{14}C]arachidonate by rabbit iris-ciliary body. Data from radio-TLC scans of extracts from incuations of iris-ciliary body homogenate developed in solvents A and P. *Error bars* indicate SD (*n* = 3)

3.3 Glutathione

Reduced glutathione (GSH) at 0.5 mM and 1 mM concentrations suppressed arachidonate metabolism by the iris-ciliary body and cornea. the formation of prostaglandin-like material, especially 6-keto-$PGF_{1\alpha}$, was inhibited, with the exception of material that co-chromatographed with unlabelled PGE_2 (Figs. 8, 9). Arachidonate metabolism by the spleen was not affected by GSH but the generation of 6-keto-$PGF_{1\alpha}$ by the lung was inhibited, whereas the formation of PGE_2 was increased.

3.4 Ebselen

In the absence of exogenously added GSH, ebselen (2–50 μM) potently inhibited arachidonate metabolism by the iris-ciliary body in a concentration dependent manner. The biosynthesis of prostaglandin-like material was inhibited with the exception of that metabolite that co-migrated with unlabelled $PGF_{2\alpha}$ which was increased and was only inhibited by the high concentrations of ebselen. The effect of 5 μM ebselen on arachidonate metabolism by the iris-ciliary body is shown in Fig. 10. In the absence of exogenous GSH, ebselen (2–50 μM) inhibited the formation of prostaglandin-like material and 12-HETE by the cornea, the extent of inhibition increasing with the concentration of ebselen. The effect of 5 μM ebselen on arachidonate metabolism by the cornea is shown in Fig. 11.

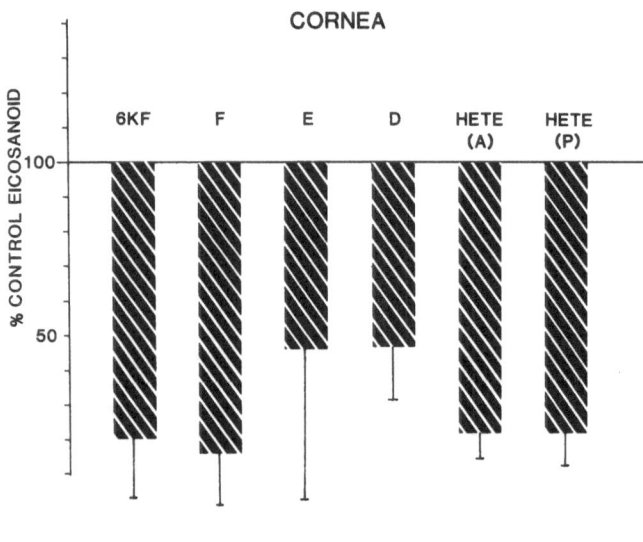

Fig. 9. Effect of reduced glutathione (1 mM) on the metabolism of [^{14}C]arachidonate by rabbit cornea homogenate. Data from radio-TLC scans of extracts developed in solvents A and P. *Error bars* indicate SD ($n = 3$)

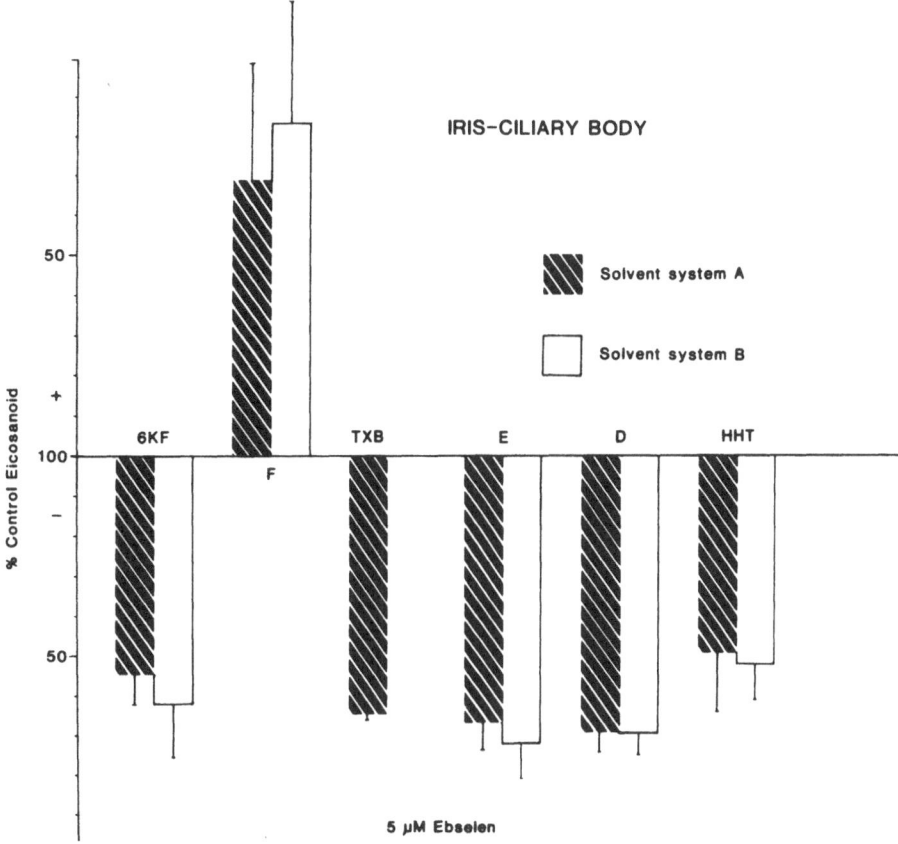

Fig. 10. Effect of ebselen (5 μM) on the metabolism of [^{14}C]arachidonate by rabbit iris-ciliary homogenate. Data from radio-TLC scans of extracts developed in solvents A and B. *Error bars* indicate SD ($n = 6$)

The cyclo-oxygenase activity by the spleen was suppressed by ebselen (2–20 μM) in the absence of exogenous GSH. However, metabolism by the lipoxygenase pathway was not significantly affected. The effect of 5 μM ebselen on arachidonate metabolism by the spleen is shown in Fig. 12. The lipoxygenase activities of the lung and especially the platelet were, however, significantly inhibited by ebselen.

4 Discussion

The effects of hydrogen peroxide upon both the ocular and non-ocular tissues examined correlated with the ability of these tissues to metabolise hydrogen peroxide. The iris-ciliary body has higher levels of GSH-peroxidase, and especially

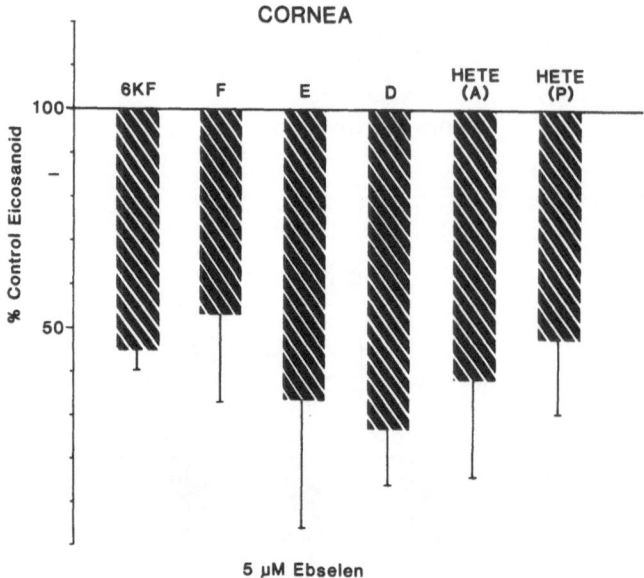

Fig. 11. Effect of ebselen (5 µM) on the metabolism of [^{14}C]arachidonate by rabbit cornea homogenate. Data from radio-TLC scans of extracts developed in solvents A and P. *Error bars* indicate SD ($n = 4$)

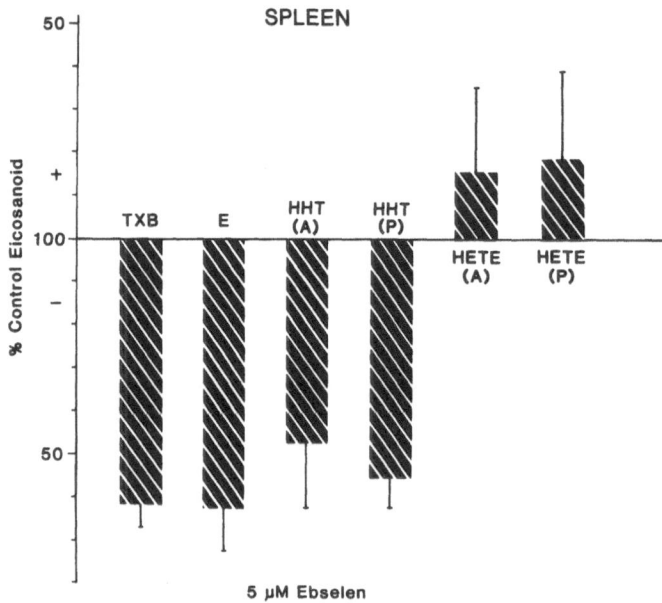

Fig. 12. Effect of ebselen (5 µM) on the metabolism of [^{14}C]arachidonate by rabbit spleen homogenate. Data from radio-TLC scans of extracts developed in solvents A and P. *Error bars* indicate SD ($n = 4$)

catalase, than the cornea (Bhuyan and Bhuyan 1977); consequently hydrogen peroxide more strongly inhibited arachidonate metabolism by the cornea than by the iris-ciliary body. The non-ocular tissues also have a high complement of these protective enzymes. This is especially true of the lung which has to combat the very high oxygen tension to which it is continually exposed. The toxicity of hydrogen peroxide is presumably mediated by hydroxyl free radicals generated via a Haber-Weiss reaction. the transition metal cation catalyst in ocular tissue may be copper as this has been shown to be present at moderate concentrations (McGahan and Bito 1983) and tissue injury resulting from inflammatory episodes might facilitate the availability of this metal. Reducing agents such as ascorbate and glutathione could recycle cuprous cations to increase oxidative insult.

The increased formation of 6-keto-$PGF_{1\alpha}$ by both the iris-ciliary body and cornea in the presence of ascorbate (Figs. 3, 4) is similar to the response of seminal vesicles and aorta treated with ascorbate (Beetens and Herman 1980). The stimulation of 6-keto-$PGF_{1\alpha}$ biosynthesis in both the iris-ciliary body and cornea by ascorbate (Figs. 3,4) is similar to that demonstrated for both seminal vesicles (Beetens and Herman 1980) and aorta (Beetens and Herman 1983). The stimulatory effect of ascorbate on eicosanoid biosynthesis was greater in the cornea than in the iris-ciliary body. This agrees with a previous observation (Polgar and Taylor 1980) of stimulated prostaglandin generation in the presence of ascorbate. An interesting observation was the enhanced formation of HHT, independent of TXB_2 biosynthesis. The biological function of HHT still remains to be clarified, especially since chemotactic and chemokinetic properties towards PMN leukocytes have been reported for this compound (Goetzl and Gorman 1978).

The stimulatory effect of ascorbate upon arachidonate metabolism may be due to the combination of both pro and anti-oxidant activity. Ascorbate efficiently scavenges oxygen free radicals (Bielski et al. 1975) and may remove the oxygen free radical that is a co-product of PG endoperoxidase activity. This free radical species is implicated in the tissue damage that accompanies PG release during the inflammatory response (Kuehl et al. 1977; Egan et al. 1978) and intoxicates both cyclo-oxygenase and especially prostacyclin synthetase. Thus removing this damaging agent would protect these activities. Ascorbate can, however, also generate hydrogen peroxide (Polgar and Taylor 1980) and low levels of hydrogen peroxide have been proposed as essential activators of cyclo-oxygenase (Hemler et al. 1979; Hemler and Lands 1980). This may be an alternative explanation for the stimulation of cyclo-oxygenase activity observed in the cornea, and this investigation has confirmed an earlier study (Taylor et al. 1982) that reported the stimulation of prostaglandin synthesis by corneal cells in the presence of ascorbate. the non-ocular tissues we examined, especially the spleen, have high endogenous ascorbate (Omaye et al. 1979; Fox et al. 1982) and in response to added exogenous ascorbate we observed no effect upon archidonate metabolism. This was probably related to the high capacity of these tissues like the iris-ciliary body, to metabolise hydrogen peroxide.

Ascorbate decreased the production of 12-HETE by the cornea which agreed with a previous study (Williams and Paterson 1986), whereas the non-ocular lipoxygenase were unaffected by ascorbate addition. It has been suggested, however, that cytochrome P-450 rather than lipoxygenase activity might generate 12-HETE in the cornea (Schwartzman et al. 1987).

GSH potently inhibited arachidonate metabolism by both ocular and non-ocular tissues with the exception of PGE_2, which was increased. It has been reported (Nugteren et al. 1966; Tai 1976) that GSH inhibits cyclo-oxygenase activity and diverts synthesis in favour of PGE_2. More recently GSH has been demonstrated to inhibit the formation of prostacyclin by both seminal vesicles (Cottee et al. 1977) and aorta (Salmon et al. 1978). GSH is the co-factor for GSH-peroxidase which together with catalase defends tissue against oxidative injury. GSH-peroxidase, unlike catalase which only detoxifies hydrogen peroxide, will reduce both organic peroxides and hydrogen peroxide. The removal of the peroxide activators of cyclo-oxygenase would result in inhibition of prostanoid biosynthesis.

The inhibition of arachidonate metabolism by both ocular and non-ocular tissues in the presence of ebselen is most probably related to its well-documented GSH-peroxidase-like activity (Muller et al. 1984; Wendel et al. 1984). This would reduce the availability of the essential hydroperoxide activator and inhibit enzymatic activity. The apparent increase in the formation of 12-HETE by the spleen may not represent a direct stimulatory effect of ebselen on the spleen lipoxygenase but a redirection of metabolism from the cyclo-oxygenase pathway to the lipoxygenase; this effect was also observed in rat spleen (Hurst et al. 1988). The rabbit lung and especially the platelet lipoxygenases were inhibited like the rat lung and human platelet activities (Hurst et al. 1988). Lung and platelet lipoxygenases have been shown to be biochemically and structurally similar (Yokoyama et al. 1983).

The ability of pro-oxidants and anti-oxidants to influence arachidonate metabolism in the eye as in other tissues is almost certainly related to the endogenous 'peroxide tone' (Hemler et al. 1979). The eye has efficient protective mechanisms to contain oxidative stress under physiological conditions. However, during such episodes such as the acute inflammatory response, in which damaging oxygen radicals may be released in concentrations that could compromise or overwhelm the anti-oxidant defenses then it is conceivable that arachidonate metabolism could be modified.

Acknowledgement. This investigation was supported by USPHS research grant No. EYO6918, the Kentucky Lions Eye Research Foundation and an unrestricted grant from Research to Prevent Blindness, Inc. We also thank Marilyn McLendon for preparation of the manuscript.

References

Beetens JR, Herman AG (1980) Enhanced formation of 6-oxo-$PGF_{1\alpha}$ by ram seminal vesicles in the presence of anti-oxidants. Br J Pharmacol 69:267P

Beetens JR, Herman AG (1983) Vitamin C increases the formation of prostacyclin by aortic rings from various species and neutralises the inhibitory effect of 15-hydroperoxy-arachidonic acid. Br J Pharmacol 80:249

Bhattacherjee P, Kulkarni S, Eakins KE (1979) Metabolism of arachidonic acid in rabbit ocular tissues. Invest Ophthalmol Vis Sci 18:172

Bhuyan KC, Bhuyan DK (1977) Regulation of hydrogen peroxide in eye humors. Effect of 3-amino-1H-1,2,4-triazole on catalase and glutathione peroxidase of rabit eye. Biochim Biophys Acta 497:641

Bielski BHJ, Richter HW, Chan PC (1975) Some properties of the ascorbate free radical. Ann NY Acad Sci 258:231

Bragt PC, Bansberg JI, Bonta IL (1980) Anti-inflammatory effects of free radical scavengers and anti-oxidants. Inflammation 4:289

Cottee F, Flower RJ, Moncada SA, Salmon JA, Vane JR (1977) Synthesis of 6-keto-$PGF_{1\alpha}$ by ram seminal vesicles. Prostaglandins 14:413

Eakins KE, Whitelock RAF, Perkins ES, Bennett A, Unger WG (1972) Release of prostaglandins in ocular inflammation. Nature 239:248

Egan RW, Gale PH, Beveridge GC, Philips GB (1978) Radical scavenging as the mechanisms for stimulation of prostaglandin Cyclo-oxygenase and depression of inflammation by lipoic acid and sodium iodide. Prostaglandins 16:861

Flatman S, Hurst JS, McDonald-Gibson RG, Jonas GEG, Slater TF (1986) Biochemical studies on a 12-lipoxygenase in human uterine cervix. Biochim Biphys Acta 883:7

Fox RR, Lam K-W, Lewen R, Lee P-F (1982) Ascorbate concentration in tissues from normal and buphthalmic rabbits. J Hered 73:109

Goetzl EJ, Gorman R (1978) Chemotactic and chemokinetic stimulation of human eosinophil and neutrophil PMN leukocytes. J Immunol 120:526

Hemler ME, Lands WEM (1980) Evidence for a peroxide-initiated free radical mechanism of prostaglandin biosynthesis. J Biol Chem 255:6253

Hemler ME, Cook HW, Lands WEM (1979) Prostaglandin synthesis can be triggered by lipid peroxides. Arch Biochem Biophys 193:340

Hurst JS, Slater TF, Lang J, Juergens G, Zollner H, Esterbauer H (1987a) Effects of the lipid peroxidation product 4-hydroxynonenal on the aggregation of human platelets. Chem Biol Interact 61:109

Hurst JS, Paterson CA, Short CS (1987b) Oxidant and anti-oxidant effects on arachidonate metabolism in ocular and non ocular tissues. Invest Ophthalmol Vis Sci [Suppl] 28:200

Hurst JS, Paterson CA, Short CS, Hughes MD (1988) Oxidant and anti-oxidant effects on arachidonate metabolism by rabbit ocular tissues, (in preparation)

Kass MA, Holmberg KE (1979) Prostaglandin and thromboxane synthesis by microsomes of rabbit ocular tissues. Invest Ophthalmol Vis Sci 17:167

Kuehl FA, Humes JL, Egan RW, Ham EA, Beveridge GC, van Arman CG (1977) Role of prostaglandin endoperoxide PGG_2 in inflammatory processes. Nature 265:170

Kulkarni PS, Rodriguez AV, Srinavasan BD (1984) Human anterior uvea synthesise lipoxygenase products from arachidonic acid. Invest Ophthalmol Vis Sci 25:221

Masuda K, Izawa Y, Mishima S (1973) Prostaglandins and uveitis: A preliminary report. Jpn J Ophthalmol 17:166

McGahan MC, Bito LZ (1983) The pathophysiology of the ocular microenvironment. I. Preliminary report on the possible involvement of copper in ocular inflammation. Curr Eye Res 2:883

Muller A, Cadenas E, Graf P, Sies H (1984) A novel biologically active seleno-organic compound. I. Glutathione peroxidase-like activity in vitro and anti-oxidant capacity of PZ 51 (ebselen). Biochem Pharmacol 33:3235

Nugteren DH, Beerthuis RK, van Dorp DA (1966) The enzymatic conversion of all cis-8,11,14 eicosatrienoic acid into PGE_1. Recl Trav Chim Pays Bas 85:405

Omaye ST, Turnbull JD, Sauberlich HE (1979) Selected methods for the determination of ascorbic acid in animal cells. Methods Enzymol 62:1

Polgar P, Taylor L (1980) Stimulation of prostaglandin synthesis by ascorbic acid via hydrogen peroxide formation. Prostaglandins 19:693

Salmon JA, Smith DR, Flower RJ, Moncada SA, Vane JR (1978) Further studies on the enzymatic conversion of prostaglandin endoperoxide into prostacyclin by porcine aortic microsomes. Biochim Biophys Acta 523:250

Schwartzman ML, Murphy RC, Abraham NG, Masferrer J, Dunn MW, McGiff JC (1987) Identification of novel corneal arachidonic acid metabolites: Characterisation of their biological activities. Invest Ophthalmol Vis Sci [Suppl] 28:328

Tai H-H (1976) Mechanisms of prostaglandin biosynthesis in rabbit kidney medulla: A rate limiting step and the differential stimulatory actions of L-adrenaline and glutathione. Biochem J 160:577

Taylor L, Menconi M, Leibowitz HM, Polgar P (1982) The effect of ascorbate, hydroperoxides and bradykinin on prostaglandin production by corneal and lens cells. Invest Ophthalmol Vis Sci 23:378

Wendel A, Fawsel M, Safayhi H, Otter R (1984) A novel biologically active seleno-organic compound. II. Activity of PZ 51 in relation to glutathione peroxidase. Biochem Pharmacol 33:3235

Williams RN, Paterson CA (1986) Modulation of corneal lipoxygenase by ascorbic acid. Exp Eye Res 43:7

Williams RN, Delamere NA, Paterson CA (1985) Generation of lipoxygenase products in avascular tissues of the eye. Exp Eye Res 41:733

Yokoyama K, Mizuno H, Mitachi T, Yoshimoto S, Yamamoto S, Pace Asciak CR (1983) Partial purification and characterisation of arachidonate-12-lipoxygenase from rat lung. Biochim Biophys Acta 750:237

Endogenous Formation of Leukotriene D_4 During Systematic Anaphylaxis in the Guinea Pig

S. Hammarström, A. Keppler, L. Örning and K. Bernström

Department of Cell Biology, Faculty of Health Sciences, University of Linköping,
581 85 Linköping, Sweden

1 Introduction

Leukotriene (LT) C_4 is a biologically active substance formed from arachidonic acid and glutathione in various cells and tissues (Samuelsson 1983) and presumed to play a major role as a mediator of allergic and anaphylactic reactions (Samuelsson 1983; Hammarström 1983). It is formed by a number of cells e.g., basophilic (MacGlashan et al. 1983) and eosinophilic leukocytes (Weller et al. 1983), circulating monocytes (Ferreri et al. 1986), mast cells (Rouzer et al. 1982), and alveolar macrophages (Razin et al. 1983).

The biological properties of LTC_4 comprise induction of airway obstruction, constriction of coronary arteries, hypotension, and plasma extravasation. Leukotriene formation during shock may therefore mediate the symptoms and cause the death of an animal in anaphylaxis (Lewis and Austen 1984; Piper 1984; Feuerstein 1985; Lefer 1986). In the present report evidence is presented for the formation of one of these metabolites in vivo as well as for endogenous leukotriene production in anesthetized guinea pigs undergoing anaphylactic shock.

2 Materials and Methods

Chemicals. [5, 6, 8, 9, 11, 12, 14, 15 − $^3H_8$]LTC_4 (Hammarström 1981), was prepared biosynthetically as described. Synthetic LTB_4, LTC_4, LTD_4, and LTE_4, were kindly provided by J. Rokach, Merck-Frosst Inc., Canada. [14, 15 − $^3H(n)$]LTD_4 (48 Ci/mmol) was obtained from Amersham. A radioimmunoassay for LTC_4, including [14, 15 − $^3H(n)$]LTC_4 (39 Ci/mmol), was purchased from New England Nuclear. γ-Glutamyl transferase from porcine kidney (EC 2.3.2.2), 4-hydroxy-2,2,6,6,-tetramethyl-piperidinooxy (HTMP), ovalbumin (grade III), and pyrilamine maleate, were purchased from Sigma. Sodium pentobarbital, diazepam (Roche), chloramphenicol (Dumex), cholic and dehydrocholic acid (Pharmacia) and dexamethasone-21-phosphate (Merck Sharp and

Nigam et al. (Eds.), Eicosanoids,
Lipid Peroxidation and Cancer
© Springer-Verlag Berlin Heidelberg 1988

Dohme) were purchased from Karolinska Apoteket (a local pharmacy). LY 171883, sodium salt was a gift from Dr. J. Fleisch of Lilly Research Laboratories, Indianapolis, USA. Solvents and chemicals were high performance liquid chromatography (HPLC) grade or *pro analysi* purity.

Male Hartley guinea pigs (8–15 weeks, 300–600 g) were obtained from Sahlins, Malmö, Sweden. The animals were fed an unrestricted standard diet purchased from EWOS; Södertälje, Sweden. For metabolic experiments, animals were pretreated with dexamethasone (50 mg/kg; i.p.) 16–24 h before bile duct cannulation. On the day of the experiment diazepam (5 mg/kg) was given i.p. 30 min before anesthesia with pentobarbital 30 mg/kg, i.p.). After performing tracheostomy and intubation (polyethylene tubing, o.d., 2.4 mm), the bile duct was cannulated (polyethylene tubing, o.d., 1.1 mm). [^{3}H$_8$]LTC$_4$, (1 µCi, dissolved in 0.4 ml 0.9% saline/ethanol 98:2 v/v) was injected into the heart or the vena cava inferior. Bile was collected for 5 h in ice-cold methanol/ water (9:1, v/v) containing 1mM HTMP and 0.5 mM EDTA (pH 7.4). Methanol was added to 80%, and 0.1 ml aliquots were removed for radioactivity determination. The remainder was centrifuged and analyzed by reverse phase(RP)-HPLC (see below). For analysis of leukotriene formation during shock, guinea pigs were sensitized 14–16 days before antigen challenge by (a) i.p. injection of 10 mg/kg of ovalbumin; and (b) i.p. injection of 3 mg/kg of ovalbumin plus 300 mg/kg of Al(OH)$_3$. Al(OH)$_3$ was added to the antigen solution 1 h before injection. The developing antibodies (Andersson 1982) are IgG in procedure (a) and IgE plus IgG in procedure (b). To minimize contribution of trauma-induced leukotriene production (Denzlinger et al. 1985) the following surgical methods were used:

1. 2 days before immunological challenge the cystic duct was ligated, the gall bladder was punctured, and the bile duct was incised. An anastomosis (Silastic 0.6 mm, tubing, o.d. connected at both ends with o.d. 1.1 mm polyethylene tubing) was inserted into the bile duct (orientation toward the liver) and into the duodenum. Upon closure of the abdominal wall, a loop of the anastomosis tubing was placed subcutaneously to allow subsequent easy access. The diazepam (5 mg/kg)/diethyl ether anesthesia was then interrupted. From 12–16 h before abdominal surgery and throughout the remainder of the experiments, the animals were given chloramphenicol (50 mg-kg-day) and cholic plus dehydrocholic acid (10 plus 2.5 mg kg-day-) orally.
2. After emptying the gall bladder, as described above, the bile duct of diazepam-treated, pentobarbital-anesthesized animals was cannulated (polyethylene tubing, o.d. 1.1 mm). The animals were kept anesthesized under a heating lamp for 3 h.

Ovalbumin, 0.2 mg/kg, was then given i.v. to pentobarbital-anesthesized animals prepared according to either surgical method 1. or 2. Rechallenge with 2 mg/kg ovalbumin was performed after 2.5 h if the initial dose was insufficient. Animals prepared according to surgical method 1. received 30 mg/kg i.v. of LY 171883 sodium salt (a LTD$_4$/E$_4$ receptor antagonist) 30 min before challenge (Fleisch et al. 1985). In surgical method 2. pyrilamine maleate (1 mg/kg i.v.) was given at the time of antigen administration. Guinea pigs without clear signs of anaphylactic reaction (respiratory distress, impaired breathing, cyanosis and micturition) were not used. Bile was collected under argon in ice-cold methanol / water (9:1, v/v) containing 1 mM HTMP and 0.5 mM EDTA (pH 7.4). Bile volumes were determined gravimetrically and the methanol concentration was adjusted to 80% v/v. After 3 h at 0 °C the samples were centrifuged for 10 min at 10000 × g and 4 °C. Aliquots of the supernatants were evaporated (at 37 °C), to dryness dissolved in 0.42 ml of methanol / water (3:7, v/v) and subjected to RP-HPLC after addition of [^{3}H$_8$]LTD$_4$, (315 pCi, 6.6 fmol) as a standard.

High-Performance Liquid Chromatography / Radioimmunoassay. RP-HPLC was performed on Radial Pak C$_{18}$ columns (5 μm particles, i.d. 8 mm) equipped with C$_{18}$ μBondapak precolumns (Waters Associates) The mobile phase consisted of methanol / water /acetic acid (64:36:0.1, v/v/v) containing 1 mM EDTA, (pH 6.2 adjusted with NH$_4$OH). The flow rate was 2 ml/min and the column recovery of LTD$_4$ was 84%. For ion-pair RP-HPLC a C$_{18}$ Nuleosil column was used. The mobile phase consisted of methanol / water / acetic acid (70:30:0.1), v/v/v) containing 2.5 mM pentanesulfonic acid (cf. Ziltener et al. 1983). The flow rate was 1 ml/min. RP-HPLC fractions for radioimmunological analysis were immediately neutralized with 1 M K$_2$CO$_3$ and stored at − 20 °C under argon for up to 5 days. An ultraviolet light absorption detector was used to monitor the 280 nm absorbance (Spectromonitor II, Laboratory Data Control) of the effluent.

The LTC$_4$ antibody used cross-reacted with leukotrienes and analogues as follows: LTC$_4$, 100%; LTD$_4$, 44% LTE$_4$, 6%; LTC$_4$ sulphone, 11%; LTD$_4$ sulphone, 8%; 5-HETE, 0.04%; and LTB$_4$, 0.003% (relative percentage, molar basis at 50% binding). The lower detection limit was 50 fmol for LTC$_4$, and 100 fmol for LTC$_4$, LTC$_4$ and LTD$_4$. Standard curves were constructed using freshly RP-HPLC purified standards. Neutralized RP-HPLC fractions (0.6–1 ml), which coeluted with the [^{3}H]LTD$_4$ standard, were evaporated to dryness and dissolved in radioimmunoassay buffer.

Enzymatic Modification of Endogenously Formed LTD$_4$. Deproteinized bile (0.4 ml, containing 315 pCi [^{3}H]LTD$_4$ as standard) collected 0–30 min after antigen challenge was evaporated to dryness, dissolved in 0.45 ml methanol / water, 3:7 v/v and fractionated by RP-HPLC (solvent 7). Aliquots (0.8 ml) of the fractions collected were used for determination of radioactivity. Fractions coeluting with the [^{3}H]LTD$_4$ standard were evaporated, dissolved in 0.05 ml 0.1 M Tris-HCL (pH 8.0) / 10 mM MgCl$_2$ / 8 mM glutathione / containing γ-glutamyl transferase (0.2 mg/ml) and incubated for 30 min at 37 °C. One volume of methanol / acetic acid (9:1, v/v) was added; the mixture was evaporated, redissolved in methanol / water (3:7 v/v) and applied to RP-HPLC. Fractions were analyzed by radioimmunoassay as described above.

3 Results

3.1 Time Course of Hepatobiliary Tritium Elimination

[$^3H_8$]LTC$_4$ was administered i.v. to bile fistulated guinea pigs. Thirty-two percent of the given radioactivity was excreted in bile during the first 15 min after injection, 50% during the first 60 min after injection and 61% after 5 h of bile collection (mean values from two experiments; variation ± 10%). The amount of radioactivity in urine collected by puncturing the bladder 5 h after administration was determined in one experiment and revealed less than 7% of the administered dose. Liver, kidney, lung, heart, and spleen contained 11.8%, 0.6%, 0.05%, 0.03%, and 0.02% of the administered radioactivity, respectively. Thus, the total recovery of radioactivity amounted in ~ 70% of the injected dose.

3.2 Identification of LTD$_4$ as a Major Metabolite of LTC$_4$ in Guinea Pig Bile

RP-HPLC analysis of extracts from bile, obtained after i.v. injection of [$^3H_8$]LTC$_4$, indicated extensive conversion to predominantly LTD$_4$: 5 min after tracer administration, 47% of the excreted radioactive material cochromatographed with synthetic LTD$_4$, and 21% with LTC$_4$. After 30 min, polar metabolites eluting close to the solvent front predominated (42%); LTD$_4$ constituted 33%; LTE$_4$, 16%; and LTC$_4$, 7.5% of the radioactivity. Later (at 2 h) 70% of the excreted material consisted of polar metabolites.

Figure 1 shows the cumulative excreted radioactivity for each component in bile vs. time: after 15 min about half of the maximally excreted radioactivity had been eliminated: 15% of the injected dose was LTD$_4$, 9% was polar metabolites, 6% was LTC$_4$, and 1% was LTE$_4$. The corresponding values after 30 min were 18% for LTD$_4$, 12% for polar metabolites, 7% for LTC$_4$, and 2.4% for LTE$_4$, and after 3 h values were, 20% for LTD$_4$, 19% for polar metabolites, 8.5% for LTC$_4$, and 4.2% for LTE$_4$ (Fig. 1). The radioactive material from the punctured urinary bladder consisted mainly of polar material as previously reported (Hammarström 1981).

4 Endogenous Formation for LTC$_4$, in Guinea Pigs

4.1 Endogenous Generation of LTD$_4$ During Anaphylactic Shock

Analyses of biliary cysteine-containing leukotrienes during anaphylactic shock were performed by RP-HPLC/radioimmunoassay in five experiments. As reported for the rat (Denzlinger et al. 1985), bile duct surgery was accompanied by appearence of endogenous cysteine-containing leukotrienes in guinea pig bile.

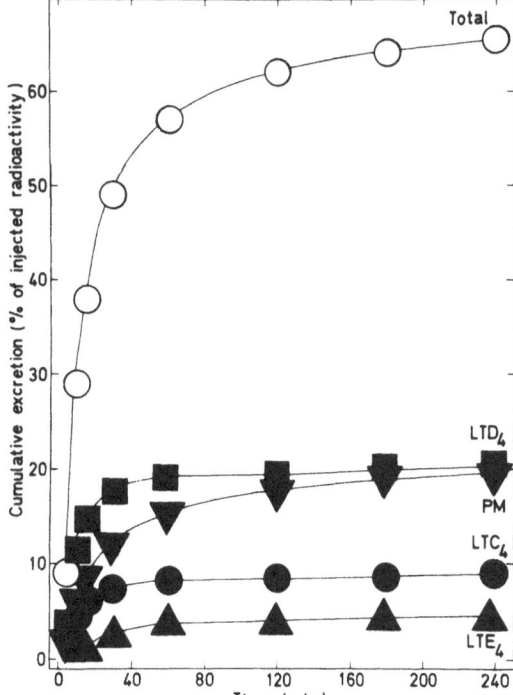

Fig. 1. Cumulative excretion of biliary [^{3}H]leukotriene metabolites. The total accumulated excretion of each metabolite is plotted as a function of time after [^{3}H$_8$]LTC$_4$ administration. *PM,* polar metabolites

The prechallenge levels of LTD$_4$ were 12 $\pm$ 2 nM in bile cannulated compared to 8.5 $\pm$ 0.5 nM in anastomosis animals (p $<$ 0.2). These values were sufficiently low not to interfere with the determination of biliary LTD$_4$ after anaphylactic challenge. Surgery method 1. frequently led to losses of animals due to wound infections, postsurgical adhesions, and ileus, whereas method 2. did not introduce these complications.

Anaphylaxis was elicited by i.v. administration of a low dose of antigen. In most animals, this led to development of clear signs of shock as evidenced by breathing difficulties, cyanosis, and micturition. Responding animals usually survived 20–60 min after challenge. If an animal survived for 2.5 h, a higher dose of antigen was given. This resulted in shock and subsequently death of the animal. Guinea pigs that did not respond even to the higher ovalbumin dose were not included in the investigation. Bile samples were collected during the entire survival period after antigen challenge and for at least 30 min before challenge. The prechallenge bile samples provided control levels of leukotrienes for each animal.

Bile samples from one experiment were subjected to RP-HPLC, and all fractions collected were analyzed by radioimmunoassay (Fig. 2). An analysis of prechallenge bile shows only low and evenly distributed immunoreactivity. Corresponding analyses of two bile samples (0–15 and 15–45 min after challenge) had a major immunoreactive component of the same retention time as the [^{3}H$_8$]LTD$_4$ standard. Integration of these chromatograms indicated biliary LTD$_4$ concentra-

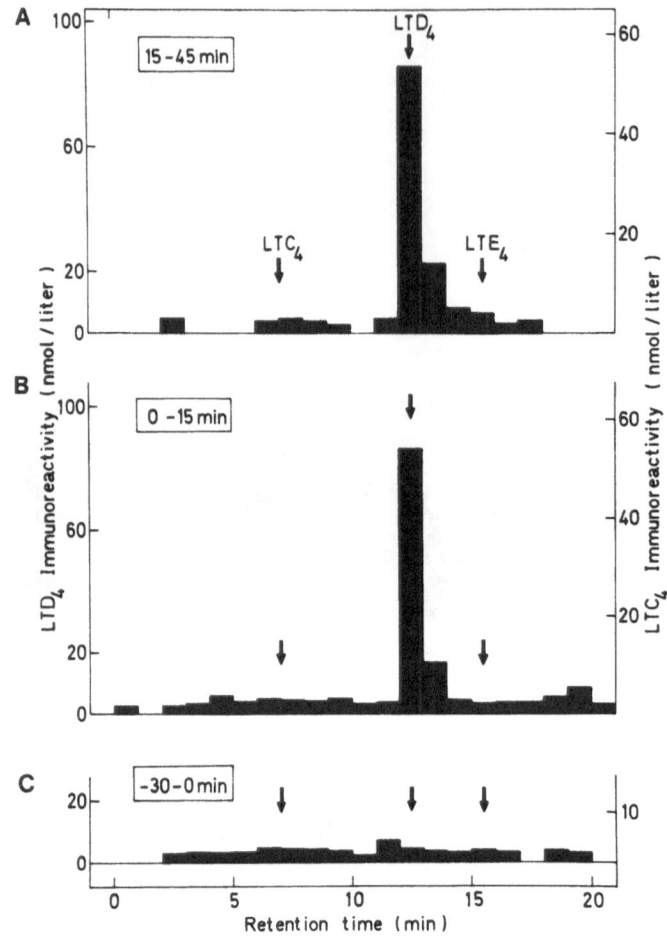

Fig. 2, A-C. Radioimmunochromatograms of biliary cysteine-conteining leukotriene in sensitized guinea pigs before and after antigen challenge. Deproteinized samples, corresponding to 0.1 ml of bile, collected 0–30 min before **(A)** and 0–15 min **(B)** or 15–45 min after immunological challenge **(C)** were analyzed by RP-HPLC/radioimmunoassay

tions of 108 nM and 116 nM, respectively. Radioimmuno-chromatographic analyses of bile collected 0–15 min after challenge gave similar results. Since both the results shown in Fig. 1 and the analyses of endogenous biliary leukotrienes in the three experiments described above indicated that LTD$_4$ was the major LTC$_4$ metabolite in guinea pig bile, radioimmunological analyses of bile samples in two subsequent experiments focused only on the LTD$_4$ region of the RP-HPLC chromatogram. The position of this region was determined using a trace amount of [^{3}H]LTD$_4$, as standard. The results of these analyses showed a basal LTD$_4$ concentration in -30 to 0-min prechallenge bile of 10 ± 1 nM (mean $\pm$ SEM, $n = 5$).

Bile collected 0–15 or 0–30 min after triggering the immunological event contained 86 ± 10 nM LTD$_4$ (mean $\pm$ SEM, $n = 5$, $P < 0.001$, Fig. 3B). The bile flow decreased to about half, from 83 ± 9 before to 38 ± 4 ml min^{-1} kg^{-1} after antigen challenge (mean $\pm$ SEM, $n = 5$ $P < 0.01$, Fig 3A), and the LTD$_4$ excretion rate rose 3.6-fold from 0.99 ± 0.16 before to 3.18 ± 0.39 pmol min^{-1} kg^{-1} after antigen challenge (mean $\pm$ SEM, $n = 5$, $P < 0.002$, Fig. 3C).

4.2 Identification of Endogenous LTD$_4$ by Enzymatic Transformations to LTC$_4$ and LTE$_4$

Part of the pooled LTD$_4$-containing fraction from RP-HPLC fractionated bile 0–30 min after the challenge were treated with an enzyme preparation containing γ-glutamyl transferase and dipeptidase. Glutathione was also added to force the former reaction in its reverse direction (cf. Hammarström 1981). Analysis of the reaction products by RP-HPLC and radioimmunoassay revealed two immunoreactive components of the same retention times as LTE$_4$ and LTC$_4$. These results provided further evidence that the endogenously excreted leukotriene was LTD$_4$.

5 Discussion

To analyze LTC$_4$ formation in vivo knowledge on the metabolism and excretion routes of this compound is required. During the past several years, the metabolism of LTC has been investigated in vitro and in vivo in several species (for review see Hammarström et al. 1985). Biliary/fecal excretion predominated in several rodents [mouse (Girard et al. 1982), guinea pig (Hammarström 1981), and rat (Denzlinger et al. 1985; Örning et al. 1986; Hagmann et al. 1986; Hagmann et

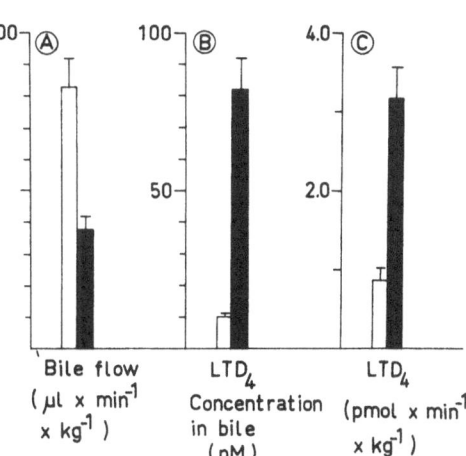

Fig. 3, A–C. Effects of antigen challenge on bile flow **(A)**, biliary LTD$_4$, concentration **(B)**, and biliary LTD$_4$ excretion rate **(C)**. *Open bars*, prechallenge values; *solid bars*, post challenge values. The *bars* represent mean values $\pm$ SEM from five experiments. The differences between values before and after antigen challenge were significant. P < 0.01 **(A)**; P < 0.001 **(B)**; and P < 0.002 **(C)**

al. 1984; Ormstad et al. 1982)] as well as in bile duct-cannulated monkey (Denz-linger et al. 1986), whereas urinary excretion was more prominent than fecal excretion in man (Örning et al. 1985).

LTC$_4$ metabolism in the guinea pig was investigated with the aim of identify-ing suitable metabolites for analyzing leukotriene formation during anaphylaxis. Animals were pretreated with dexamethasone to prevent leukotriene formation in response to the surgical manipulations by inhibiting precursor fatty acid re-lease. This improved the experimental situation markedly. The structures of two fecal LTC$_4$ metabolites in the rat have been found to be N-acetyl-LTE$_4$ and N-acetyl-trans-LTE$_4$ (Örning et al. 1986). LTE$_4$ was identified as a urinary LTC$_4$ metabolite in man (Örning et al. 1985) and a biliary metabolite in the monkey (Denzlinger et al. 1986). The present results demonstrated that LTD$_4$ is a major biliary LTC$_4$ metabolite in the guinea pig.

The respiratory, cardiovascular, and microciruculatory effects of LTC$_4$ and LTD$_4$ mentioned above suggest that they can produce the major consequences of acute systemic anaphylaxis (Lewis and Austen 1984; Piper 1984; Feuerstein 1985; Lefer 1986). It has also been documented that inhibitors of leukotriene formation can prevent several anaphylactic responses (Aehringhaus et al. 1983; Hedqvist et al. 1984; Dahlen et al. 1985). The purpose of this investigation was to develop a suitable animal model which would enable the demonstration of in vivo leukotriene synthesis during systemic anaphylaxis. Due to the hepatobiliary elimination of LTC$_4$ in the guinea pig, bile was chosen for these measurements. To avoid interference by trauma-induced leukotriene formation (see above), a bilio-duodenal anastomosis was implanted 2 days before ovalbumin challenge. The analytical results were satisfactory, but the death rate was high due to infec-tions, postsurgical adhesions and ileus. This necessitated pretreatment with chlo-ramphenicol and bile flow stimulating agents (cholic and dehydrocholic acid). An alternative approach, consisting of bile duct surgery followed by a 3-h inter-val before antigen challenge, proved to be more successful with a failure rate (lethality during and after the operation) close to zero. Pyrilamine maleate treat-ment in this method increased the survival time after challenge suggesting that histamine receptor interaction contributed to the lethal effect of antigen chal-lenge. Bile, collected before and after i.v. challenge with ovalbumin, was ana-lyzed by RP-HPLC and radioimmunoassay. LTD$_4$ was identified as the major immunoreactive component in anaphylactic guinea pig bile. the analyses showed an 8.6-fold increase in LTD$_4$ concentration in bile after challenge compared to bile before the challenge. Antigen provocation was also accompanied by de-creased bile flow (to about half) as observed in rats undergoing endotoxin shock (Hagman et al. 1984). In spite of the reduced bile flow, the LTD$_4$ excretion rate had increased 3.6-fold after antigen challenge of sensitized guinea pigs. The ob-servation that LTD$_4$ is the predominant endogenous leukotriene metabolite in guinea pig bile was in good agreement with the metabolic tracer studies per-formed as part of this investigation and in obvious contrast to results obtained in rats and monkeys. In the latter species N-acetyl-LTE$_4$ (Örning et al. 1986; Hag-mann et al. 1986) and LTE$_4$ (Denzlinger et al. 1986), respectively, were the dom-inating LTC$_4$ metabolites. It is interesting that LTD$_4$ predominates in the guinea pig since LTD$_4$ is more active than LTC$_4$, N-acetyl-LTE$_4$, and LTE$_4$ on several biological parameters (Lefer 1985; Lewis et al. 1981) and because guinea pigs are especially sensitive to exogenous cysteine-containing leukotrienes (Feuerstein 1984) as well as to conditions that induce endogenous synthesis of these compounds such as trauma and anaphylaxis. Intravenous administration of LTC_4 or LTD_4 (1 nmol/kg) evoked widespread plasma extravasation in the guinea pig (marker, Evans blue) (Hua et al. 1985) and a 10–fold higher dose produced shock-like states in artifically ventilated guinea pigs (Lux et al. 1983). In spontaneously breathing guinea pigs, 2 nmol of LTD_4 per kg i.v. can be lethal (S.-E. Dahlen, personal communication), Intra-atrial injection of 0.1 nmol of LTC_4 per kg produced marked cardiovascular effects in man (Kaijser et al. to be published).

The present analyses demonstrate that LTC_4 is endogenously produced in guinea pigs after immunological challenge and suggest that the amounts formed during the initial 30-min period may approach 1 nmol/kg. This figure is based on the LTD_4 excretion rate (129.6 nmol 30 min^{-1} kg^{-1}), the HPLC recovery of LTD_4 (84%), and the excretion of i.v. administered [³H₈]LTC_4 as biliary LTD_4 (18% during 30 min).

Acknowledgement. (This work was supported by grants from the Swedish Medical Research Council; project 03X-5914).

References

Aehringhaus U, Peskar BA, Wittenberg HR, Wölbling RH (1983) Effect of inhibition of synthesis and receptor antagonism of SRS-A in cardiac anaphylaxis. Br J Pharmacol 80:73-80

Andersson P (1982) Effects of inhibitors of anaphylactic mediators in two models of bronchial anaphylaxis in anaesthetized guinea-pigs. Br J Pharmacol 77:301-307

Dahlen SE, Palmertz UJ, Hedqvist P (1985) Methylprednisolone inhibition of leukotriene-dependent airway anaphylaxis in the guinea pig in vivo. Agents and Actions 17:310-311

Denzlinger C, Rapp S, Hagmann W, Keppler D (1985) Leukotrienes as mediators in tissue trauma. Science 230:330-332

Denzlinger C, Guhlman A, Scheuber PH, Wilker D, Hammer DK, Keppler D (1986) Metabolism and analysis of cysteinyl leukotrienes in the monkey. J Biol Chem 261:15601-15606

Ferreri NR, Howland WC, Spiegelberg HL (1986) Release of leukotrienes C₄ and B₄ and prostaglandin E₄ from human monocytes stimulated with aggregated IgG, IgA and IgE. J Immunol 136:4188-4193

Feuerstein G (1984) Leukotrienes and the cardiovascular system. Prostaglandins 27:781-802

Feuerstein G (1985) Autonomic pharmacology of leukotrienes. J Auton Pharmacol 5:149-168

Fleisch JH, Rinkema LE, Haisch KD, Swanson-Bean D, Goodson T, Ho PPK, Marshal WS (1985) LY 171883, 1-⟨2-hydroxy-3-propyl-4⟨-(1H-tetrazol-5-yl)butoxy⟩phenyl⟩ethanone, an orally active leukotriene D₄ antagonist. J Phrmacol Exp Ther 233:148-157

Girard Y, Larue M, Jones T, Rokach J (1982) Synthesis of the sulfones of leukotrienes C₄, D₄ and E₄. Tetrahedron Lett 23:1023-1026

Gustafsson B (1959) Lightweight stainless steel systems for rearing germ-free animals. Ann NY Acad Sci 78:17-28

Hagmann W, Denzlinger C, Keppler D (1984) Role of peptide leukotrienes and their hepatobiliary elimination in endotoxin action. Circ Shock 14:223-235

Hagmann W, Denzlinger C, Rapp S, Weckbecker G, Keppler D (1986) Identification of the major endogenous leukotriene metabolite in the bile of rats as N-acetyl leukotriene E₄. Prostaglandins 31:239–251

Hammarström S (1981) Metabolism of leukotriene C₃ in the guinea pig. J Biol Chem 256:9573–9578

Hammarström S (1983) Leukotrienes. Ann Rev Biochem 52:355–377

Hammarström S, Örning L, Bernström K (1985) Metabolism of leukotrienes. Mol Cell Biochem 69:7–16

Hedqvist P, Dahlen SE, Palmertz U (1984) Leukotriene-dependent airway anaphylaxis in guinea pigs. Prostaglandins 28:605–608

Hua XY, Dahlen SE, Lundberg JM, Hammarström S, Hedqvist P (1985) Leukotrienes C₄, D₄ and E₄ cause widespread and extensive plasma extravasation in the guinea pig. Naunyn Schmiedebers Arch Pharmacol 330:136–141

Lefer AM (1985) Eicosanoids as mediators of ischemia and shock. Fed Proc 44:275–280

Lefer AM (1986) Leukotrienes as mediators of ischemia and shock. Biochem Pharmacol 35:123–127

Lewis RA, Austen KF (1984) The biologically active leukotrienes. J Clin Invest 73:889–897

Lewis RA, Drazen JA, Austen KF, Toda M, Brion F, Marfat A, Corey EJ (1981) Contractile activities of structural analogs of leukotrienes C and D: role of the polar substituents. Proc Natl Acad Sci USA 78:4579–4583

Lux WE Jr., Feuerstein G, Faden AI (1983) Thyrotropin-releasing hormone reverses the hypotension and bradycardia produced by leukotriene D₄ in unanaesthetized guinea pigs. Prost Leukotr Med 10:301–307

MacGlashan DW Jr., Scleimer RP, Peters SP, Schulman ES, Adams GK, Sobotka AK, Newball HH, Lichtenstein LM (1983) Comparative studies of human basophils and mast cells. Fed Proc 42:2504–2509

Ormstad K, Uehara N, Orrenius S, Örning L, Hammarström S (1982) Uptake and metabolism of leukotriene C₃ by isolated rat organs and cells. Biochem Biophys Res Comm 104:1434–1440

Örning L, Kaijser L, Hammarström S (1985) In vivo metabolism of leukotriene C₄ in man. Biochem Biophys Res Commun 130:214–220

Örning L, Norin E, Gustafsson B, Hammarström S (1986) In vivo metabolism of leukotriene C₄ in germ-free and conventional rats J Biol Chem 261:766–771

Piper PJ (1984) Formation and actions of leukotrienes. Physiol Rev 64:743–761

Razin E, Mencia-Huerta JM, Stevens RL, Lewis RA, Liu FT, Corey EJ, Austen KF (1983) IgE mediated release of leukotriene C₄, chondroitin sulfate E proteoglycan, β-hexosaminidase, and histamine from cultured bone marrow-derived mouse mast cells. J Exp Med 157:189–201

Rouzer CA, Scott WA, Hamil AL, Liu FT, Katz DHW, Cohn ZA (1982) Secretion of leukotriene C and other arachidonic acid metabolites by macrophages challenged with immunoglobulin E immune complexes. J Exp Med 156:1077–1086

Samuelsson B (1983) Leukotrienes: mediators of immediate hypersensitivity reactions and inflammation. Science 220:568–575

Weller PF, Lee CW, Foster DW, Corey EJ, Austen KF, Lewis RA (1983) Generation and metabolism of 5-lipoxygenase pathway leukotrienes by human eosinophils: predominant production of leukotriene C₄. Proc Natl Acad Sci USA 80:7626–7630

Ziltener HJ, Chavaillaz PA, Jörg A (1983) Leukotriene formation by eosinophil leukocytes. Analysis with ion-pair high pressure liquid chromatography and effect of the respiratory burst. Hoppe Seylers Z Physiol Chem 364:1029–1037

Theoretical Calculations of the Structures of Some Prostaglandins and Inhibitors of Lipoxygenase

J. Barker and C. Thomson

NFCR Project, Department of Chemistry, University of St. Andrews,
St. Andrews KY16 9ST, UK

1 Introduction

The methods of computational chemistry have now developed to the point where calculations of the structure and properties of molecules of biological importance can yield important information complementary to that obtained in experimental studies (Richards 1983). There are many recent reviews of these exciting developments (Naráy-Szabó 1986; Beveridge and Jorgensen 1986), and the advent of sophisticated colour graphics displays has seen a mushrooming of applications to a wide range of problems, especially in pharmacology, where computer assisted drug design is of increasing importance (Venkataraghvan and Feldman 1985).

In the study of the various stages in cancer induction we have previously applied these methods to the elucidation of the mechanisms involved in the decomposition of N,N^1-dimethylnitrosamine (Thomson and Reynolds 1986), and also to the mechanism of action of the enzyme glyoxalase-I and the design of inhibitors of this enzyme (Thomson et al. 1985, Brandt et al. 1986).

We have recently become interested in tumour promotion (Thomson 1986; Cuthbertson and Thomson 1987), and in particular the role of lipid peroxidation and the molecules of the arachidonic acid cascade in this part of the carcinogenic process.

There is a great deal of experimental evidence on the role of prostaglandins and related molecules in cancer (Fischer and Slaga 1985) but there have been very few theoretical investigations of these relatively large molecules using the most up-to-date methodology. We report the results of our preliminary studies in this paper. Full details will be reported elsewhere.

One of the difficulties in the study of the products of the arachidonic cascade is the transient nature of the various molecules. Structural data on these molecules is available for some of the most stable of the species, but in order to fully understand the mechanistic aspects of the various pathways, we need structural and property data on *all* the intermediates. This information can now be reliably obtained from theoretical calculations.

Nigam et al. (Eds.), Eicosanoids,
Lipid Peroxidation and Cancer
© Springer-Verlag Berlin Heidelberg 1988

2 Theoretical Methods and Computational Aspects

We refer the reader to some excellent reviews for details of the methods used (Richards 1983; Clark 1985) but briefly summarise here the important points.

It is now feasible to locate the energy minima on the potential energy surface of moderately sized polyatomic molecules and the resulting structure is usually in very good agreement with available experimental structural data (Clark 1985; Hehre et al. 1986; Dewar 1985). Therefore, we have considerable confidence in the reliability of *predicted* structures from theory: particularly important in the case of unstable intermediates.

The accuracy of the results is, of course, dependent on the method used. For the largest molecules, our strategy is to use one of the empirical molecular mechanics methods to obtain a fully optimized structure, in particular, the AMBER programme (Kollman et al. 1986) which is then used as input for a semi-empirical self-consistent field method, which is of course a proper quantum mechanical method. We have used the MNDO method (Dewar and Thiel 1977) as implemented in the MOPAC programme (Stewart 1987). Both the structures obtained by the molecular mechanics and the MNDO are usually in very good agreement with experiment. In addition, the wave function obtained with the latter method can be used to calculate other useful molecular properties.

Of these, the molecular electrostatic potential (MEP) around the molecule has proved to be of very great value in the study of the interactions between enzymes and their substrates and inhibitors (Politzer and Truhlar 1981). This property is very conveniently displayed on a colour graphics terminal, and the different regions of potential are colour coded for ease of interpretation.

The calculations reported in the present paper were carried out on the University VAX 11/785 and our own MICROVAX-2 GPX work station, using the programmes AMBER and MOPAC V3.1, which contains the option for carrying out MNDO calculations. The display of the structures and MEP maps was carried out using either the ChemX programme (Chemical Design, Oxford) or our own graphics package (Edge 1987; Higgins, unpublished).

We have also calculated more accurate ab-initio wave functions for the optimized geometry (Hehre et al. 1986) but these results will be reported at a later date.

3 Previous Structural Information and Theoretical Work

There has not been any previous work on these molecules at the level of theory we describe here; there have been a few CNDO calculations (Kothekar and Kalia 1985), but not with geometry optimization. There have also been some early molecular mechanics calculations (Murakami and Akahori 1974, 1977a, b, 1979).

X-ray structures have been determined for several of the prostaglandins, including PGE_2, $PFG_{2\alpha}$ and $PGF_{2\beta}$ (de Titta et al. 1980; Langs et al. 1977), but not for the other molecules dealt with below. We emphasise, however, that complete geometry optimization is carried out in all cases, and the agreement with experiment is discussed below.

In the present paper we will deal first primarily with the calculations on PGE_2 and $PGF_{2\beta}$ for which there is reliable crystal structure data (de Titta et al. 1980) and second with the structure of some lipoxygenase inhibitors.

4 Lipoxygenase and Tumour Promotion

This paper is the first in a series dealing with the structure of a large number of molecules of importance in the arachidonic acid cascade, and in the present paper we concentrate on the molecules of importance in the 5-lipoxygenase pathway (Needleham et al. 1986) because of its importance in the study of tumour promotion, and in cancer in general.

During the last few years, many different studies have shown that inhibitors of 5-lipoxygenase can inhibit tumour promotion by the phorbol esters and related molecules (Furstenberger and Marks 1985; Marks and Furstenberger 1985; Fischer et al. 1982). Unfortunately, there is not yet any structural or sequence information available on the active site of the enzyme, but a comparison of the structure of the substrate and inhibitors of the enzyme might shed some light on the structural and electonic features of the molecules which might facilitate the design of more effective inhibitors.

5 Results and Discussion

Arachidonic acid is oxidised first to the 5-hydroperoxyeicosa-tetraenoic acid (HPETE). This molecule loses an oxygen atom to give 5-S hydroxy-eicosa-5, 8, 11, 14 tetraenoic acid (HETE). Similar molecules oxidised at the 12- position are produced by 12-lipoxygenase. (Needleham et al. 1986).

The mechanistic aspects of the potential energy surfaces of these reactions will be investigated later: at the present time we report only the structure of the molecules involved, and those of the most important inhibitors.

The results of a geometry optimization yield, for the MNDO method, not only the geometrical parameters but also the heat of formation ΔH_f^θ (298k), the molecular dipole moment, and the vibrational frequencies. Table 1 gives the computed values of ΔH_f^θ, the total energy and the dipole moments for the lowest energy conformation of the molecules studies in this work.

Table 1. Heats of formation, total energy and dipole moment for MNDO optimised geometry

Molecule	ΔH_f^{θ} (298 k)[a]	E[b]	μ[c]
Arachidonic acid	− 82.09	−3655.98	1.73
5-HPETE	− 97.20	−4297.54	0.60
5-HETE	− 122.24	−3978.17	2.25
PGE$_2$	− 227.40	−4623.63	6.796
PGF$_{2\beta}$	− 244.45	−4652.71	2.29
12-HPETE	− 96.62	−4297.51	1.79
12-HETE	− 118.94	−3978.03	1.30
PDG$_2$	− 231.98	−4623.83	1.85
Thromboxane	− 273.72	−4974.42	3.38

[a] Values in kcal mol^{-1}
[b] Values in eV
[c] Values in Debye (D)

5.1 Structure of Arachidonic Acid

It is clear that the long alkyl chain makes the molecules very flexible, but the minimum energy configuration is in the expected U-shape (Fig. 1). The MEP is more or less neutral over most of the molecule, except for the clearly distinct carboxyl group. (Fig. 1)

There is no crystal structure data available, but solution NMR studies are consistent with the molecule existing in an equilibrium between the U-form and more open chain forms (Gunstone 1974).

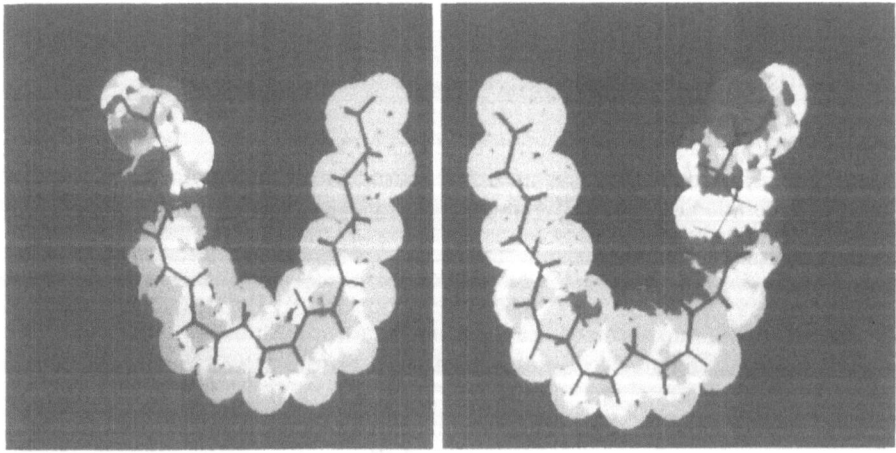

Fig. 1. MEP on van der Waals' surface of arachidonic acid

5.2 The Structures of 5-HPETE and 5-HETE

The overall shape is similar to that of AA, but in the case of 5-HPETE, the two chains are somewhat closer. Table 2, which lists some of the computed bond lengths and angles, shows how similar the structural parameters are for these two molecules. The MEP are illustrated in Fig. 1 and 3.

Table 2. Selected calculated bond lengths and interbond angles in HPETE and HETE

	HPETE	HETE
Bond lengths		
O1 –C2	1.232	1.232
O3 –C2	1.357	1.357
C12–O17	1.433	1.404
C15–C18	1.348	1.348
C20–C22	1.350	1.350
O17–O55	1.292	–
O17–H55	–	0.947
Bond angles		
O1 –C2 –O3	118.2	118.1
C12–O17–H55	–	112.4
C12–O17–O55	115.8	–

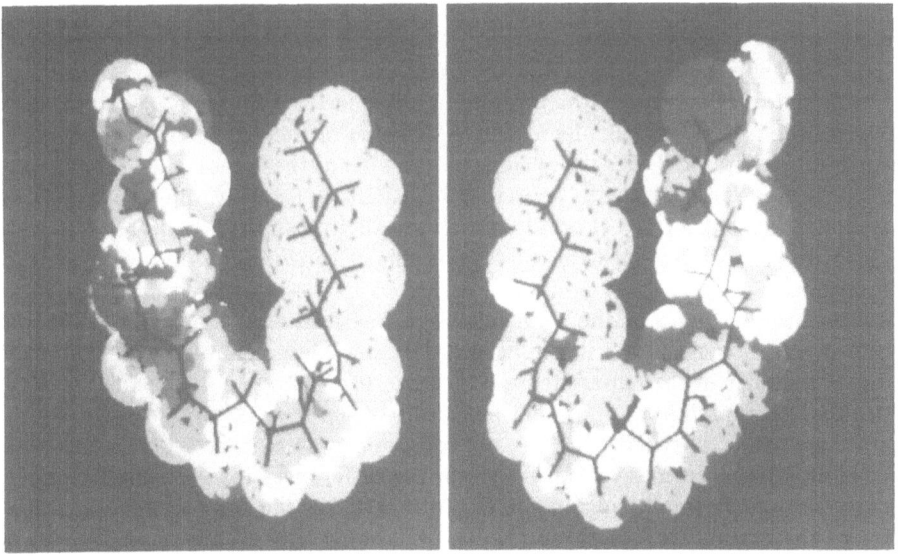

Fig. 2. MEP on van der Waals' surface of 5-HPETE

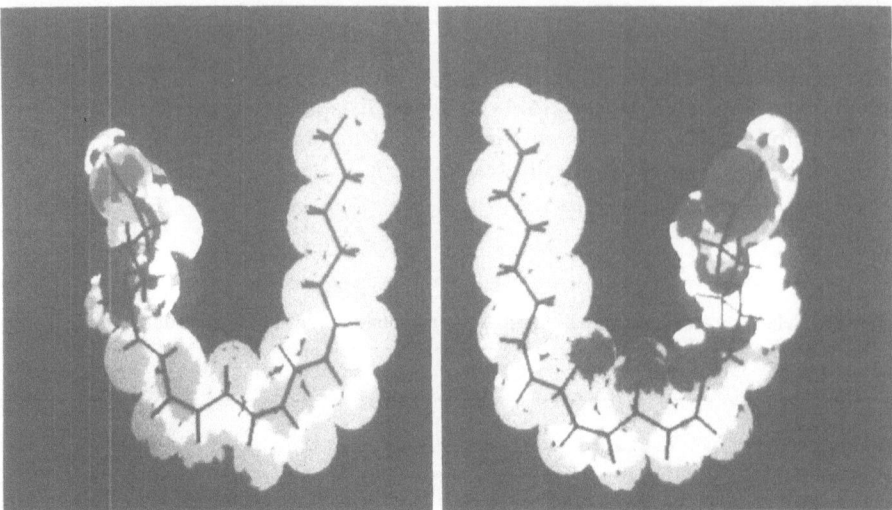

Fig. 3. MEP on van der Waals' surface of 5-HETE

5.3 The Structures of 12-HPETE and 12-HETE

These two molecules are distinctly different in appearance, and of lower stability than the 5-isomers. Further discussion of these molecules will be reported at a later date.

5.4 Prostaglandins PGE$_2$ and PGF$_2$

The X-ray structure of these two molecules have been determined (de Titta et al. 1980), and our computed structures can be compared with them in detail. Table 3 gives the X-ray and calculated values of some of the key internal coordinates for both PGE$_2$ and PGF$_{2\beta}$. Considering the size of these molecules, the overall agreement is very good, with bond lengths usually in error by approximately 0.05 (too large except for $C = O$ bonds, which are slightly shorter than observed), angles within approximately 5°, and torsion angles to within 10–15°, with the correct sign.

The overall shape of the molecule is reproduced well, the main difference between the experimental structure and the calculated is that the two chains are closer together in the experimental structure. It is always possible that we have not located the global minimum, but so far we have not found a lower energy structure. The computed MEP are not reproduced here, since a black and white version is less informative, but in colour the differences between the two molecules are rather small, since the OH group MEP is the only significant difference. Further analysis of these results, together with full colour MEP, will be reported later. We emphasise that these are the first calculations which give a reliable wave function from which we can compute the MEP.

Table 3. Comparison of observed and calculated values of the internal coordinates for PGE_2 and $PGF_{2\beta}$

	PGE₂		PGF₂β	
	X ray	Calc	X ray	Calc
Bond lengths				
C1 –O12	1.196	1.226	1.192	1.230
C1 –O13	1.256	1.365	1.302	1.360
C8 –C9	1.489	1.546	1.548	1.578
C9 –O9	1.235	1.222	1.428	1.398
C9 –C10	1.464	1.531	1.527	1.566
C10–C11	1.511	1.559	1.532	1.557
C11–O11	1.405	1.395	1.440	1.396
C5 –C6	1.348	1.347	1.306	1.348
C13–C14	1.288	1.347	1.316	1.347
C15–O15	1.446	1.403	1.451	1.405
Angles				
C4 –C5 –C6	124.9	130.4	127.0	128.6
C7 –C8 –C9	117.3	116.0	114.6	116.1
C8 –C9 –O9	122.1	124.7	112.6	109.8
C10–C11–O11	113.6	110.0	109.0	116.0
C14–C15–O15	107.8	107.0	106.5	110.6
Torsion angles				
C5 –C6 –C7 –C8	– 126.7	– 131.9	140.8	126.2
C7 –C8 –C9 –O9	– 33.3	– 41.1	– 90.0	– 92.4
C9 –C10–C11–O11	– 151.6	– 140.4	– 146.5	– 144.9
C10–C11–C12–C13	163.9	148.2	160.4	156.1

6 Lipoxygenase Inhibitors

Since products of the lipoxygenase pathway are involved in a variety of pathological processes (Fischer and Slaga 1985), lipoxygenase inhibitors have been found to be of use in these conditions; in particular they inhibit the second stage of tumour promotion.

Among these compounds are some flavones, such as quercetin, previously studied by us in connection with the inhibition of glyoxalase-I (Brandt et al. 1986), ETYA (5, 8, 11, 14 eicosatetraynoic acid), NDGA (nordihydroguaretic acid), AA861 (2,35 trimethyl-6-(12-hydroxy 5, 10 dodecadiyniol)1, 4 benzoquinone), BW755C (3-amino-1-[m-(trifluoromethyl)-phenyl]-2-pyrazoline), and phenidone (1-phenyl-3-pyrazolidinone) (Fischer et al. 1982). With such a diversity of chemical structures, it seemed to us that it was possible that similarities would be found in their MEP.

Table 4 lists the computed ΔH_f^θ and dipole moments of the inhibitors we have examined. This work is not yet complete, and at this time we will comment only briefly on the results. The inhibitors phenidone, BW755C and NDGA all have at least one similar part of the MEP which is similar, and which is associated either with 2OH groups or 2 N atoms. Fitting of the molecules using the fitting algo-

Table 4. Heats of formation, total energies and dipole moment of 5-lipoxygenase inhibitors

Molecule	ΔH_f^θ (298 k)	E	μ
Phenidone	9.46	-2027.86	3.03
BW755C	-73.73	-3478.87	5.06
ETYA	32.89	-3537.67	1.83
NDGA	-154.56	-3902.54	3.70

rithm in CHEMX gives a reasonable fit, and the relevant section of the molecule superimposes quite closely on the —COOH group of arachidonic acid. The MEP in this region are quite similar.

In summary, the MEP of these molecules show that they probably interact with the active site of the enzyme where the —COOH group of arachidonic acid is bound. A more quantitative analysis of the MEP in this region (in progress)

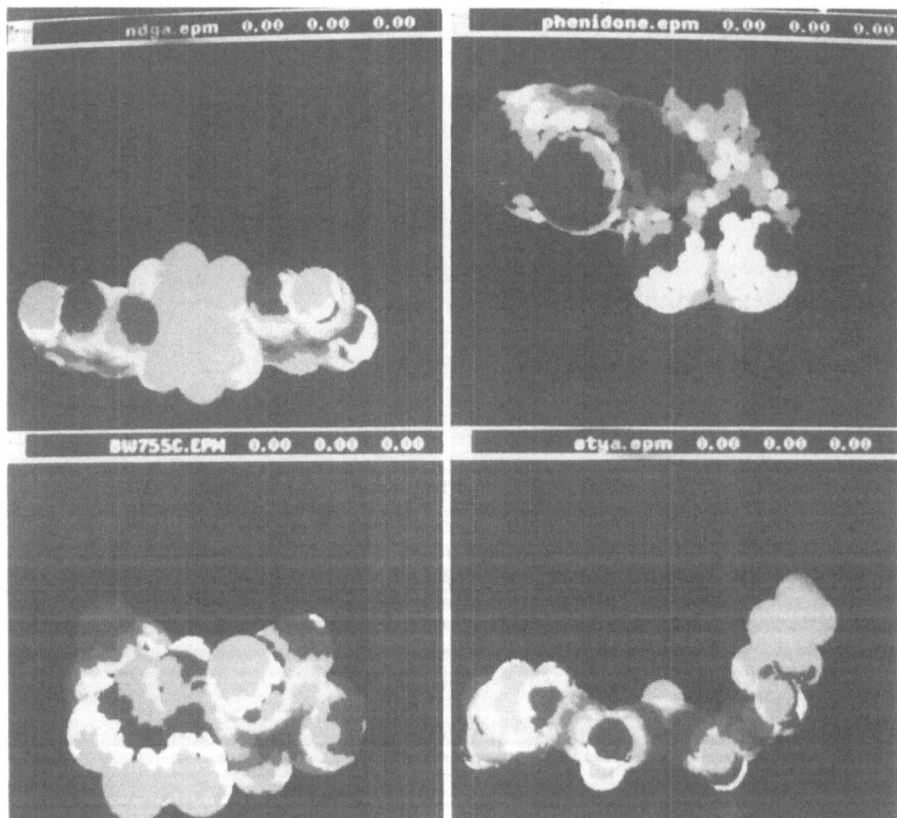

Fig. 4. MEP of some 5-lipoxygenase inhibitors on van der Waals' surface

should enable one to suggest other more specific inhibitors, as we did for glyoxalase-1 (Brandt et al. 1986). MEP are given in Fig. 4.

7 Conclusion

In this paper, we have presented the first results of complete geometry optimization calculations, using the semi-empirical MNDO method, of some prostaglandins, arachidonic acid, and several inhibitors of 5-lipoxygenase. The good agreement of the calculated structures with experiment, where known, gives us good reason to expect that a detailed analysis of the MEP, to be presented elsewhere, will help elucidate how the inhibitors mimic the substrate molecules, and we are confident that future theoretical work on these important molecules will be most rewarding.

*Acknowledgements.*The authors would like to express their deep gratitude to Digital Equipment Corporation (DEC) for the donation of the MICROVAX-2/ GPX work station, and to the National Foundation for Cancer Research, Bethesda, USA for continued financial report.

References

Beveridge RB, Jorgensen WL (1986) Computer simulation of chemical and biomolecular systems. Ann Acad Sci 482
Brandt RB, Laux JE, Yates SW, Thomson C, Edge C (1986) Inhibition of glyoxalase. I. in vitro by coumarin and coumarin derivatives. Int J Quantum Chem S13:155-165I
Clark T (1985) A handbook of computational chemistry. Wiley, New York
Cuthbertson AF, Thomson C (1987) J Mol Graph 5:92-96
De Titta GT,, Langs DA, Edmonds JW, Duac WL (1980) Prostadienoic acids PGE_2 and $PGR_{2\beta}$: Crystallographic studies of conformational transmission and receptor recognition. Acta Chrystallogr 836:638-645
Dewar MJS (1985) Quantum mechanical molecular models. J Phys Chem 89:2145-2150
Dewar MJS Thiel WJ (1977) Ground states of molecules 38: The MNDO method. Approximations and parameters. J Am Chem Soc 99:4899-4907
Edge C (1987) Thesis, University of St Andrews
Fischer SM, Slaga TJ Eds) (1985) Arachidonic acid metabolism and tumour promotion. Nijhoff, Boston
Fischer SM, Mills GD, Slaga TM (1982) Inhibition of mouse skin tumour promotion by several inhibitors of arachidonic acid metabolism. Carcinogenesis 3:1243-1245
Furstenberger G, Marks F (1985) In: Fischer SM, Slaga TJ (eds) Arachidonic acid metabolism and tumour promotion. Nijhoff, Boston
Gunstone FD (1974) The structural analysis of fatty acids and esters by nuclear magnetic resonance. University of Amsterdam
Hehre WJ, Radom L, Schleyer P, Pople JA (1986) Ab-initio molecular orbital theory. Wiley, New York
Kollman P, Weiner PK, Coldwell J, Singh UC (1986) Amber V3 0. UCSF, San Francisco
Kothekar V, Kalia K (1985) Solution conformation and electrostatic potential distribution of prostaglandins and thromboxanes and their relation to specificity. FEBS Lett 193:99-104

Langs DA, Erman M, de Titta GT (1977) Conformations of prostaglandin $F_{2\alpha}$ and recognition of prostaglandins by their receptors. Science 197:1003–1005

Marks F, Furstenberger G (1985) In: Sies H (ed) Oxidative stress. Academic, New York

Murakami A, Akahori Y (1974) Conformational analysis of prostaglandins. I. Theoretical calculation of the conformation of prostaglandin $F_{1\beta}$ derivatives. Chem Pharm Bull (Tokyo) 22:1133–1139

Murakami A, Akahori Y (1977a) Conformational analysis of Prostaglandins. III. Study on active sites and conformation-action relationships. Chem Pharm Bull (Tokyo) 25:3155–3162

Murakami A, Akahori Y (1977b) Conformational analysis of prostaglandins. II. Most probable conformation of prostaglandins A, B, e and F. Chem Pharm Bull (Tokyo) 25:2870–2874

Murakami A, Akahori Y (1979) Conformational analysis of prostaglandins. IV. Relationship between melting point and calculated conformational energy of prostaglandins. Chem. Pharm Bull (Tokyo) 27:548–550

Naráy-Szabo G (ed) (1986) Theoretical chemistry of biological systems. Elsevier, Amsterdam

Needleham P, Turk J, Jakschik BA, Morrison AR, Lefkowith JB (1986) Arachidonic acid metabolism. Ann Rev Biochem 55:69–102

Politzer P, Truhlar DG (1981) Chemical applications of atomic and molecular electrostatic potentials. Plenum, New York

Richards WG (1983) Quantum pharmacology, 2nd ed. Butterworth, London

Stewart JJP (1987) MOPAC V3.1. USAF Academy, Colorado Springs

Thomson C (1986) Some recent developments concerning the mechanisms of tumour promotion. Int J Quantum Chem 813:297–305

Thomson C, Reynolds CA (1986) A theoretical study of N-nitrosamine metabolites: Possible alkylating species in carcinogenesis by N, N-dimethyl, nitrosamine. Int J Quantum Chem 30:751–762

Thomson C, Edge C, Brandt R (1985) Structure and motion: Membranes nucleic acids and proteins. In: Clementi E, et al. (eds) Adenine, New York p 375

Venkataraghvan V, Feldman RJ (1975) Macromolecular structure and specificity: Computer assisted modelling and applications. Ann NY Acad Sci 439

Lipid Peroxidation and Cancer I

Lipid Peroxidation and Cell Division in Normal and Tumour Tissues

T. F. Slater

Department of Biology and Biochemistry, Brunel University, Uxbridge, Middlesex UB8 3PH, UK

Free radical intermediates have been implicated in various aspects of carcinogenesis, such as in examples of initiation (see Pullman et al. 1980; Floyd 1982; McBrien and Slater 1982) and in promotion (Kensler and Trusch 1984). In Addition, free radical-mediated disturbances have been reported in a number of established cancers; it is this aspect of the inter-relationships of free radical reactions and cancer that will be considered here.

In studies commenced in the mid-1960s it was shown (Slater and Cook 1970) that frozen powdered samples of normal human uterine cervix contained a relatively large amount of a free radical species that gave a very strong signal when examined by electron spin resonance (esr) spectroscopy. Similar frozen powdered samples prepared from invasive cancer of the cervix were found to give very much reduced esr signals. The very strong esr signal observed with frozen powdered samples of normal cervix, with a g value close to the free electron spin position, is not observed significantly in frozen intact samples of normal cervix (Benedetto et al. 1981); in these samples there is, however, a strong signal in the region of g = 2.11–2.15. Evidence has been provided (Benedetto 1982) to support the view that grinding samples of frozen intact cervix under liquid nitrogen to form a fine powder is accompanied by a loss of the g = 2.11–2.15 signal and the appearance of the g = 2.004 signal. The latter signal was identified as a peroxyl-species, using S-, X-, and Q-band esr (Tomasi et al. 1984). Peroxyl-radicals are usually oxidising species, and the work outlined above led to studies of the thiol: disulphide changes in normal and abnormal cervix that are reported elsewhere in this book (Nöhammer et al. this volume; Benedetto et al. this volume; Principe et al., this volume).

Peroxyl radicals are important intermediates of lipid peroxidation, a free radical-mediated chain reaction (for review see Slater 1975; Comporti 1985). Although lipid peroxidation may include peroxidative damage to all types of lipid, including cholesterol, this discussion will be restricted to changes that affect polyunsaturated fatty acids (PUFAHs). Normally, the first reaction in lipid peroxidation is the abstraction of a hydrogen-atom by a reactive free radical species, $R^{\bullet}$. In intact cells and tissues the reactive free radical $R^{\bullet}$ may arise from physiological substances such as H_2O_2, from electron-donation reactions to suitable acceptors such as O_2, or from reactions catalysed by transition metal ions including Fe^{2+}, Fe^{3+}, Cu^{2+}. Under conditions in vitro the process of lipid peroxidation is usually studied (Slater and Cheeseman 1987) after artificial stimulation:

Nigam et al. (Eds.), Eicosanoids,
Lipid Peroxidation and Cancer
© Springer-Verlag Berlin Heidelberg 1988

for example, with γ-irradiation, with Fe^{2+}-ascorbate, with iron-chelates such as Fe^{2+}-ADP and an NADPH-flavoprotein, with CCl_4 and the NADPH-cytochrome P_{450} electron transport chain, with cumene hydroperoxide and so on. When lipid peroxidation is stimulated in such ways in tissue homogenates or intracellular fractions the PUFAHs that are mainly affected are the highly unsaturated acids, arachidonate and decosahexaenoate. From the above brief discussion it is evident that lipid peroxidation is usually studied after stimulation but may also occur naturally, and may be either essentially non-enzymic or enzymic in character. It should also be emphasised in this context that a number of enzyme-catalysed peroxidative reactions are of very considerable biological importance and are under close metabolic control; these include prostaglandin-G_2 synthase, and lipoxygenase reactions.

Lipid peroxidation of PUFAHs such as arachidonate results in the formation of a considerable variety of products (Table 1). What is not always appreciated is that some of these products have significant biological effects at very low concentrations; in consequence, even apparently very low levels of lipid peroxidation may result in substantial cellular responses. For example, Lands and co-workers (Hemler et al. 1979) have shown that submicromolar concentrations of lipid hydroperoxides are a necessary stimulus for prostaglandin synthesis. Several independent studies have shown that nanomolar concentrations of epoxy-derivatives of arachidonate have striking effects on the release of various hormones (Capdevila et al. 1983; Snyder et al. 1983); the 4-hydroxy-alkenals have many interesting biological properties (see Esterbauer 1985) including effects on chemotaxis at nanomolar concentration: these are dose-response effects that are consistent with physiological mediators both at intra- and extra- cellular levels (see Slater 1987). It should be stressed that the formation of these very active products of lipid peroxidation can enable a highly localised free radical reaction (in a biomembrane, for example) to result in metabolic perturbations at large distances from the original initiation step due to diffusion of the secondary and tertiary products (Slater 1976). Lipid peroxidation is very often a very damaging and degradative process; in the rancidity of foods for example. If lipid peroxidation overwhelms the normally efficient cellular defence systems then it can cause cytotoxicity and cell death (for a discussion of free radical-mediated cell damage, cause or consequence see Slater 1988). However, the established biological effects of very low concentrations of products of lipid peroxidation allows the contention that small rates of lipid peroxidation in normal cells may have important physiological consequences. One such consequence could be the

Table 1. Products of lipid peroxidation

Dienes
Lipid hydroperoxides
Epoxides
Hydroxy-alkenals
Alkanes
Chemiluminescence
$L^{\bullet}$, $LOO^{\bullet}$, $LO^{\bullet}$ (esr spin trap)

effect of 4-hydroxy-alkenals, such as 4-hydroxy-nonenal on DNA-synthesis and on thiol-dependent reactions (Schauenstein et al. 1977). In this way, the endogenous production of low levels of 4-hydroxy-alkenals could act as a 'coarse control' on DNA synthesis; in contrast, inhibition of lipid peroxidation might then be expected to be associated with a stimulation of cell division (Slater 1973).

It has been known for a long time that many liver tumours undergo lipid peroxidation far less readily than normal liver when stimulated in various ways (see Cheeseman et al. 1984, 1986a). A list of liver tumours that have been reported to exhibit a low rate of lipid peroxidation compared with normal liver is given in Table 2. Some data for the Novikoff and Yoshida tumours are given in Table 3. It can be seen that there are significant decreases in the content of PUFAHs, undetectable levels of cytochrome P_{450} and very much reduced rates of lipid peroxidation under different stimulated conditions. The low rate of lipid peroxidation in the microsomal fractions isolated from liver tumours may be due to (a) a decreased content of the substrate PUFAHs; (b) a decreased content of the NADPH-flavoprotein that can regenerate Fe^{2+} from Fe^{3+}, and may also donate electrons to O_2 to yield $O_2^{\cdot-}$ and H_2O_2; (c) a decreased content of cytochrome P_{450}, which can function in free radical initiation reactions as a mixed function oxidase, and also as a peroxidase acting on lipid hydroperoxides to generate peroxyl-radicals; (d) an increased content of antioxidants that can inhibit initiations or act as chain breakers. It is clear from the data in Table 3 that there are significant decreases in PUFAHs and in cytochrome P_{450}, but what

Table 2. Liver tumours with depressed lipid peroxidation

Novikoff	Morris 44
Yoshida	Morris 3924A
Aflatoxin	Morris 9618A
Ethionine	D 23
Butter yellow	D 30
Diethylnitrosamine	D 192A

Table 3. Cytochrome P_{450} and polyunsaturated fatty acids in normal rat liver microsomes, and in microsomes prepared from Novikoff and Yoshida tumour cells

Measurement	Normal	Novikoff	Yoshida
Cytochrome P_{450} pmol/mg protein	710 ± 20	ND	ND
$C_{20:4}$ % total fatty acids	19.1 ± 2.4	10.6 ± 0.3	9.6 ± 0.8
$C_{22:6}$ % total fatty acids	6.1 ± 0.3	1.4 ± 0.3	5.3 ± 0.4
$C_{18:1}/C_{18:0}$	0.39	1.03	1.32

ND, Not detectable
(From Cheeseman et al. 1986a, 1988)

about the content of antioxidants? Certainly, there appears to be an increased content of antioxidant in Novikoff and Yoshida tumours since exposure to γ-radiation, which generates reactive free radicals via direct radiolysis of water, does not provoke significant lipid peroxidation even though substantial amounts of PUFAHs are present.

Antioxidants can be classified (see Burton et al. 1983) as preventative or chain-breaking; a list of the more important members of these two classes of antioxidant in liver tissue is given in Table 4. There have been many studies on antioxidant content of animal and human tumours. For example, it has been known for a long time (see Greenstein and Andervont 1943) that catalase is often very much reduced in activity; a recent study confirming this in human liver tumours is by Corrocher et al. (1986). These authors have also reported a lower level of glutathione peroxidase activity and glutathione. Moreover, there are several reports of a decreased activity of superoxide dismutase and glutathione reductase in liver tumours (Pinto and Bartley 1973; Peskin et al. 1977; Oberley and Buettner 1979; Tisdale and Mahmoud 1983). However, in lipid extracts of Novikoff and Yoshida tumours there is a substantial *increase* in chain-breaking antioxidant activity, and this has been identified with an increased content of α-tocopherol (Cheeseman et al. 1986a, 1988); see Table 5. Thus these liver tumour cells accumulate a much higher intracellular concentration of α-tocopherol than is found in normal hepatocytes, and this results (together with other factors mentioned previously) in a greatly reduced rate of lipid peroxidation.

Table 4. Major antioxidants in liver and other tissues

Preventative antioxidants
  Some metal chelators
  Superoxide dismutase
  Catalase
  Glutathione peroxidases
  Glutathione transferases

Chain-breaking antioxidants
  Glutathione
  Ascorbate
  Urate
  Ubiquinone
  β-Carotene
  α-Tocopherol

Table 5. α-Tocopherol in normal rat liver microsomes and in microsomes prepared from Novikoff and Yoshida tumour cells

Measurement	Normal	Novikoff	Yoshida
α-tocopherol : lipid nmol/mg	1.67 ± 0.31	4.87 ± 0.37	3.06 ± 0.42
α-tocopherol : bis-allylic methylenes $\times 10^4$	5.2	33	38.6

(From Cheeseman et al. 1986a, 1988)

What is the biological significance of this? One possible explanation is that the increased content of antioxidant, and the associated low rate of lipid peroxidation, ensures that genetic material is at least partially protected from free radical-mediated damage during a time when it is more susceptible to damage during DNA-synthesis. Another possibility is one already mentioned earlier: products of lipid peroxidation inhibit DNA synthesis, so that if this is a physiologically important reaction then it would become necessary to inhibit lipid peroxidation prior to the onset of DNA synthesis.

In order to study these questions further we turned to the regenerating rat liver model; studies on this are described in more detail by Cheesman et al. in this book. During liver regeneration there are cycles of DNA-synthesis and mitosis, and microsomal membrane fractions prepared from regenerating livers are much less suscepthible to stimulated lipid peroxidation at times of maximum thymidine kinase activity than at other periods of the cell cycle (see Cheeseman et al. 1986b, and this volume). The depressed response to stimulation of lipid peroxidation in membrane fractions prepared from regenerating liver appears partly due to an increased input of α-tocopherol, and to an increased content of a GSH-dependent reaction. As a result there is a cyclical variation in microsomal antioxidant activity. In consequence, in this model, there appears to be a physiological role for lipid peroxidation that may be inter-related with the stimulus for DNA-synthesis.

Acknowledgements. We are grateful to the Cancer Research Campaign and to the Association for International Cancer Research for financial support.

References

Benedetto C (1982) Biochemical studies on human uterine cancer. In: McBrien DCH, Slater TF (eds) Free radicals lipid peroxidation and cancer. Academic, London pp 27-45

Benedetto C, Bocci A, Dianzani MU, Chiringhello B, Slater TF, Tomasi A, Vannini V (1981) Electron spin resonance studies on normal human uterus and cervix, and on benign and malignant uterine tumours. Cancer Res 41:2936-2942

Burton GW, Cheeseman KH, Doba T, Ingold KU, Slater TF (1983) Vitamin E as an antioxidant in vitro and in vivo. Ciba 101:4-14

Capdevila J, Chacos N, Falck JR, Manna S, Negro-Vilar A, Ojeda SR (1983) Novel hypothalamic arachidonate products stimulate somatostatin release from the median eminence. Endocrinology 113:421-423

Cheeseman KH, Burton GW, Ingold KU, Slater TF (1984) Lipid peroxidation and lipid antioxidants in normal and tumour cells. Toxicol Pathol 12:235-239

Cheeseman KH, Collins M, Proudfoot K, Slater TF, Burton GW, Webb AC and Ingold KU (1986a) Studies on lipid peroxidation in normal and tumour tissues. The Novikoff rat liver tumour. Biochem J 235:507-514

Cheeseman KH, Collins M, Maddix S, Milia A, Proudfoot K, Slater TF, Burton GW, et al. (1986b) FEBS Lett 209:191-196

Cheeseman KH, Emery S, Maddix SP, Slater TF, Burton GW and Ingold KU (1988) Studies on lipid peroxidation in normal and tumour tissues. The Yoshida rat liver tumour. Biochem J 250:247-252

Comporti M (1985) Biology of disease. Lipid peroxidation and cellular damage in toxic liver injury. Lab Invest 53:599-623

Corrocher R, Casaril M, Bellisola G, Gabrielli GB, Nicoli N, Guidi GC, de Sandre G (1986) Severe impairment of antioxidant system in human hepatoma. Cancer 58:1658-1662

Esterbauer H (1985) Lipid peroxidation products: formation, chemical properties and biological activities. In: Poli G, Cheeseman KH, Dianzani MU, Slater TF (ed) Free radicals in liver injury. IRL, Oxford, pp 29-47

Floyd RA (ed) (1982) Free radicals and Cancer. Dekker, New York

Greenstein JP, Andervont HB (1943) The liver catalase activity of tumor-bearing mice and the effect of spontaneous regression and of removal of certain tumors. JNCI, 2:345-355

Hemler ME, Cook HW, Lands WEM (1979) Prostaglandin synthesis can be triggered by lipid peroxides. Arch Biochem Biophys 193:340-345

Kensler TW, Trush MA (1984) Role of oxygen radicals in tumor promotion. Environ Mutagen 6:593-616

McBrien DCH, Slater TS (eds) (1982) Free radicals, lipid peroxidation and cancer. Academic, London

Oberley LW, Buettner GR (1979) Role of superoxide dismutase in cancer: a review. Cancer Res 39:1141-1149

Peskin AV, Keon YM, Zbarsky IB, Konstantinov AK (1977) Superoxide dismutase and glutathione peroxidase activities in tumors. FEBS Lett 78:41-45

Pinto RE, Bartley W (1973) Glutathione reductase and glutathione peroxidase activities in hepatomous livers of rats treated with diethyl nitrosamine. FEBS Lett 32:307-309

Pullman B, Ts'o POP, Gelboin H (eds) (1980) Carcinogenesis: Fundamental mechanisms and environmental effects. Reidel, Dordrecht

Schauenstein E, Esterbauer H, Zollner H (1977) Aldehydes in biological Systems. Pion, London

Slater TF (1973) Barriers and class distinctions in biochemistry: adrift in a hostile world. Inaugural lecture, Brunel University, Biochemistry Department, Uxbridge

Slater TF (1975) The role of lipid peroxidation in liver injury. In: Keppler D (ed) Pathogenesis and mechanisms of liver cell necrosis. MTP Press, Lancaster

Slater TF (1976) Biochemical pathology in microtime. In: Dianzani MU, Ugazio G, Sena LM (eds) Recent advances in biochemical pathology: Toxic liver injury. Minerva Medica, Turin pp 99-108

Slater TF (1987) Lipid peroxidation and intercellular messengers in relation to cell injury. Agents Actions 22:333-334

Slater TF (1988) Free radical disturbances and tissue damage: cause or consequence? In: Rice-Evans C, Halliwell B (eds) Free radicals: a search for a new methodology. Richlieu, (in press)

Slater TF, Cheeseman KH (1987) Lipid peroxidation. In: Benedetto C, McDonald-Gibson RG, Nigam S, Slater TF (eds) Prostaglandins and related substances: A practical approach. IRL, Oxford, pp 243-258

Slater TF, Cook JWR (1970) Electron spin resonance studies on normal and malignant human tissue, and in normal and damaged rat liver. In: Evans DMD (ed) Cytology Automation. Livingstone, London, pp. 108-120

Snyder GD, Capdevila J, Chacos N, Manna S, Falck JR (1983) Action of luteinising hormone-releasing hormone: involvement of novel arachidonate acid metabolites. Proc Natl Acad Sci USA 80:3504-3507

Tisdale MJ, Mahmoud MB (1983) Activities of free radical metabolising enzymes in tumours. Br J Cancer 47:809-812

Tomasi A, Benedetto C, Nilges M, Slater TF, Swartz HM, Symons MCR (1984) Studies on human uterine cervix and rat cervix using S-, X- and Q- band electron spin resonance spectroscopy. Biochem J 224:431-436

The Involvement of Free Radicals and Lipid Peroxidation in Carcinogenesis

T. A. Connors

Medical Research Council Toxicology Unit, Woodmansterne Road, Carshalton, Surrey SM5 4EF, UK

1 Introduction

Associations between environmental factors and human cancer have been known for many years. The high incidence of scrotal cancer in chimney sweeps was recorded in the eighteenth century while in the nineteenth century a link had already been observed between lung cancer and the mining of certain ores and between bladder cancer and the large scale synthesis of aniline dyes. A relationship between ionising radiation and cancer was proposed soon after the discovery of Xrays, while for almost 80 years scientists have used viruses as a means of inducing cancers in laboratory animals.

2 Chemical Carcinogens

Knowledge of the mechanism of carcinogenesis has developed over a long period with contributions from different disciplines. Early and mainly descriptive studies established that discrete chemicals in coal tars were carcinogenic and a clear role has now been established for electrophilic reactants formed from organic chemicals as initiators of the carcinogenic process. Studies on alkylating agents were crucial experiments in drawing attention to the key role of DNA in cancer induction. Nitrogen mustards were known to be carcinogenic and mutagenic and early studies on mechanisms showed that they reacted covalently with cellular constituents via a carbenium ion intermediate (Fig. 1). The species is

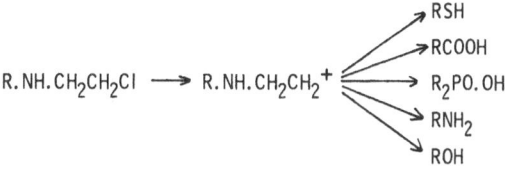

Fig. 1. The mechanism of alkylation of cellular molecules by nitrogen mustard occurring by the formation of a transient carbenium ion

Nigam et al. (Eds.), Eicosanoids,
Lipid Peroxidation and Cancer
© Springer-Verlag Berlin Heidelberg 1988

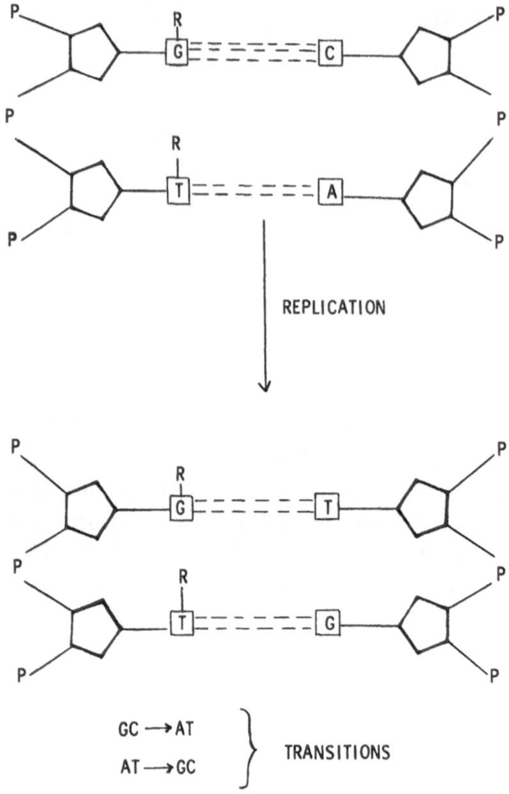

P
R
G
C
P
P
P
R
T
A
P
P
REPLICATION
P
P
R
G
T
P
P
R
T
G
P
P
GC → AT
AT → GC
TRANSITIONS

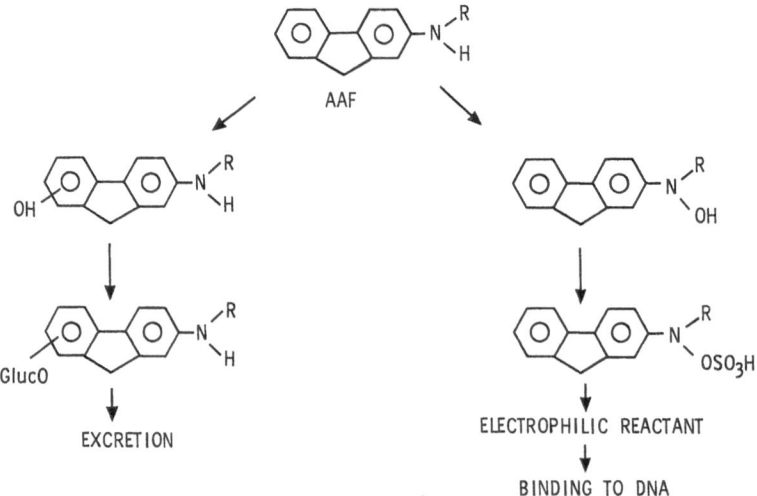

Fig. 2. Point mutation arising as a result of replication before repair of alkylated bases

R
N
H
AAF
R
N
H
OH
R
N
OH
R
N
H
GlucO
EXCRETION
R
N
OSO3H
ELECTROPHILIC REACTANT
BINDING TO DNA

Fig. 3. Metabolism of acetylaminofluorene (AAF)

particularly reactive to ionised sulphur and oxygen moieties and uncharged amino groups. Many experiments suggested an essential role for covalent binding to DNA as the mechanism of cytotoxicity. That such a reaction might also be mutagenic was demonstrated initially in bacteria using simple alkylating agents (Loveless 1969). Alkylation of O^6 guanine for example is likely to interfere with hydrogen bonding of the base pairs (Fig. 2). If replication occurred before repair, mispairing could occur and a further round of replication would result in a 'fixed' mutation represented by a GC to AT transition. Experiments of this type strengthened the hypothesis put forward very early on, that cancer originated from a somatic cell mutation. These experiments were however just a laboratory curiosity since a large number of chemicals were known with the ability to cause cancer in laboratory animals (e.g. benzpyrene) but which did not contain any electrophilic centres. An important advance pioneered by the Millers (Miller 1970) was to show that chemically inert carcinogens could be metabolised in vivo to electrophilic reactants. Acetylaminofluorene (AAF, Fig. 3) is, for example, largely detoxified by ring hydroxylation and conjugation. A small proportion however is N-hydroxylated and conjugated with sulphate to form an unstable species which breaks down to an electrophile with DNA binding properties. The majority of chemical carcinogens are now known to be metabolised by similar pathways and thus have the ability to mutate DNA by a number of mechanisms.

3 Viral Carcinogenesis and Oncogenes

While mutation was seen to be an early and crucial event, primarily as a result of studies on chemical carcinogens, these experiments gave no indication of the nature of the gene or genes, mutation of which was essential for carcinogenesis. This information came from two sources — studies on oncogenic RNA viruses and DNA transfection experiments where cells are persuaded to incorporate DNA fragments administered externally. RNA viruses have a simple genome consisting of the information required for the synthesis of the viral core and coat proteins and reverse transcriptase flanked by promoting LTR sequences. On infection the few molecules of the reverse transcriptase that the virus contains are activated and convert the simple RNA genome into a DNA copy which is integrated into the host DNA and actively transcribed. Rapidly transforming RNA viruses which are not normally replication competent rely on helper viruses for replication, but can nevertheless rapidly induce tumours in the appropriate host. Studies on the Rous sarcoma virus revealed the presence of a single viral gene (src) responsible for cell transformation. The demonstration that src was initially of cellular origin (Stephelin et al. 1976) and had been acquired from the chicken genome by a process of recombination led to the concept that the genome of higher animals contains a number of genes referred to as proto-oncogenes which are presumably essential for the rapid cell proliferation and differentiation required during development but are repressed in the mature cell.

The complementary studies on DNA transfection demonstrated that DNA from malignant cells could transform cell lines to the malignant state (Krontiris and Cooper 1981) and the view that an essential stage in cancer was activation of a cellular proto-oncogene was strengthened. A fairly large number of cellular oncogenes have now been described many of which are also carried by tumour viruses. The future challenge is to determine how the protein products of oncogenes can transform cells. The limited information so far suggests that they are involved in cell proliferation and differentiation and are of a number of distinct types (Table 1). Some oncogenic proteins are related to growth factors or their receptors and the initiation of secondary messenger responses. Others are found in the nucleus and play an important role in the cell cycle probably by regulating m-RNA transcription. The current evidence suggests that cells acquire the full malignant phenotype only after the activation of two or more oncogenes. The middle T or large T oncogenes of the polyoma virus for instance only cause tumours if they act together (Rassoulzadegan et al. 1982). Recent studies of transgenic mice show that expression of c-myc and V-Ha-ras are essential for early tumour induction (Sinn et al. 1987). Since the tumours that arise are clonogenic then other factors are also necessary for full transformation. Although point mutation as described above is one mechanism which is known to occur in some malignancies, there are clearly other events which might lead to inappropriate gene activation (Table 2). Mutations can of course occur by other mecha-

Table 1. Oncogene products

GTP/GDP binding proteins
  Inside surface of plasma membrane: modulates c-AMP pathway? ras family
Nuclear proteins
  Regulation of m-RNA transcription: myc myb fos
Growth factors
  Products related to polypeptide growth factors such a PDGF: sis
Protein kinases
  Regulation of phosphorylation of proteins involved in secondary messenger pathway
    Serine/threonine kinase: mos, rat
    Tyrosine kinases – large numbers of oncogene products; abl, fes, fms, erbB, tsr

---

Table 2. Potentially carcinogenic events

Point mutation
Chromosomal translocation
Gene amplification
Rapidly transforming RNA viruses
Viral integration
Gene deletion
Chromosomal rearrangement

---

nisms such as faulty repair or continued DNA replication in the presence of covalently bound molecules. Chromosome translocations and gene amplification have also been associated with some human cancers. Besides rapidly transforming RNA viruses, other viruses including DNA viruses may alter oncogene expression in a number of ways. They might after insertion into cellular DNA cause activation by a powerful transcription enhancement of oncogenes or by insertion of a viral promoter close to an oncogene sequence. One might also expect that agents that damage DNA may cause altered expression due to gene disruption or deletion or by causing chromosomal rearrangements. An important finding is that the chromosomal changes seen after irradiation or free radical induced damage are quite distinct from the changes seen after exposure to cytotoxic and carcinogenic chemicals. The implication is that free radicals may play an important role in carcinogenesis which is quite different from the mainly initiating, DNA covalent binding properties of chemical carcinogens.

4 Role of Free Radicals in Carcinogenesis

Experiments showing that chemical carcinogens may be activated by the generation of free radical intermediates are numerous and have usually involved in vitro systems. Prostaglandin synthetase for example can convert benzpyrene to an intermediate on the activation pathway by a free radical mechanism (Marnett 1981). Non-enzymic mechanisms involving peroxyl radicals generated during lipid peroxidation (Kraus and Eling 1984) play a role in the metabolism of a procarcinogen to a carcinogen and may be particularly important in certain tissues (Gower and Wills 1987). However there is no firm evidence for the direct addition in vivo of an organic radical to DNA and so their role in initiation of cancer is obscure. The large organic free radicals could generate peroxide and ultimately the hydroxyl radical and there is a great deal of evidence that hydroxyl radicals cause single strand breaks in DNA. Hydroxyl radicals can certainly interact with DNA bases and possibly induce error prone repair mechanisms resulting in mutations. The problem about ascribing a role for hydroxyl radicals is their transient nature (half-life 10^{-9}s) which makes it unlikely that sufficient concentrations could be generated close enough to DNA to cause significant damage. The problem is overcome by a substance such as the cigarette tar radical described by Pryor (1987). This free radical, whose existence at present is inferred from ESR studies, is a chemical containing relatively long lived semiquinone radicals and which can associate with DNA. The semiquinone groups generate superoxide and peroxide which in the presence of chelated metals forms the OH˙ radical. Since the site of production is close to DNA then the chemical is very effective in damaging DNA. Alternatively it has been suggested that the more stable peroxyl free radicals may be involved in carcinogenesis (Marnett 1987).

The current evidence would suggest however that if free radicals play an essential role in transforming cells then this is most likely to be a post-initiation or promotional event. It has been known for many years that malignancy is a multistage process. The initiation event is itself complex since it requires alkylation at an appropriate site in DNA and cell replication before repair has taken place. The cell biologist recognises a number of distinct events in cell transformation such as the acquisition of 'immortality' followed by partial independance from serum growth factors and anchorage independence before the cells can form malignant tumours in suitable hosts. A number of systems exist, for example a simple skin painting model, which can separate cancer into initiating and promoting stages (Yuspa 1987). An 'incomplete' carcinogen, which is very often a chemical which at a higher dose is carcinogenic, is applied to the skin and is non-carcinogenic. Application of a promoter, which must be after the complete carcinogen, for an appropriate number of doses can lead to a high cancer incidence although the promoter itself is essentially non-carcinogenic (Fig. 4). This enables one to study separately promotional events although, as with initiation, it is likely that promotional events are multistage since the known promoters have a myriad of biological effects. There is growing evidence that free radicals play an essential role in the promotion of initiated cells. The evidence has been elegantly reviewed by Ames (1983) and Cerutti (1985) where numerous agents that induce a pro-oxidant state in the cell are shown to be powerful promoters with their effects being reversed by anti-oxidants. Since most of these agents are poor mutagens and powerful clastogens it has been proposed that promotional events may involve specific gene rearrangements which allow expression of the previously mutated DNA. It is of interest that most powerful promoters have three properties in common, namely the ability to generate free radicals in cells, to stimulate mitogenesis and to cause chromosomal aberrations. It is tempting to assume that hydroxyl radicals cause single strand breaks which are normally very readily repaired since cells can apparently tolerate a large number of free radical induced breaks without cytotoxicity (Cantoni et al. 1987). However if replication takes place before repair then gene rearrangements may occur which activate mutated oncogenes. If two or more oncogenes must be expressed and in a defined sequence then the need for several applications of the promoter after the initiator can be explained as well as the reversibility of the promoter if sufficient doses are not administered. It has been shown by Parsons (1988) that free

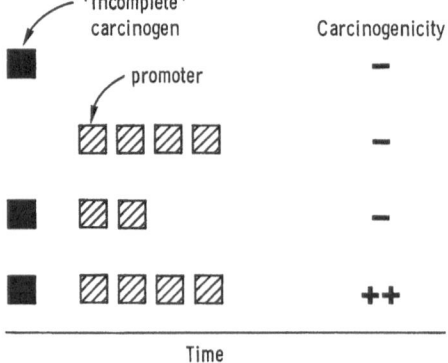

Time

Fig. 4. Two-stage model for carcinogenesis. The promoter must always be applied after the 'incomplete' carcinogen and for a minimal period

radical generating systems can cause DNA single strand breaks in cell culture. If these cells are synchronised they can be shown to be extraordinarily sensitive to free radicals for a short period of the S-phase. The ability of promoters to induce proliferation may therefore also be important in that it ensures that cells enter a highly sensitive phase during the production of free radicals.

The scheme in Fig. 5 might represent essential stages of carcinogenesis. Cells are continually being exposed to electrophilic reactants which have the potential to cause a mutagenic event. However there are adequate repair mechanisms and unless division takes place before a crucial alkylation is removed covalent alkylation of DNA is not particularly harmful. The promotional stage is essentially a chromsomal rearrangement induced by the hydroxyl radical followed by mitogenesis. This might be a specific property of OH˙ radicals since it has already been mentioned that the chromosomal damage caused by ionising radiation and free radicals is qualitatively different from that seen by electrophilic reagents.

In the laboratory scientists have chosen as models for cancer chemicals which at high dose levels cause a high incidence of tumours after a short latent period. These complete carcinogens would presumably have all the necessary properties shown in Fig. 5. All commonly used complete carcinogens are known to be electrophilic reactants per se or be capable of forming them in vivo. At the high dose levels used most of them are cytotoxic and would be expected to cause a compensatory proliferative response. There is also accumulating evidence that these carcinogens may form free radical intermediates as well as electrophilic reactants on metabolism. In some cases the evidence is circumstantial, for example the ability of anti-oxidants to protect against aflatoxin B_1 induced liver cancer (Mandel et al. 1988) or the rapid oxidative stress seen soon after nitrosamine administration (Ahotupa et al. 1987) but in other cases much more direct. An OH˙ radical attack on DNA will result in the formation of thymine glycol, for example, and evidence has been obtained that this product is formed by benzpyrene and aflatoxin B_1 (Cerutti 1985; Ide et al. 1983). More recently and using a monoclonal antibody, thymine glycol has been detected in a dose dependent manner in cells exposed to both benzpyrene and N-hydroxy-2-naphthylamine (Leadon 1987).

Studies on complete carcinogens which compress a number of essential carcinogenic events into a short period are quite artificial and almost certainly do not mimic the human situation except perhaps where patients are treated with alky-

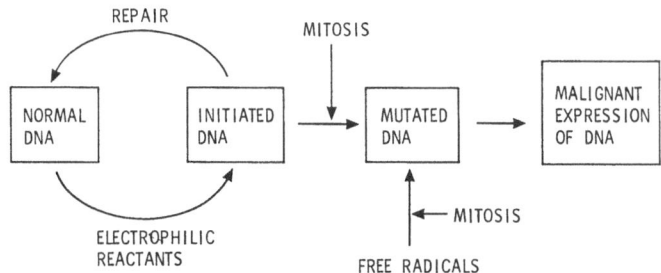

Fig. 5. Two-stage model for carcinogenesis

lating agents or radiation as part of cancer therapy. In humans estimation of the levels of alkylated haemoglobin in blood suggest that one is continually being exposed to potentially mutagenic events since levels of alkylated haemoglobin are related to the level of intracellular DNA alkylation (Bailey et al. 1987). The cells have adequate means of removing these alkylations and a more critical event is probably the division of cells before repair has taken place, perhaps by an alteration of the hormonal or growth factor environment. While complete carcinogens form both electrophilic reactants and free radicals at high dose levels, at the extremely low dose levels of human exposure it may be that the promoting hydroxyl radical is not formed and that this arises by a different sequence of events. This would explain the generally long latent period of cancer induction if initiating and promoting events, both of which are multistage in nature, were the result of unrelated incidents. More attention has been paid recently to promotional events and it is of interest that Parke and his colleagues (Ioannides et al. 1984) have suggested that the majority of chemical carcinogens are activated by the P448 type of enzyme (cytochrome P_1-450 in the mouse and P450c in the rat). Since there is some evidence that the P448 enzyme is readily uncoupled (Parkinson et al. 1986) and may effectively generate superoxide, then by producing both electrophilic reactants and free radicals it may have a crucial role in carcinogenesis. In humans persistent induction of this enzyme by food constituents or contaminants might be important. A number of chemicals are known which are poor substrates of cytochrome P448 but have long biological half-lives as a result of which they induce the enzyme over a long period of time. TCDD and DDT have these properties and are potent promoting agents (Poland et al. 1982; Pitot et al. 1980; Rajanopo, this volume). It has recently been shown that hexachlorobenzene which causes a persistent induction of $P_1$450 in mice is a powerful hepatocarcinogen if the mice are pretreated with iron. Evidence is now being sought that any chemical which persistently induces $P_1$450 and maintains a chronic oxidative state will be a liver carcinogen in mice pretreated with iron (Smith A. 1982, personal communication).

Attempts to remove carcinogens from the environment rely in part on the knowledge obtained from animal carcinogenicity experiments or short term tests which primarily detect chemicals which act as initiating carcinogens. Since this is only the first and arguably the least important stage of carcinogenesis it would seem important if one is to prevent cancer or design intervention programmes as described by Bruce (1987), to pay more attention to later post initiation events. It would appear that much would be gained from the detailed study of the role of free radicals in these later stages.

References

Ahotupa M, Bussachini-Griot V, Bereziat JC, Camus AM, Bartsch H (1987) Rapid oxidative stress induced by N-nitrosamines. Biochem Biophys Res Commun 146:1047–1054
Ames BN (1983) Dietary carcinogen and anticarcinogens. Science 221:1256–1264

Bailey E, Farmer PB, Shuker DEG (1987) Estimation of exposure to alkylating carcinogens by the GC-MS determination of adducts to haemoglobin and nucleic acid bases in urine. Arch Toxicol 60:187-191

Bruce WR (1987) Hypotheses for the origin of colon cancer. Cancer Res 47:4237-4242

Cantoni O, Murray D, Meyn RE (1987) Induction and repair of DNA single strand breaks in EM9 mutant CHO cells treated with hydrogen peroxide. Chem Biol Interact 63:29-38

Cerutti PA (1985) Prooxidant states and tumor promotion. Science 227:375-381

Gower JD, Wills ED (1987) The oxidation of benzo(a)pyrene-7,8-dihyrodiol mediated by lipid peroxidation in the rat intestine and the effect of dietary lipids. Chem Biol Interact 63:63-74

Ide M-L, Kaneko M, Cerutti PA (1983) Benzo(a)pyrene and ascorbate-CuSO$_4$ induce DNA damage in human cells by indirect action. In: McBrien DCH, Slater TF (eds) Protective agents in cancer. Academic, London, pp 125-136

Ioannides C, Lum PY, Parke DV (1984) Cytochrome P448 and the activation of toxic chemicals and carcinogens. Xenobiotica 14:119-137

Kraus RS, Eling TE (1984) Arachidonic dependent cooxidation. Biochem Pharmacol 33:3319-3324

Krontiris GT, Cooper GM (1981) Transforming activity of human tumor DNA's. Proc Natl Acad Sci USA 78:1181-1184

Leadon SA (1987) Production of thymine glycols in DNA by radiation and chemical carcinogens as detected by monoclonal antibody. Br J Cancer [Suppl 8] 55:113-117

Loveless A (1969) Possible relevance of 0-6 alkylation of deoxyguanosine to the mutagenicity and carcinogenicity of nitrosamines and nitrosamides. Nature 223:206-207

Mandel HG, Manson MM, Judah DJ, Simpson JL, Green JA, Forrester LM, Wolf CR, Neal GE (1988) A metabolic basis for the protective effect of the antioxidant ethoxyquin on aflatoxin B$_1$ hepatocarcinogenesis in rat. Cancer Res (in press)

Marnett LJ (1981) Polycyclic aromatic hydrocarbon oxidation during prostaglandin biosynthesis. Life Sci 29:531-546

Marnett LJ (1987) Peroxyl free radicals: potential mediators of tumour initiation and promotion. Carcinogenesis 8:1365-1373

Miller JA (1970) Carcinogenesis by chemicals an overview. Cancer Res 30:559-576

Parkinson A, Thomas PE, Ryan DE, Gorsky LD, Smively JE, Sayer JM, Jerina DM, Levin W (1986) Mechanism of inactivation of rat liver microsomal cytochrome P-450c by 2-bromo-4'-nitroacetophenone. J Biol Chem 261:11487-11495

Parsons PG (1988) Potency, selectivity and cell cycle dependence of catechols in human tumour cells in vitro. Biochem Pharmacol 37:1711-1715

Pitot HC, Goldsworthy T, Campbell HA, Poland A (1980) Quantitative evaluation of the promotion by 2,3,7,8-tetrachloridbenzo-p-dioxin of hepatocarcinogenesis from diethylnitrosamine. Cancer Res 40:3616-3620

Poland A, Palen D, Glover E (1982) Tumour promotion by TCDD in skin of MRS/J hairless mice. Nature 300:271-274

Pryor WA (1987) Cigarette smoke and the involvement of free radical reactions in chemical carcinogenesis. Br J Cancer [Suppl 8] 55:19-23

Rassoulzadegan M, Cowie A, Carr A, Glaichenhaus N, Kamen R, Cuzin F (1982) The roles of individual polyoma virus early proteins in oncogenic transformation. Nature 300:713-718

Sinn E, Muller W, Pattengale P, Tepler I, Wallace R, Leder P (1987) Coexpression of MMT/v-Ha-ras and MMTY/cOmyc genes in transfenic mice. Synergistic action of oncogenes in vivo. Cell 49:465-475

Stephelin D, Varmus HE, Bishop JM, Vogt PK (1976) DNA related to the transforming gene(s) of avian sarcoma viruses is present in normal avian DNA. Nature 260:170-173

Yuspa SH (1987) Tumour promotion. In: Fortner JG, Rhoads JE (eds) Accomplishments in cancer research 1986. Lipincott, Philadelphia, pp 169-180

Diene Conjugation and Peroxidation in Precancer

T. L. Dormandy and D. G. Wickens

Department of Chemical Pathology, Whittington Hospital, Highgate Hill, London N19 5NF, UK

1 Introduction

Lipid peroxidation has had a bad press. The reasons for this go back to the pioneering work of Farmer and his group in the 1930s (Farmer and Sundralingham 1942; Farmer et al. 1942, 1943; Farmer 1942, 1946; Bolland and Koch 1945; Farmer 1946). In a series of publications they established free-radical-mediated chain autoxidation as the cause of rancidification and of related breakdown processes in fats and oils. Nothing can detract from the greatness of their achievement or indeed the rightness of their findings. But, quite unwittingly, they set in motion two misleading trends. One of the key events in the autoxidation chains they described is the interaction of lipid free radicals with molecular oxygen to form peroxy free radicals and then peroxides. In the type of material they studied this is the inevitable sequel of lipid free-radical generation. Very quickly in everyday usage the free-radical mediated breakdown of lipids became synonymous with peroxidation. This is particularly true today in areas of biomedical research: indeed, it finds expression in the title of the present Symposium. It can no longer be supported or justified by experimental evidence. Second, in medicine, lipid peroxidation, the cause of rancidification in food chemistry and of structural breakdown in biochemistry, has become closely associated with disease. This is a distortion of reality.

2 Diene Conjugation as a Peroxidation Marker

A decade before Farmer and his group embarked on their studies Gillam et al. (1931) described the development of a light-absorption peak around 230–235nm when natural fats are stored or subjected to alkali isomerisation. A few years later Edisbury et al. (1933) showed that the peak is the function of a diene-conjugated bond sequence, i.e. a bond sequence in which two double bonds are separated by one single bond (—C=C—C=C—). Since in the bulk of naturally

Nigam et al. (Eds.), Eicosanoids,
Lipid Peroxidation and Cancer
© Springer-Verlag Berlin Heidelberg 1988

occurring polyunsaturated lipids the double bonds are separated by two single bonds (—C=C—C—C=C—) the diene-conjugated configuration could be regarded as "anomalous"; it was attributed by Edisbury et al. (1935) to the formation of cyclic breakdown products. Farmer and his colleagues corrected this view (Farmer et al. 1942, 1943; Farmer 1942; Bolland and Koch 1945). They showed that free-radical attack on a polyunsaturated lipid involves a shift in the position of the double bonds: when a hydrogen atom is abstracted from the methylenic group between double bonds the double bonds "close-up" (by way of a hybrid radical) and become diene conjugated. The diene-conjugated configuration is retained by the peroxides formed and by most of the oxygen-containing peroxidation products. Farmer's concept was modified and extended by a number of workers – Bergstrom (1945) in particular established an order of precedence among more-or-less stable isomers – but in essentials it has remained unchallenged. Its practical importance lay in the simplicity of measuring a characteristic uv-light absorption peak. The technique was rapidly adopted by industry as a sensitive indicator of oxidative damage to fats, oils and food; and it remains widely used today.

3 Diene Conjugation in Biomedical Research

The recognition in the 1950s that free radicals may be as important in biological as in industrial chemistry created a similar need for specific "markers" and diene conjugation again offered a promising approach. Over the next 25 years, despite technical difficulties and some inconsistencies between different methodologies, the measurement helped to establish the role of free radicals in a wide range of biochemical and pathological processes, e.g. carbon tetrachloride toxicity (Recknagel and Ghoshal 1966a-c; Rao and Recknagel 1968; Di Luzio and Hartman 1969; Benedetti et al. 1974, 1977), ethanol toxicity (Di Luzio 1964; MacDonald 1973; Shaw et al. 1981), iron overload (Bacon et al. 1983) and ozone toxicity (Goldstein et al. 1969; Kyei-Aboagye et al. 1973; Mustafa et al. 1973).

The chief limitation of the method did not become apparent until its tentative introduction into clinical chemistry (Di Luzio 1972; Shaw et al. 1981; Wickens et al. 1981; Lunec et al. 1981, 1982). Although the spectroscopic results were not inconsistent with evidence based on other indirect "free-radical markers", they raised a number of perplexing questions. Peroxides and peroxidation products are almost undetectable in fresh human tissues and tissue fluids. Diene-conjugated material is not. Diene conjugation, moreover, was too non-specific a measurement for clinical use. The absorption peak at 234 nm undoubtedly measured a characteristic bond sequence, but, in human serum and tissues, this sequence could be a feature of a large number and variety of molecules. The uncertainties led to studies aimed at establishing the "sources" of diene conjugation. The results were unexpected.

4 Octadeca-9,11-dienoic Acid (18:2(9,11))

Cawood et al. (1983, 1984) showed that over 95% of diene conjugation in human serum and tissues could be ascribed to a single fatty-acid residue present in esterified form in all the main lipid classes (triglycerides, phospholipids, cholesteryl esters) as well as in free fatty acids. More surprisingly this fatty acid was shown *not* to be a peroxide or a peroxidation breakdown product. It was identified as a diene-conjugated isomer of linoleic acid, octadeca-9*cis*,11*trans*-dienoic acid (Iversen et al. 1984; Smith et al. 1988). A series of comparatively simple high-performance liquid chromatographic (HPLC) methods were developed for measuring the concentration of the compound in different lipid classes (Iversen et al. 1985). These methods also measured the concentration of the non-diene-conjugated parent compound (linoleic acid: octadeca-9*cis*,12*cis*-dienoic acid: 18:2(9,12)). The two could be related to each other in the form of a molar ratio (MR):

$$MR = \frac{\text{octadeca-9,11-dienoic acid} \times 100}{\text{octadeca-9,12-dienoic acid}}$$

Clinical studies showed that both in absolute terms and as a fraction of 18:2(9,12) the concentration of 18:2(9,11) varied independently in different lipids, one of several reasons why measurements of "total diene conjugation" had often proved difficult to interpret. Initial studies in human serum and bile focused on phospholipids and revealed significant abnormalities in a number of pathological states: chronic alcoholism (Fink et al. 1985, 1986; Szebeni et al. 1986), paraquat poisoning (Crump et al. 1985), pre-eclamptic toxaemia of pregnancy (Erskine et al. 1985) and primary biliary cirrhosis (Braganza and Day 1987).

5 18:2(9,11) in Cervical Precancer

Two main considerations led to the application of the new methodology to cancer and precancer of the human cervix. First, free-radical mechanisms have long been suspected of playing a part in malignant change (Emanuel 1976). Second, although theoretically cervical cytology offers, today, the best scope for effective prophylaxis against cancer it hardly does so in practise. (The practical difficulties arise partly from the inherent observer dependency of morphological diagnosis and partly from the apparent impossibility of automating this prohibitively expensive and time-consuming microscopic technique.) Initial studies on biopsy material showed a highly significant separation between normal and precancerous (CIN) epithelium (Griffin et al. 1987). The method was then adapted to exfoliated cells, the same type of material as used for the Papanicolaou technique (Tay et al. 1987). Since the 18:2(9,12) concentration does not differ significantly in normal and precancerous tissue (Griffin et al. 1987), it could be used as a "built-in" reference measurement, differences in the 18:2(9,11) concentra-

Table 1. Mean (SD) molar ratios of 18:2(9,11) to 18:2(9,12) in exfoliated cells.

Diagnosis	n	Molar ratio	Significance[a]
Normal	40	1.80 (0.47)	–
HPV infection	13	2.33 (0.68)	$p < 0.01$
CIN 1	40	2.89 (1.31)	$p < 0.0001$
CIN 2	38	3.06 (1.42)	$p < 0.0001$
CIN 3	81	3.32 (1.26)	$p < 0.0001$
Invasive cancer	3	5.25	–

[a] Mann-Whitney test against normal

tion being expressed as a molar ratio. Table 1 shows the results published in a preliminary report (Singer et al. 1987). These findings are highly encouraging: they point to a sensitivity at least comparable to the Papanicolaou technique. The specificity of the method remains to be established, a more complex and difficult task. There is no doubt that the method is capable of complete automation and is non-observer dependent. Very preliminary results from cancer in other sites suggest that an increase in 18:2(9,11) in malignant cells may not be confined to the cervical epithelium.

6 The Survival Value of Peroxidation

These findings raise many questions which still need to be answered. We would, however, point to one apparent contradiction which may have wider implications.

There is now considerable evidence to suggest that peroxidation is significantly diminished in malignant tissue. We would mention in particular the attenuation of the peroxy ESR signal in carcinoma of cervix reported by Benedetto et al. (1981) and the apparent resistance of malignant tissue to oxidative stress (Burton et al. 1983). On the face of it this may seem difficult to reconcile with the accumulation of a free-radical product in cervical precancer. The contradiction is, of course, only apparent. The 18:2(9,11) is not a peroxide, indeed it may reflect a block in the "normal" peroxidative sequel of a free-radical attack on lipids. This prompts us to consider briefly and at a frankly speculative level the possible survival value of lipid peroxidation.

Except in a few exceptional instances lipid peroxidation is inextricably linked to cell damage and cell death. The relationship is reciprocal. Lipid peroxidation on any scale is incompatible with cell survival and cell damage due to any cause favours lipid peroxidation. Indeed, it seems not unreasonable to suggest that lipid peroxidation may be an essential as well as an invariable step in the self-destroying mechanism of all cells. That such a mechanism exists cannot be seriously doubted. If it did not, cell turnover would come to a halt. (Theoretically it would be possible for cell turnover to be governed by cell production: it just happens that this is not how nature operates. In almost every instance it is the rate of cell destruction which governs cell production.) If lipid peroxidation is in fact an essential component of this built-in cytotoxic mechanism then it has immense survival value even under normal circumstances. Its survival value in cancer may be even greater.

In the course of any scientific meeting related to cancer much time and thought is usually devoted to carcinogenesis, i.e. to mechanisms which initiate and promote malignant change. The more menacing the picture becomes – and no such meeting passes without new carcinogenic dangers being revealed – the more one wonders how we survive even as long as we do. Part of the answer must surely be twofold. First, malignant change is essentially a form of cell damage. Second, the vast majority of cells damaged in this way are eliminated by the self-destroying mechanism built into every cell.

It is of course almost certainly possible for a powerful carcinogenic mechanism to overcome this natural cytotoxic process. But it is equally possible that the great majority of clinical cancers (at least in certain sites) develop because the cytotoxic process is weakened or suppressed. If this is so then the accumulation of 18:2(9,11) and similar compounds, indicating a block in free-radical mediated peroxidation, may reflect not carcinogenesis but the failure of normal cytotoxic activity. Perhaps such speculation has practical relevance. Not only are we conditioned by our work to think of peroxidation as inherently damaging and dangerous and, therefore, in the most generalised terms, a potential cause of disease: we are also increasingly exposed to a torrent of promotional literature dedicated to the idea of "protecting" us against this danger. Before we accept this message uncritically we should reflect that damage, destruction and death at the cellular level is a sine qua non of our health and survival as an organism and perhaps our main protection against cancer.

References

Bacon BR, Tavill AS, Brittenham GM, Park CH, Recknagel RO (1983) Hepatic lipid peroxidation in vivo in rats with chronic iron overload. J Clin Invest 71:429–439

Benedetti A, Ferrali M, Chieli E, Comporti M (1974) A study of the relationships between carbon tetrachloride-induced lipid peroxidation and liver damage in rats pretreated with vitamin E. Chem Biol Interact 9:117–134

Benedetti A, Casini AF, Ferrali M, Comporti M (1977) Early alterations induced by carbon tetrachloride in the lipids of the membranes of the endoplasmic reticulum of the liver cell. I. Separation and partial characterisation of altered lipids. Chem Biol Interact 17:151–166

Benedetto C, Bocci A, Dianzani MU, Ghiringhello B, Slater TF, Tomasi A, Vannini V (1981) Electron spin resonance studies on normal human uterus and cervix and on benign and malignant uterine tumours. Cancer Res 41:2936–2942

Bergstrom S (1945) The autoxidation of methyl ester of linoleic acid. Ark Kem Mineral Geol 21A:1–18

Bolland JL, Koch HP (1945) The course of autoxidation reactions in polyisoprenes and allied compounds. IX. The primary thermal oxidation products of ethyl linoleate. J Chem Soc 1945:445–447

Braganza JM, Day JP (1987) Serum octadeca-9,11-dienoic acid concentrations in primary biliary cirrhosis. Lancet 1:987–988

Burton GW, Cheeseman KH, Ingold KU, Slater TF (1983) Lipid antioxidants and products of lipid peroxidation as potential tumour protective agents. Biochem Soc Trans 11:261–262

Cawood P, Wickens DG, Iversen SA, Braganza JM, Dormandy TL (1983) The nature of diene conjugation in human serum, bile and duodenal fluid. FEBS Lett 162:239–243

Cawood P, Iversen SA, Dormandy TL (1984) The nature of diene conjugation in biological fluids. In: Bors W, Saran M, Tait D (eds) Oxygen free radicals in chemistry and biology. de Gruyter, Berlin, pp 355–358

Crump BJ, Thurnham DI, Situnayake RD, Davis M (1985) Free radicals and alcoholism. Lancet 2:955–956

Di Luzio NR (1964) Prevention of acute ethanol-induced fatty liver by the simultaneous administration of antioxidants. Life Sci 3:113–118

Di Luzio NR (1972) Protective effect of vitamin E on plasma lipid dienes in man. J Agric Food Chem 20:486–490

Di Luzio NR, Hartman AD (1969) The effect of ethanol and carbon tetrachloride administration on hepatic lipid-soluble antioxidants. Exp Mol Pathol 11:38–52

Edisbury JR, Morton RA, Lovern JA (1933) Absorption spectra in relation to the constituents of fish oils. Biochem J 27:1451–1460

Edisbury JR, Morton RA, Lovern JA (1935) The absorption spectra of fatty acids from fish-liver oils. Biochem J 29:899–908

Emanuel NM (1976) Free radicals and the action of inhibitors of radical processes under pathological states and ageing in living organisms and in man. Q Rev Biophys 9:283–308

Erskine KJ, Iversen SA, Davies R (1985) An altered ratio of 18:2(9,11) to 18:2(9,12) linoleic acid in plasma phospholipids as a possible predictor of pre-eclampsia. Lancet 1:554–555

Farmer EH (1942) Ionic and radical mechanisms in olefinic systems, with special reference to processes of double-bond displacement, vulcanisation and photo-gelling. Trans Faraday Soc 38:356–361

Farmer EH (1946) Peroxidation in relation to olefinic structure. Trans Faraday Soc 42:228–236

Farmer EH, Sundralingham A (1942) The course of oxidation reactions in polyisoprenes and allied compounds. I. The structure and reactive tendencies of the peroxides of simple olefins. J Chem Soc 1942:121–139

Farmer EH, Bloomfield GF, Sundralingham A, Sutton DA (1942) The course and mechanism of autoxidation reactions in olefinic and polyolefinic substances including rubber. Trans Faraday Soc 38:348–356

Farmer EH, Koch HP, Sutton DA (1943) The course of autoxidation reactions in polyisoprenes and allied compounds. VII. Rearrangement of double bonds during autoxidation. J Chem Soc 1943:541–547

Fink R, Clemens MR, Marjot DH, Patsalos P, Cawood P, Norden AG, Iversen SA, Dormandy TL (1985) Increased free-radical activity in alcoholics. Lancet 2:291–294

Fink R, Clemens MR, Marjot DH, Patsalos P, Cawood P, Iversen SA, Dormandy TL (1986) Plasma markers of lipid peroxidation in alcoholic patients. In: Rice-Evans C (ed) Free radicals, cell damage and disease. Richelieu, London, pp 127–131

Gillam AE, Heilbron IM, Hilditch TP, Morton RA (1931) Spectrographic data of natural fats and their fatty acids in relation to vitamin A. Biochem J 25:30–38

Goldstein BD, Lodi C, Collinson C, Balchum OJ (1969) Ozone and lipid peroxidation. Arch Environ Health 18:631–635

Griffin JFA, Wickens DG, Tay SK, Singer A, Dormandy TL (1987) Recognition of cervical neoplasia by the estimation of a free radical reaction product (octadeca-9,11-dienoic acid) in biopsy material. Clin Chim Acta 163:143–148

Iversen SA, Cawood P, Madigan MJ, Lawson AM, Dormandy TL (1984) Identification of a diene conjugated component of human lipid as octadeca-9,11-dienoic acid. FEBS Lett 171:320–324

Iversen SA, Cawood P, Dormandy TL (1985) A method for the measurement of a diene-conjugated derivative of linoleic acid, 18:2(9,11), in serum phospholipid, and possible origins. Ann Clin Biochem 22:137–140

Kyei-Aboagye K, Hazucha M, Wyszogrodski I, Rubenstein D, Avery ME (1973) The effect of ozone exposure in vivo on the appearance of lung tissue lipids in the endobronchial lavage of rabbits. Biochem Biophys Res Commun 54:907–913

Lunec J, Halloran SP, White AG, Dormandy TL (1981) Free-radical oxidation (peroxidation) products in serum and synovial fluid in rheumatoid arthritis. J Rheumatol 8:233–245

Lunec J, Wickens DG, Graff TL, Dormandy TL (1982) Copper, free radicals and rheumatoid arthritis. In: Sorenson JRJ (ed) Inflammatory diseases and copper. Humana, New Jersey, pp 231–242

MacDonald CM (1973) The effects of ethanol on hepatic lipid peroxidation and on the activities of glutathione reductase and peroxidase. FEBS Lett 35:227–230

Mustafa MG, de Lucia AJ, York GK, Arth C, Cross CE (1973) Ozone interaction with the rodent lung. II. Effects on oxygen consumption of mitochondria. J Lab Clin Med 82:357–365

Rao KS, Recknagel RO (1968) Early onset of lipoperoxidation in rat liver after carbon tetrachloride administration. Exp Mol Pathol 9:271–278

Recknagel RO, Ghoshal AK (1966a) Lipoperoxidation as a vector in carbon tetrachloride hepatotoxicity. Lab Invest 15:132–148

Recknagel RO, Ghoshal AK (1966b) New data on the question of lipoperoxidation in carbon tetrachloride poisoning. Exp Mol Pathol 5:108–117

Recknagel RO, Ghoshal AK (1966c) Quantitative estimation of peroxidative degeneration of rat liver microsomal and mitochondrial lipids after carbon tetrachloride poisoning. Exp Mol Pathol 5:413–426

Shaw S, Jayatilleke E, Ross WA, Gordon ER, Lieber CS (1981) Ethanol-induced lipid peroxidation: potentiation by long-term alcohol feeding and attenuation by methionine. J Lab Clin Med 98:417–424

Singer A, Tay SK, Griffin JFA, Wickens DG, Dormandy TL (1987) Diagnosis of cervical neoplasia by the estimation of octadeca-9,11-dienoic acid. Lancet 1:537–539

Smith GN, Taj M, Braganza JM (1988) Free Radic Biol Med, (submitted)

Szebeni J, Eskelson C, Sampliner R, Hartmann B, Griffin J, Dormandy T, Watson RR (1986) Plasma fatty acid pattern including diene-conjugated linoleic acid in ethanol users and patients with ethanol related liver disease. Alcoholism Clin Exp Res 10:647–650

Tay SK, Singer A, Griffin JFA, Wickens DG, Dormandy TL (1987) Recognition of cervical neoplasia by the estimation of a free radical reaction product (octadeca-9,11-dienoic acid) in exfoliated cells. Clin Chim Acta 163:149–152

Wickens DG, Wilkins MH, Lunec J, Ball G, Dormandy TL (1981) Free-radical oxidation (peroxidation) products in plasma in normal and abnormal pregnancy. Ann Clin Biochem 18:158–162

Oxygen Radical Regulation of Vascular Reactivity

G. Thomas and P. W. Ramwell

Georgetown University Medical Center, Department of Physiology and Biophysics, Washington, D.C. 20007, USA

1 Introduction

Lipid peroxides are implicated in the pathophysiology of many disease processes including cancer, atherosclerosis and inflammation (Frankel 1984). Several mechanisms have been suggested for these peroxide mediated events. These include loss of membrane integrity, liberation of lysosomal enzymes and initiation of destructive free radical mediated reactions (Bridges et al. 1983; Mak et al. 1983). However, the actual mechanism is still unknown. Enhanced prostaglandin synthesis occurs in human tumours. This may relate to the stimulation of prostaglandin H (PGH) synthetase by lipid peroxides (Hemler and Lands 1980). On the other hand lipid peroxides, by inhibiting prostacyclin (PGI_2) synthesis, may affect the tone of vascular smooth muscle. In addition to lipid peroxides, vascular endothelium may also interact with hydrogen peroxide (H_2O_2) derived from polymorphonuclear leukocytes (PMN). The PMN derived H_2O_2 stimulates PGI_2 production by endothelial cells at sites of inflammation (Harlan and Callahan 1984). The integrity of vascular endothelium is obligatory for many vasoactive compounds to elicit relaxation (Furchgott and Zawadzki 1980). This endothelium dependent relaxation is abolished by superoxide anion ($O_2^{\cdot -}$) (Rubanyi and Vanhoutte 1986). Little is known of the vascular activity of peroxides and therefore we have investigated the effect of several peroxides on the tone of isolated blood vessels with and without endothelium. We report here the mechanism of vascular reactivity of several peroxides and their interaction with $O_2^{\cdot -}$.

2 Materials and Methods

The method of Crawford et al. (1978) was used to prepare 15-HPETE, and 15-HETE was obtained by subsequent reduction with glutathione (Coene et al. 1986). All the other reagents and enzymes were purchased from Sigma Chemical

Nigam et al. (Eds.), Eicosanoids,
Lipid Peroxidation and Cancer
© Springer-Verlag Berlin Heidelberg 1988

Co. (St. Louis, Missouri, USA). Aortic rings (3–4 mm) were prepared from Sprague Dawley rats as suggested previously (Thomas and Ramwell 1986). The rings were suspended in 5 ml baths containing Krebs-bicarbonate buffer equilibrated with 95% O_2 and 5% CO_2 (pH 7.4) and warmed to 37°C. The tension generated was measured isometrically using a Harvard force transducer. The different peroxides were added to the bath in cumulative doses after the rings were pre-contracted with $PGF_{2\alpha}$ (1×10^{-5}M). In experiments with inhibitors, the reagents were added 5 min prior to the addition of $PGF_{2\alpha}$.

3 Results and Discussion

Previous studies have shown that the integrity of vascular endothelium is required for many vasodilating agents such as acetylcholine to elicit relaxation (Furchgott and Zawadzki 1980). Here, we have investigated the response of isolated ring segments of rat thoracic aorta to several peroxides. Fig. 1 shows the effect of H_2O_2 on aortic rings pre-contracted with $PGF_{2\alpha}$ ($1 \cdot 10^{-5}$M). Hydrogen peroxide dose-dependently relaxed the vessels to the same degree whether endo-

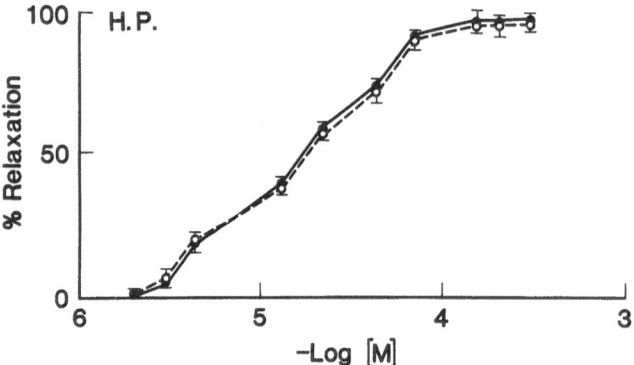

Fig. 1. Relaxing effect of hydrogen peroxide (*H.P.*) in rat aortic rings pre-contracted with $PGF_{2\alpha}$ (1×10^{-5} M), with (•—•) and without (o---o) endothelium. Each point is the mean ± SEM of seven determinations

Table 1. Relaxation potencies of various hydroperoxides in rat aorta pre-contracted with $PGF_{2\alpha}$ (1×10^{-5} M).

Compound	Dose for 50% relaxation ($\times 10^{-5}$ M)	Dose for 100% relaxation ($\times 10^{-5}$ M)
Hydrogen peroxide	2.0	16.0
Tert-butyl hydroperoxide	1.7	16.0
Cumene hydroperoxide	23.0	120.0
15-HPETE	3.6	15.0

thelium was present or not. This is in contrast to the situation observed with compounds like acetylcholine for which an intact endothelium is obligatory. Ta-

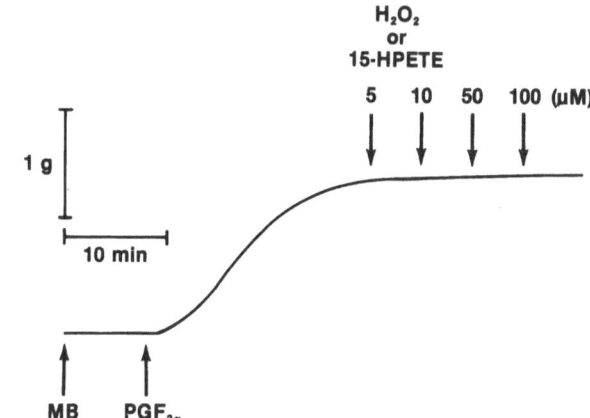

Fig. 2. Inhibitory effect of methylene blue (*MB*; 2×10^{-5} M) on the relaxation effect of hydrogen peroxide (H$_2$O$_2$) or 15-hydroperoxy eicosatetraenoic acid (*15-HPETE*)

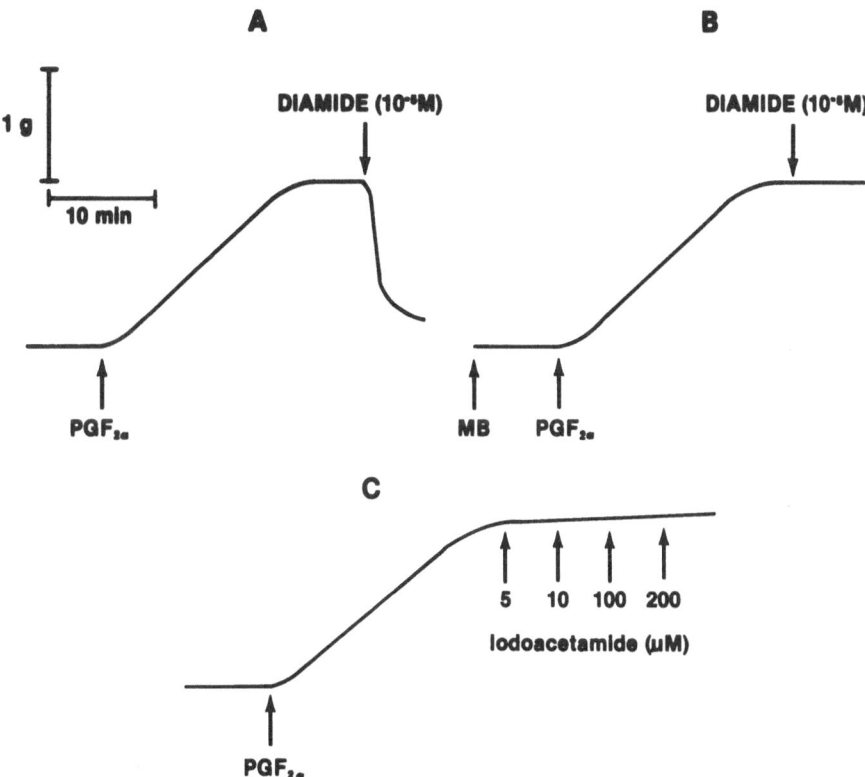

Fig. 3. A Relaxation effect of -SH oxidizing agent, diamide (1×10^{-5} M). **B** Effect of methylene blue (*MB*; 1×10^{-5} M) on diamide-elicited relaxation. **C** Effect of -SH alkylating agent, iodoacetamide. The rat aortic rings were contracted with PGF$_{2\alpha}$ (1×10^{-5} M)

ble 1 summarizes the relative potencies of hydrogen peroxide, tert-butyl hydro-peroxide, cumene hydroperoxide and 15-HPETE. There were not significant dif-ferences among the different hydroperoxides in their relaxation responses. Our studies were done in presence of either indomethacin ($2 \cdot 10^{-5}$M) or eicosatetray-noic acid, ETYA ($1 \cdot 10^{-5}$M). This indicates that the relaxation elicited by the peroxides was not mediated by the release of cyclo-oxygenase or lipoxygenase products. Previous reports have shown that endothelium dependent relaxation is inhibited by methylene blue, an inhibitor of soluble guanylate cyclase (Rapoport et al. 1983). Our results indicate that methylene blue inhibited the relaxation response to the peroxides (Fig. 2). Fatty acid peroxides and H_2O_2 are known to activate soluble guanylate cyclase (Graff et al. 1978; White et al. 1976). These data support the idea that peroxides elicit vascular relaxation by directly activat-ing soluble guanylate cyclase. Activation of guanylate cyclase depends on the redox state of the system. Thiol reductants like dithiothreitol or glutathione pre-vent activation of soluble guanylate cyclase and oxidants activate the enzyme (Graff et al. 1978). To test this hypothesis, we used diamide, an intracellular -SH oxidizing agent. Fig. 3 (A, B) shows that diamide ($1 \cdot 10^{-5}$M) relaxed vessels pre-contracted with $PGF_{2\alpha}$ and this relaxation is abolished by methylene blue. On the other hand, iodoacetamide, a thiol alkylating agent had no effect (Fig. 3C). Thus activation of soluble guanylate cyclase by the peroxides is probably related to the oxidation of intracellular thiol groups. This supports the previous report by Graff et al. (1978) on the activation of guanylate cyclase by endoperoxides and fatty acid hydroperoxides.

We have shown recently that in addition to 15-HPETE, other dihydroperox-ides like 5,15-diHPETE and 8,15-diHPETE also elicit relaxation of several iso-lated vascular tissues (D'Alarco et al. 1987). The hydroperoxides are further me-tabolized by the endothelial cells to their corresponding hydroxy fatty acids (Hopkins et al. 1984). However, studies (Forstermann and Neufang 1984) have shown that mono- and di-hydroxy fatty acids do not possess vasodilating prop-erties.

Direct and indirect evidence indicates that endothelium dependent relaxation is inhibited by $O_2^{\cdot -}$ (Rubanyi and Vanhoutte 1986). Moncada et al. (1986) have shown that several $O_2^{\cdot -}$ generating systems like hydroquinone, pyrogallol and Fe^{++}ions inhibit the vascular relaxation elicited by the endothelium derived re-laxing factor. This inhibitory action is reversed in presence of superoxide dismu-tase, but not by catalase. this implies $O_2^{\cdot -}$ destroys EDRF.

We have used a $O_2^{\cdot -}$ generating system xanthine oxidase (0.025 U/ml)-xan-thine (0.1 mM). Initial studies have indicated that this combination completely abolishes endothelium dependent relaxation to acetylcholine. Our results show that H_2O_2 elicited relaxation is not affected by the $O_2^{\cdot -}$ generating system (Fig. 4A, B). Hydroxyl radical, OH$^{\cdot}$, is an activator of soluble guanylate cyclase (Mit-tal and Murad 1977). Interaction of H_2O_2 with $O_2^{\cdot -}$ generates OH$^{\cdot}$ radicals (Pet-ers and Foote 1976). In our study, mannitol (10 mM), a scavenger of OH$^{\cdot}$, had no effect on H_2O_2 mediated relaxation, but the relaxation was abolished by ca-talase (300 U/ml) (Fig. 4C). These results suggest that peroxides like H_2O_2 elicit relaxation by directly activating soluble guanylate cyclase in smooth muscle

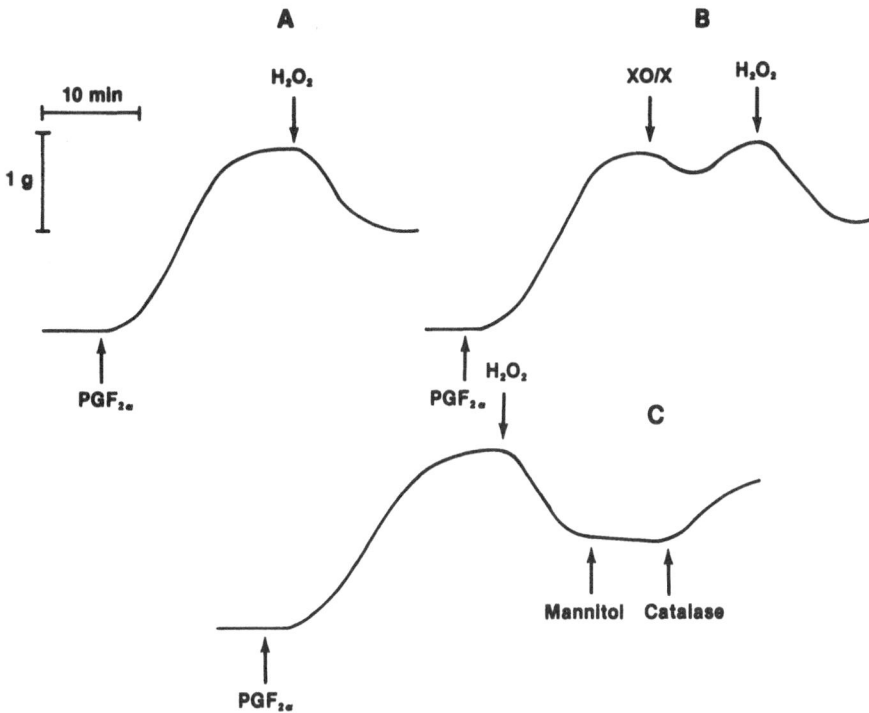

Fig. 4. **A** Relaxation effect of H_2O_2 (1×10^{-5} M) on rat aortic ring. **B** Effect of xanthine oxidase (0.025 U/ml)/xanthine (0.1 mM) (XO/X). **C** Effect of mannitol (10 mM) and catalase (300 U/ml) on H_2O_2-elicited relaxation. The rings were contracted with $PGF_{2\alpha}$ (1×10^{-5} M)

and it is not mediated by any oxygen radicals. Unlike EDRF, peroxide elicited relaxation is not inhibited by $O_2^{\cdot-}$.

In addition to their effect on blood vessels, the hydroperoxides may also have an effect on platelets. It has been shown that hydroperoxides such as 15-HPETE, 12-HPETE, and 9- and 13-hydroperoxy octadecadienoic acid (9- and 13-HPODE), inhibit arachidonic acid induced platelet aggregation and secretion (Bult et al. 1987). Similar to our studies with blood vessels, these effects of hydroperoxides on platelets are non-specific since benzoyl peroxide, a structurally dissimilar peroxide also has inhibitory effects on platelets (Bult et al. 1987).

These paradoxical effects of hydroperoxides may explain some of the observed vascular abnormalities and bleeding episodes in myeloproliferative diseases (Schafer 1984). Interestingly, 13-HPODE, a major lipoxygenase metabolite formed by leukocytes and endothelial cells prevents platelet adhesion to the endothelial surface (Buchanan et al. 1985).

Lipid hydro-peroxides are elevated in many pathological conditions and are reported to cause vascular injury (Goto 1982). Normally, in most tissues hydroperoxide levels are strictly regulated by the glutathione/glutathione peroxidase system. However, in many diseases (e.g. diabetes) and in nutritional deficiencies (e.g. selenium deficiency), the activity of glutathione peroxide/glutathione system will be reduced with concomittant increase in lipid peroxide levels. this may lead to the deleterious effects of hydroperoxides. It has been shown that acute as well as chronic injection of 15-HPETE to rabbits elicits severe intimal damage in the aorta (Hirai et al. 1982). In our experiments, the different peroxides did not affect the functional intergrity of the vascular endothelium. Exposure of the vessels to peroxides did not affect acetylcholine elicited endothelium dependent relaxation. This may be due to the smaller dose and shorter duration of exposure of the vessels to hydroperoxides in our experiments. We did observe an attenuation of the relaxation response to acetylcholine after the vessels were preincubated with hydroperoxides for longer than 60 min (data not shown). this damaging effect may be due to the generation of secondary free radical intermediates from hydroperoxides.

In conclusion, we report here that peroxides such as H_2O_2, 15-HPETE, 8.15-diHPETE and 4,15-diHPETE elicit vascular relaxation. This effect may be mediated through the activation of soluble guanylate cyclase by the peroxides which oxidize essential thiol groups in the enzyme. It is unlikely that this effect relates to the generation of other radicals such as $O_2^{\cdot-}$ or $OH^{\cdot}$.

References

Bridges SJW, Benford DJ, Hubbard SA (1983) Mechanisms of toxic injury. Ann NY Acad Sci 407:42–63

Buchanan MR, Haas TA, Lagarde M, Guichardant M (1985) 13-Hydroxy octa decadienoic acid is the vessel wall chemorepellant factor LOX. J Biol Chem 260:16056–16059

Bult H, Coene M-C, Claeys SM, Laekmann GM, Herman AG (1987) Hydroxy-hydroperoxy derivatives of linoleic and arachidonic acid suppress the activation of rabbit blood platelets. Prostaglandin Thromboxane Lukotriene Res 17:224–228

Coene M-C, Bult H, Claeys M, Laekmann GM, Herman AG (1986) Inhibition of rabbit platelet activation by lipoxygenase products of arachidonic acid and linoleic acid. Thromb Res 42:205–214

Crawford CG, van Alpen GWHM, Cook HW, Lands WEM (1978) The effect of precursors and product analogs of prostaglandin cyclooxygenase upon iris sphincter muscle. Life Sci 23:1255k–1262

D'Alarco M, Corey EJ, Cunard C, Ramwell PW, Uotila P, Vargas R, Wroblewska B (1987) The vasodilation induced by hydroperoxy metabolites of arachidonic acid in the rat mesenteric and pulmonary circulation. Br J Pharmacol 91:627–632

Forstermann U, Neufang B (1984) The endothelium dependent relaxation of rabbit aorta: effects of antioxidants and hydroxlated eicosatetraenoic acids. Br J Pharmacol 82:765–767

Frankel EN (1984) Lipid oxidation: Mechanisms, products and biological significance. JAOCS 61:1908–1916

Furchgott RF, Zawadski JV (1980) The obligatory role of endothelial cells in the relaxation of arterial smooth muscle by acetylcholine. Nature 288:373–376

Goto Y (1982) Lipid peroxides as a cause of vascular disease. In: Yagi K (ed) Lipid peroxides in biology and medicine. Academic, New York, pp 295–303

Graff G, Stephenson JH, Glass DB, Haddox MK, Goldberg ND (1978) Activation of soluble splenic cell guanylate cyclase by prostaglandin endoperoxides and fatty acid hydroperoxides. J Biol Chem 253:7662–7676

Harlan JM, Callahan KS (1984) Role of hydrogen peroxide in the neutrophil mediated release of prostacyclin from cultured endothelial cells. J Clin Invest 74:442–448

Hemler ME, Lands WEM (1980) Evidence for a peroxide initiated free radical mechanism of prostaglandin biosynthesis. J Biol Chem 255:6253-6262

Hirai S, Okamoto K, Morimatsu M (1982) Lipid peroxides in the aging process. In: Yagi K (ed) Lipid peroxides in biology and medicine. Academic, New York, pp 305-315

Hopkins NK, Oglesby TD, Bundy GL, Gormann RR (1984) Biosynthesis and metabolism of 15-hydroperoxy-5, 8, 11, 13-eicosatetraenoic acid by human umbilical vein endothelial cells. J Biol Chem 259:14048-14053

Mak IT, Misra HP, Weglicki WB (1983) Temporal relationship of free radical induced lipid peroxidation and loss of latent enzyme activity in highly enriched hepatic lysosomes. J Biol Chem 258:13733-13737

Mittal CK, Murad F (1977) Activation of guanylate cyclase by superoxide dismutase and hydroxyl radical: A physiological regulator of guanosine 3', 5'-monophosphate formation. Proc Natl Acad Sci USA 74:4360-4364

Moncada S, Palmer RM, Gryglewski RJ (1986) Mechanism of action of some inhibitors of endothelium derived relaxing factor. Proc Natl Acad Sci USA 83:9164-9168

Peters JW, Foote CS (1976) Chemistry of superoxide ion. Its reaction with hydroperoxides. J Am Chem Soc 98:873-875

Rapoport RM, Draznin MB, Murad F (1983) Endothelium dependent relaxation in rat aorta may be mediated through cyclic GMP-dependent protein phosphorylation. Nature 306:174-176

Rubanyi GM, Vanhoutte PM (1986) Superoxide anions and hyperoxia inactivate endothelium derived relaxing factor. Am J Physiol 250:H822-H827

Schafer AI (1984) Bleeding and thrombosis in the myeloproliferative disorders. Blood 64:1-12

Thomas G, Ramwell PW (1986) Induction of vascular relaxation by hydroperoxides. Biochem Biphys Res Commun 139:102-108

White AA, Crawford KM, Patt CS, Lad PJ (1976) Activation of soluble guanylate cyclase from rat lung by hydrogen peroxide. J Biol Chem 251:7304-7312

Arachidonic Acid-Induced Activation of NADPH Oxidase in Membranes of Human Neutrophils and HL-60 Leukemic Cells is Regulated by Guanine Nucleotide-Binding Proteins and is Independent of Ca^{2+} and Protein Kinase C

R. Seifert[1], W. Rosenthal[1], C. Schächtele[2] and G. Schultz[1]

1 Introduction

Neutrophils possess a plasma membrane-bound NADPH oxidase, which catalyzes superoxide ($O_2^{\cdot-}$) formation using NADPH as electron donor (Rossi 1986). $O_2^{\cdot-}$ formation can be activated by the chemotactic peptide, N-formyl-L-methionyl-L-leucyl-L-phenylalanine (FMLP). Several mechanisms have been suggested to be involved in the activation of NADPH oxidase as depicted in Fig. 1.

a) Binding of FMLP to plasma membrane receptors activates phospholipase C-catalyzed polyphosphoinositide degradation to diacylglycerol and inositol 1,4,5-trisphosphate, which involves a pertussis-toxin-sensitive guanine nucleotide-binding protein (G-protein) (Ohta et al. 1985). Diacylglycerol like the

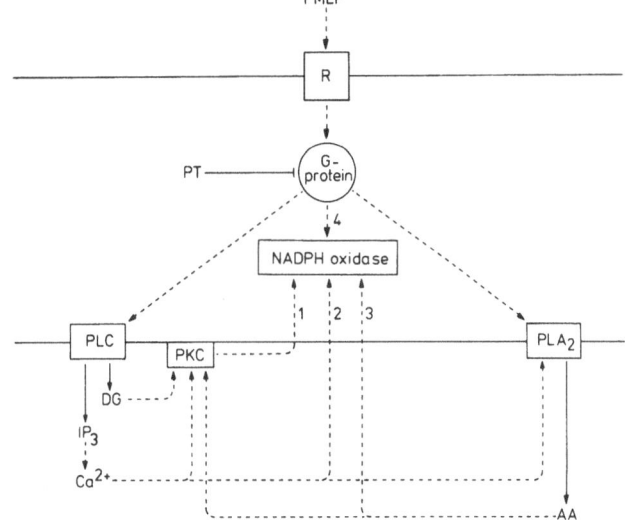

Fig. 1. Possible mechanisms involved in FMLP-induced activation of NADPH oxidase. *R*, Receptor; *PT*, pertussis toxin; *PLC*, phospholipase C; *PLA₂*, phospholipase A₂; *DG*, diacylglycerol; *IP₃*, inositol 1,4,5-trisphosphate. Numbers indicate possible mechanisms of NADPH oxidase stimulation

[1] Institut für Pharmakologie, Freie Universität Berlin, Thielallee 69–73, 1000 Berlin 33, FRG
[2] Goedecke Forschungsinstitut, Biochemische Pharmakologie, Mooswaldallee 1–9
  7800 Freiburg, FRG

Nigam et al. (Eds.), Eicosanoids,
Lipid Peroxidation and Cancer
© Springer-Verlag Berlin Heidelberg 1988

tumourpromoting phorbol ester, 4β-phorbol 12-myristate 13-acetate (PMA) activates phospholipid/Ca^{2+} – dependent protein kinase C (PKC) by increasing the apparent affinity of the enzyme for Ca^{2+} (Nishizuka 1984). As PMA and cell-permeable diacylglycerols are potent activators of NADPH oxidase, it is assumed that PKC mediates activation of $O_2^{\cdot-}$ formation (Rossi 1986).

b) The FLMP-induced release of inositol 1,4,5-trisphosphate leads to intracellular Ca^{2+} mobilization. As the Ca^{2+} ionophore, A 23 187, induces $O_2^{\cdot-}$ formation and as Ca^{2+}-depleted neutrophils do not respond to FMLP, NADPH oxidase activation has been suggested to be a Ca^{2+}-dependent process (Rossi 1986).

c) In neutrophils, phospholipase C activation is accompanied by phospholipase A_2-mediated release of arachidonic acid (AA, C 20:4 *cis*), a process which also appears to involve a G-protein (Okajima and Ui 1984; Ohta et al. 1985). As AA induces $O_2^{\cdot-}$ formation in intact neutrophils (Bromberg and Pick 1983; Badwey et al. 1984) and in cell-free systems (Bromberg and Pick 1984; Curnutte et al. 1987), it was suggested that AA or AA metabolites may serve as intracellular messengers for activation of NADPH oxidase.

d) Activation of NADPH oxidase by chemotactic peptides may involve a direct G-protein effect as pertussis-toxin inhibits activation of $O_2^{\cdot-}$ formation by FMLP in Ca^{2+}-depleted and PMA-primed neutrophils, where NADPH oxidase is activated independently of phospholipase C and presumably of phospholipase A_2 (Grzeskowiak et al. 1986).

We investigated these four possibilities of NADPH oxidase regulation using membranes from human neutrophils and HL-60 leukemic cells and present evidence that the enzyme is regulated by G-proteins, independently of Ca^{2+} and PKC (Seifert et al. 1986, 1988; Seifert and Schultz 1987a, b).

2 Results

2.1 Reversible Activation of NADPH Oxidase by Fatty Acids

AA, other *cis*- and *trans*-unsaturated fatty acids and SDS (sodium dodecyl sulfate) activated NADPH oxidase in membranes of human neutrophils in the presence of Mg^{2+} and a cytosolic factor (CF) (Fig. 2). Arachidic acid (C 20:0), 5,8, 11,14-eicosatetraynoic acid (ETYA) and esters of unsaturated fatty acids did not activate the enzyme. The $O_2^{\cdot-}$-generation rates obtained with a maximally effective concentration of AA (50 μM) were in the same order of magnitude as the rates obtained with other unsaturated fatty acids and SDS.

In order to evaluate the role of AA metabolites in the activation of NADPH oxidase, the effect of ETYA, a potent inhibitor of lipoxygenases and cyclo-oxygenase at concentrations below 10 μM (Aharony et al. 1984), was studied (Fig. 3). ETYA at concentrations higher than 30 μM inhibited $O_2^{\cdot-}$ formation elicited by AA and linoleic acid (C 18:2 *cis*), which are substrates for lipoxygenases and

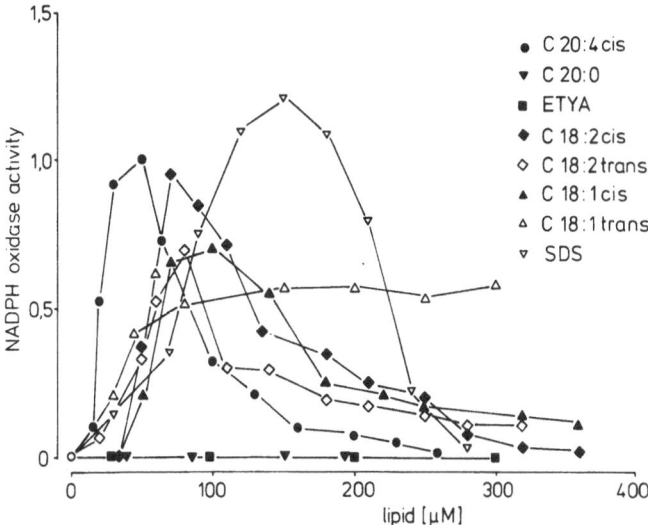

Fig. 2. Activation of NADPH oxidase in membranes of human neutrophils by fatty acids and SDS. NADPH oxidase activity is given in arbitrary units (enzyme activity measured in the presence of 50 μM AA = 1). For experimental details see Seifert and Schultz (1987a)

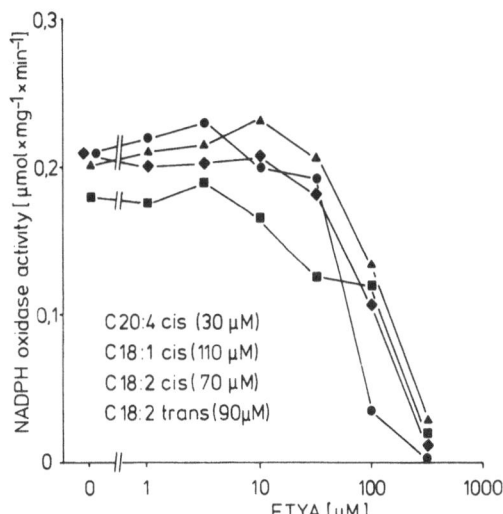

Fig. 3. Inhibition of fatty acid-induced NADPH oxidase activation in neutrophil membranes by ETYA. Experimental procedures are given in Seifert and Schultz (1987a)

cyclo-oxygenase, and $O_2^{\cdot-}$ formation induced by oleic (C 18:1 *cis*) and linolelaidic acid (C 18:2 *trans*), both not being substrates for these enzymes (Kinsella et al. 1981; Needleman et al. 1986). The cyclo-oxygenase inhibitors, diclofenac and piroxicam, at concentrations of up to 100 μM, and sodium salicylate, at concentrations of up to 10 mM (Flower 1974), did not inhibit AA-induced $O_2^{\cdot-}$

formation in neutrophil membranes. In addition, *bis* (tert.-butyl)peroxide (0.1–1000 μM) did not activate NADPH oxidase, and soybean lipoxygenase was no substitute for CF. These results indicate that metabolites of unsaturated fatty acids are not obligatory intermediates in activation of NADPH oxidase by fatty acids and that CF is neither a lipoxygenase nor cyclo-oxygenase. The inhibitory effect of ETYA on $O_2^{\cdot-}$ formation may be explained by competitive antagonism of ETYA and unsaturated fatty acids at sites which are not localized on AA-metabolizing enzymes. It is also unlikely that unsaturated fatty acids activate NADPH oxidase by increasing membrane fluidity as *trans*-unsaturated fatty acids do not increase the fluidity of plasma membranes (Karnowski et al. 1982).

The interaction of membranes, CF and AA was studied by omitting NADPH from the reaction mixture; after a preincubation, the reaction was started by the addition of NADPH. In HL-60 membranes, activation of NADPH oxidase was a very rapid process, reaching its maximum already after a 30 sec preincubation. NADPH oxidase activation was maximal when all components were present during the 3 min preincubation (Table 1). Addition of AA together with NADPH reduced the $O_2^{\cdot-}$ generation rate by 40%. The addition of membranes after preincubation of CF and AA almost completely prevented turnover of the enzyme, whereas the omission of CF reduced the $O_2^{\cdot-}$ generation rate only by 18%. In each case, bovine serum albumin (BSA), which binds AA quantitatively (Badwey et al. 1984), completely prevented $O_2^{\cdot-}$ formation. BSA rapidly terminated AA-induced $O_2^{\cdot-}$ formation in HL-60 membranes. The activated NADPH oxidase of HL-60 membranes, sedimented by centrifugation and resuspended in buffer, did not generate $O_2^{\cdot-}$ until both AA and CF were re-added. These data indicate that AA probably interacts with HL-60 membranes and not with CF, and that the activation process of NADPH oxidase is reversible.

Table 1. Activation of NADPH oxidase in membranes of HL-60 cells by arachidonic acid and its reversibility by BSA

Constituents of the activation mixture	Addition at $t = 0$ min	NADPH oxidase activity $(\text{nmol} \times \text{mg}^{-1} \times \text{min}^{-1})$
CF, M, AA	NADPH	16.9 ± 2.4
CF, M, AA	NADPH + BSA	0
CF, M	NADPH + AA	10.0 ± 2.2
CF, M	NADPH + AA + BSA	0
CF, AA	NADPH + M	2.5 ± 1.0
CF, AA	NADPH + M + BSA	0
M, AA	NADPH + CF	13.8 ± 2.0
M, AA	NADPH + CF + BSA	0

HL-60 membranes (M), CF, AA (200 μM) NADPH (500 μM) and BSA (33.3 μM) were added to reaction mixtures as indicated. Data are the mean ± SEM of three experiments.

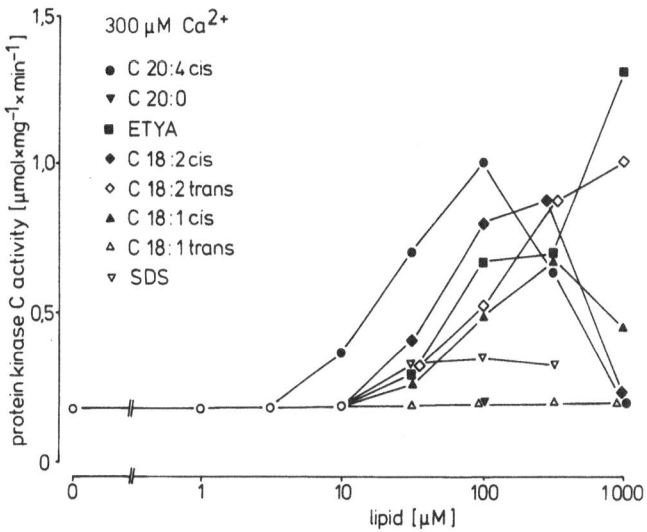

Fig. 4. Activation of protein kinase C by fatty acids and SDS. For experimental details see Seifert et al. (1988)

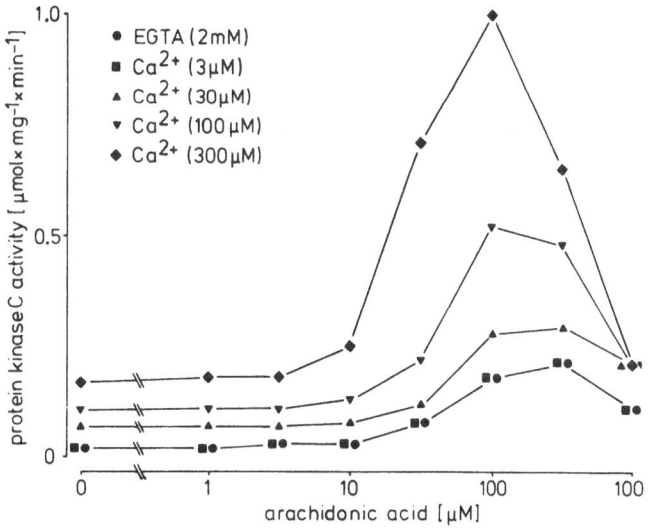

Fig. 5. Ca^{2+} dependency of protein kinase C activation by arachidonic acid. For experimental details see Seifert et al. (1988)

2.2 The Role of Ca^{2+} and Protein Kinase C in Activation of NADPH Oxidase by Fatty Acids

In order to explore the role of PKC in fatty acid-induced $O_2^{\cdot-}$ generation, we studied the effects of fatty acids on PKC purified from rat brain. In the presence of 300 μM Ca^{2+}, *cis*-unsaturated fatty acids (AA, linoleic and oleic acid) and linolelaidic acid activated the enzyme to similar extents with maximal stimulation occurring at concentrations between 30 μM and 1 mM (Fig. 4). In contrast, arachidic and elaidic acid (C 18:1 *trans*) were inactive and SDS activated PKC only slightly. ETYA, at concentrations above 10 μM, strongly activated PKC. At similar concentrations, ETYA has been reported to inhibit glucose transport (Tsunawaki and Nathan 1986) and soluble guanylate cyclase (Böhme et al. 1983) and to enhance formation of 5-hydroxyeicosatetraenoic acid (Bokoch and Reed 1981). AA activated PKC independently of Ca^{2+} employed at concentrations up to 10 μM; at higher concentrations, Ca^{2+} potentiated the effects of AA (Fig. 5). Using hydrophobic interaction chromatography, it was recently shown that Ca^{2+} increased the hydrophobicity of PKC (Walsh et al. 1984). Thus, Ca^{2+} may facilitate hydrophobic interaction between AA and PKC, thereby increasing the enzyme activity. AA-induced activation of PKC was potentiated by 1,2-dioleoyl-*rac*-glycerol (DO) (Fig. 6). PKC activation by AA (100 μM) and DO (10 μg/ml) was five times higher than that obtained with either lipid alone. Synergistic enzyme activation was also seen with PMA and in the presence of phospholipids. As certain hormonal stimuli, including chemotactic peptides, cause release of diacylglycerol and AA from phospholipids by parallel activation of, pre-

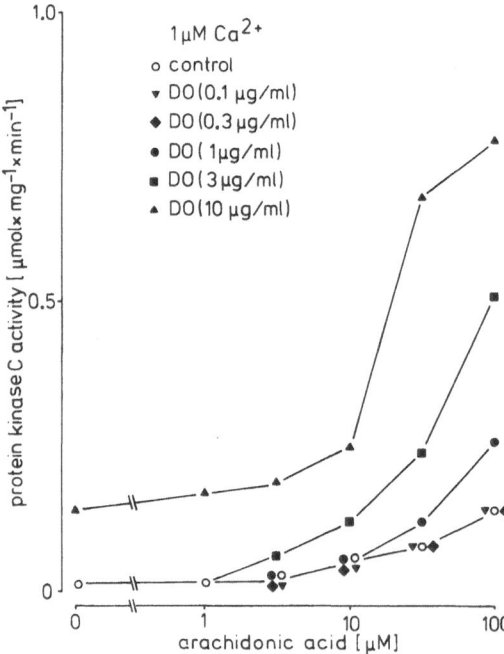

Fig. 6. Synergistic activation of protein kinase C by arachidonic acid and diacylglycerol. For experimental details see Seifert et al. (1988)

sumably, G-protein-regulated phospholipases C and A_2 (Ohta et al. 1985; Burch et al. 1986), AA and diacylglycerol may be involved synergistically in hormonal stimulation of PKC (see Fig. 1).

Based on these data, it appears that NADPH oxidase activation by fatty acids is independent of Ca^{2+} and PKC and that CF is different from PKC, as ETYA is a potent activator of PKC but inhibits $O_2^{\cdot-}$ formation. Elaidic acid and SDS strongly activated $O_2^{\cdot-}$ formation but not PKC. Fatty acids activated $O_2^{\cdot-}$ formation in the absence of exogenous Ca^{2+}, and chelation of endogenous Ca^{2+} by 10 mM EGTA did not affect $O_2^{\cdot-}$ formation. Ca^{2+} at concentrations above 100 μM inhibited AA-induced $O_2^{\cdot-}$ formation; the inhibition was almost complete with Ca^{2+} at a concentration of 1 mM. In contrast, AA-induced PKC activation was potentiated by Ca^{2+} at concentrations above 10 μM. $O_2^{\cdot-}$ formation elicited by fatty acids and SDS was not inhibited by 1-(5-isoquinolinesulfonyl)-2-methylpiperazine (H-7), a potent inhibitor of PKC. Purified PKC was no substitute for CF and did not enhance $O_2^{\cdot-}$ formation in the presence of CF. The removal of ATP from the reaction mixtures by preincubation with hexokinase and glucose or dialysis of CF did not abolish the stimulatory effect of AA.

2.3 Involvement of G-Proteins in the Regulation of NADPH Oxidase

The stable GTP-analogue, guanosine 5'-O-(3-thiotriphosphate) (GTPγS), a potent activator of G-proteins (Gilman 1987) stimulated $O_2^{\cdot-}$ formation elicited by AA in HL-60 membranes about three-fold. Guanylyl imidodiphosphate (GppNHp), another stable guanine nucleotide, and NaF plus $Al(SO_4)_3$ (presumably through AlF_4^-) stimulated $O_2^{\cdot-}$ formation about three-fold, whereas GTP, ATP and adenylyl imidodiphosphate (AppNHp) were inactive (Table 2). GDP (1 mM) completely prevented basal $O_2^{\cdot-}$ formation and concentration-dependently terminated basal and GTPγS-stimulated $O_2^{\cdot-}$ formation (Fig. 7). Effects similar to those obtained with GDP were observed with the stable GDP-analogue, guanosine 5'-O-(2-thiodiphosphate) (GDPβS), wheres ADP and UDP

Table 2. Influence of guanine and adenine nucleotides and fluoride on NADPH oxidase activity in HL-60 membranes

Addition	NADPH oxidase activity ($nmol \times mg^{-1} \times min^{-1}$)
none	30.5 ± 3.7
ATP (100 μM)	29.8 ± 2.5
AppNHp (100 μM)	32.6 ± 3.0
ATPγS (100 μM)	80.7 ± 5.7
GTP (10 μM)	31.5 ± 3.2
GppNHp (10 μM)	102.5 ± 8.7
GTPγS (10 μM)	100.9 ± 5.5
NaF (20 mM) + $Al_2(SO_4)_3$ (5 μM)	98.7 ± 6.0

Nucleotides, fluoride or solvent were added to reaction mixtures containing AA (200 μM). Data are the mean ± SEM of three experiments.

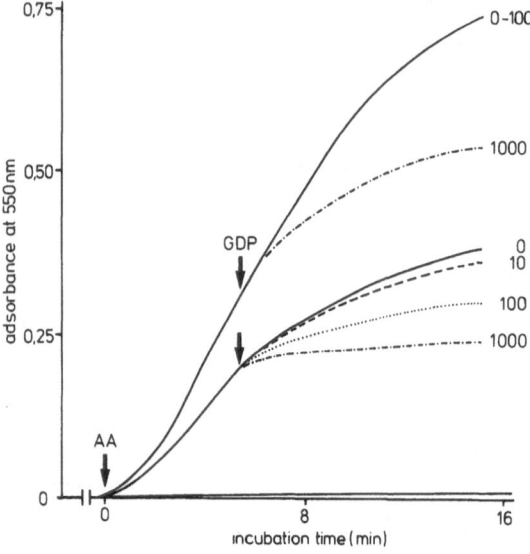

Fig. 7. Activation of $O_2^{\cdot-}$ formation in HL-60 membranes by GTPγS and its deactivation by GDP. Assays contained either GDP (1 mM, *lower trace*), solvent (*middle trace*) or GTPγS (10 μM, *upper trace*). Six minutes after AA (200 μM), GDP at the concentrations indicated was added to the reaction mixtures. For experimental details see Seifert and Schultz (1987b)

were inactive. These results indicate that GDP competes with G-protein-activating nucleotides, i.e. with endogenous GTP and less effectively with exogenous GTPγS, thus promoting inactivation of G-proteins (Gilman 1987) and consequent deactivation of NADPH oxidase. The results also implicate that the induction of basal $O_2^{\cdot-}$ generation rates does not only require AA and CF but also active G-proteins.

Adenosine 5'-O-(3-thiotriphosphate) (ATPγS) stimulated $O_2^{\cdot-}$ formation in HL-60 membranes more than two-fold (Table 2). This can possibly be explained by the following mechanism: the thiophosphate group of ATPγS can be transferred to phosphate acceptors by kinase-mediated reactions (Yi-Chi Sun et al. 1980). HL-60 membranes possess nucleoside diphosphokinase (NDPK) activity catalyzing thiophosphorylation of endogenous GDP to GTPγS which, in turn, activates G-proteins and hence NADPH oxidase. This interpretation is supported by the finding that UDP and ADP, inhibitors of NDPK (Kimura and Shimada 1983) inhibits the stimulatory effect of ATPγS but not that of GTPγS. NDPK activity is absolutely dependent of Mg^{2+} (Parks and Agarwal 1973) and chelation of Mg^{2+} by EDTA abolishes the stimulatory effect of ATPγS but not that of GTPγS. Phosphorylation of endogenous GDP to GTP by creatine kinase and creatine phosphate completely prevented the thiophosphorylation of GDP to GTPγS by ATPγS.

3 Discussion

Studies in cell-free systems greatly contributed to our understanding of NADPH oxidase regulation (Fig. 8); however, many problems remain to be solved.

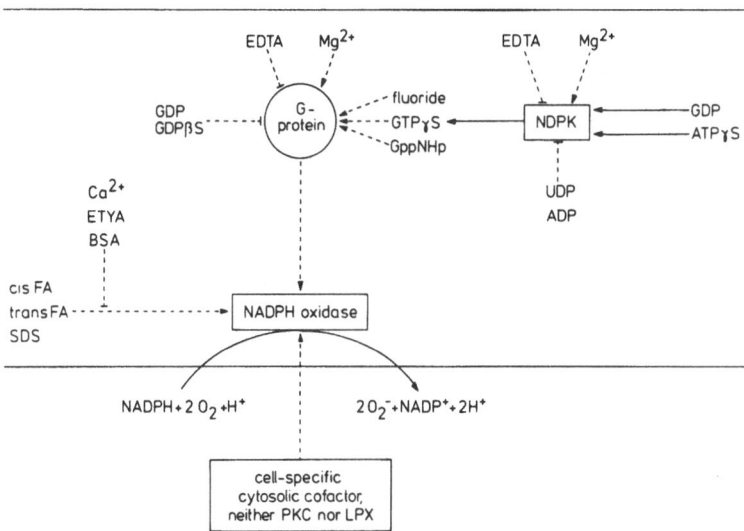

Fig. 8. Regulation of NADPH oxidase in membranes of human neutrophils and HL-60 human leukemic cells. *Cis-FA, cis*-unsaturated fatty acids; *trans-FA, trans*-unsaturated fatty acids; *LPX*, lipoxygenase

a) Neutrophils possess a phospholipase A_2 that can be activated by chemotactic peptides (Bormann et al. 1984; Okajima and Ui 1984). Inhibition of phospholipase A_2 by the phospholipase A_2-inhibitor, p-bromophenacyl bromide, or of G-proteins by pertussis toxin suppresses both $O_2^{\cdot -}$ formation and release of AA (Bromberg and Pick 1983; Okajima and Ui 1984; Ohta et al. 1985). AA reversibly activates NADPH oxidase in intact cells and in cell-free systems and, therefore, mimics reversible NADPH oxidase activation by chemotactic peptides. In intact cells the FMLP-induced enzyme activation is antagonized by competitive FMLP-antagonists (Jesaitis et al. 1986), and enzyme activation induced by AA is reversed by BSA (Badwey et al. 1984). In cell-free systems AA-induced $O_2^{\cdot -}$ formation is antagonized by Ca^{2+}, BSA, ETYA, GDP and GDPβS. However, as the release of AA is not an obligatory prerequisite for receptor-operated activation of NADPH oxidase (Tsunawaki and Nathan 1986) the question remains open as to whether AA acts as intracellular messenger to trigger activation of NADPH oxidase in intact cells.

b) The stimulatory effects of AA on NADPH oxidase in cell-free systems are independent of lipoxygenase, cyclo-oxygenase, PKC and membrane fluidity. AA may interact directly and reversibly with membrane components of NADPH oxidase. As the esters of unsaturated fatty acids failed to induce $O_2^{\cdot -}$ formation, it may be tempting to speculate that AA may activate NADPH oxidase by virtue of its function as an ionic surfactant, binding to a component of NADPH oxidase and leading to a change in enzyme charge and conformation (Bromberg and Pick 1985).

c) It remains to be clarified, whether NADPH oxidase, in analogy to adenylyl cyclase, is under the direct control of G-proteins (Gilman 1987), In addition, it is still not known whether the enzyme is regulated by the same G-protein (G_n) that has been suggested to be involved in FMLP-induced activation of phospholipase C (Gierschik et al. 1987) or by G_i, the G-protein involved in the inhibition of adenylyl cyclase, or by a distinct G-protein specific for NADPH oxidase.

d) Until now, it has been possible to activate phospholipase C and phospholipase A_2 but not NADPH oxidase in cell-free systems using FMLP (Bormann et al. 1984; Smith et al. 1985). The reason for this difference is unknown but it has been suggested that the FMLP-induced $O_2^{\cdot -}$ formation depends on an intact cytoskeleton (Jesaitis et al. 1986) which is destroyed during cell disruption.

e) Finally, the identity of CF involved in the activation of NADPH oxidase remains to be resolved. CF appears to be a phagocyte-specific, heat-labile and dialysis-resistent macromolecule, which is apparently different from lipoxygenases, caclo-oxygenase and PKC. The identification of CF may be facilitated by the recent finding that CF is absent in the cytosol of certain patients with chronic granulomatous disease (Curnutte et al. 1987) and in the cytosol of undifferentiated HL-60 cells.

Acknowledgements. The authors thank Mrs. R. Krüger for help in the preparation of the manuscript. The technical assistance of Miss M. Wulfern, Miss K. Dorst and Mrs. E. Glaß is greatly appreciated. The authors are grateful to Dr. K.-H. Jakobs (Pharmakologisches Institut der Universität Heidelberg) for supplying HL-60 cells. This work was supported by a grant of the Deutsche Forschungsgemeinschaft and of the Fonds der Chemischen Industrie.

References

Aharony D, Smith JB, Silver MJ (1984) Platelet lipoxygenase: Inhibitors. In: Chakrin LW, Bailey DM (eds) The leukotrienes. Academic, New York, pp 116–123

Badwey JA, Curnutte JT, Robinson JA, Berde CB, Karnovsky MJ, Karnovsky ML (1984) Effects of free fatty acids on release of superoxide and on change of shape by human neutrophils. J Biol Chem 259:7870–7877

Böhme E, Gerzer R, Grossmann G, Herz J, Mülsch A, Spies C, Schultz G (1983) Regulation of soluble guanylate cyclase activity. In: Dumont JE, Nunez J, Denton RM (eds) Hormones and cell regulation, vol 7. Elsevier, Amsterdam, pp 147–161

Bokoch GM, Reed PW (1981) Evidence for inhibition of leukotriene A_4 synthesis by 5,8,11, 14-eicosatetraynoic acid in guinea pig polymorphonuclear leukocytes. J Biol Chem 256:4156–4159

Bormann BJ, Huang C-K, Mackin WM, Becker EL (1984) Receptor-mediated activation of phospholipase A_2 in rabbit neutrophil plasma membrane. Proc Natl Acad Sci USA 81:767–770

Bromberg Y, Pick E (1983) Unsaturated fatty acids as second messengers of superoxide generation by macrophages. Cell Immunol 79:240–252

Bromberg Y, Pick E (1984) Unsaturated fatty acids stimulate NADPH-dependent superoxide production by cell-free system derived from macrophages. Cell Immunol 88:213–221

Bromberg Y, Pick E (1985) Activation of NADPH-dependent superoxide production in a cell-free system by sodium dodecyl sulfate. J Biol Chem 260:13539–13545

Burch RM, Luini A, Axelrod J (1986) Phospholipase A_2 and phospholipase C are activated by distinct GTP-binding proteins in response to α_1-adrenergic stimulation in FRTL5 thyroid cells. Proc Natl Acad Sci USA 83:7201–7205

Curnutte JT, Kuver R, Scott PJ (1987) Activation of neutrophil NADPH oxidase in a cell-free system. J Biol Chem 262:5563–5569

Flower RJ (1974) Drugs which inhibit prostaglandin biosynthesis. Pharmacol Rev 26:33–67

Gierschick P, Sidiropoulos D, Spiegel A, Jakobs KH (1987) Purification and immunochemical characterization of the major pertussis-toxin-sensitive guanine-nucleotide-binding protein of bovine-neutrophil membranes. Eur J Biochem 165:185–194

Gilman AG (1987) G Proteins: Transducers of receptor-generated signals. Annu Rev Biochem 56:615–649

Grzeskowiak M, della Bianca V, Cassatella MA, Rossi F (1986) Complete dissociation between the activation of phosphoinositide turnover and of NADPH oxidase by formyl-methionyl-leucyl-phenylalanine in human neutrophils depleted of Ca^{2+} and primed with subthreshold doses of phorbol 12, myristate 13, acetate. Biochem Biophys Res Commun 135:785–794

Jesaitis AJ, Tolley JO, Allen RA (1986) Receptor-cytoskeleton interactions and membrane traffic may regulate chemoattractant-induced superoxide production in human granulocytes. J Biol Chem 261:13662–13669

Karnowski MJ, Kleinfeld AM, Hovver RL, Klausner RD (1982) The concept of lipid domains in membranes. J Cell Biol 94:1–6

Kimura N, Shimada N (1983) GDP does not mediate but rather inhibits hormonal signals to adenylate cyclase. J Biol Chem 258:2278–2283

Kinsella JE, Bruckner G, Mai J, Shimp J (1981) Metabolism of trans fatty acids with emphasis on the effects of trans, trans-octadecadienoate on lipid composition, essential fatty acid, and prostaglandins: An overview. Am J Clin Nutr 34:2307–2318

Needleman P, Turk J, Jakschik BA, Morrison AR, Lefkowith JB (1986) Arachidonic acid metabolism. Annu Rev Biochem 55:69–102

Nishizuka Y (1984) The role of protein kinase C in cell surface signal transduction and tumor promotion. Nature 308:693–698

Ohta H, Okajima F, Ui M (1985) Inhibition by islet-activating protein of a chemotactic peptide-induced early breakdown of inositol phospholipids and Ca^{2+} mobilization in guinea pig neutrophils. J Biol Chem 260:15771–15780

Okajima F, Ui M (1984) ADP-ribosylation of the specific membrane protein by islet-activating protein, pertussis toxin, associated with inhibition of a chemotactic peptide-induced arachidonate release in neutrophils. A possible role of the toxin substrate in Ca^{2+}-mobilizing biosignaling. J Biol Chem 259:13863–13871

Parks RE, Agarwal RP (1973) Nucleoside diphosphokinases. In: Boyer PD (ed) The enzymes, vol 8A, 3rd edn. Academic, New York, pp 307–333

Rossi F (1986) The O_2^--forming NADPH oxidase of the phagocytes: nature, mechanisms of activation and function. Biochim Biophys Acta 853:65–89

Seifert R, Schultz G (1987a) Fatty acid-induced activation of NADPH oxidase in plasma membranes of human neutrophils depends on neutrophil cytosol and is potentiated by stable guanine nucleotides. Eur J Biochem 162:563–569

Seifert R, Schultz G (1987b) Reversible activation of NADPH oxidase in membranes of HL-60 human leukemic cells. Biochem Biophys Res Commun 146:1296–1302

Seifert R, Rosenthal W, Schultz G (1986) Guanine nucleotides stimulate NADPH oxidase in membranes of human neutrophils. FEBS Lett d205:161–165

Seifert R, Schächtele C, Rosenthal W, Schultz G (1988) Activation of protein kinase C by cis- and trans-fatty acids and its potentiation by diacylglycerol. Biochem Biophys Res Commun 154:20–26

Smith CD, Lane BC, Kusaka I, Verghese MW, Snyderman R (1985) Chemoattractant receptor-induced hydrolysis of phosphatidylinositol 4,5-bisphosphate in human polymorphonuclear leukocyte membranes. J Biol Chem 260:5875–5878

Tsunawaki S, Nathan CF (1986) Release of arachidonate and reduction of oxygen. J Biol Chem 261:11563–11570

Walsh MP, Valentine KA, Ngai PK, Carruthers CA, Hollenberg MD (1984) Ca^{2+}-dependent hydrophobic-interaction chromatography. Biochem J 224:117–127

Yi-Chi Sun I, Johnson EM, Allfrey VG (1980) Affinity purification of newly phosphorylated protein molecules. J Biol Chem 255:742–747

Lipid Peroxidation and Cancer II

Biochemistry of Oxidative Stress: Recent Experimental Work

H. Sies

Institut für Physiologische Chemie I, Universität Düsseldorf, Moorenstraße 5, 4000 Düsseldorf 1, FRG

1 Introduction

Reactive oxygen species have been shown to occur as a normal part of aerobic life, and they can lead to toxicity in cells and organs (for reviews, see Chance et al. 1979; Sies 1985, 1986). This is of importance in pathophysiological conditions such as carcinogenesis, inflammation, radiation damage, and also in drug and xenobiotic toxicity. Reactive oxygen species are also generated in the process of redox cycling, as used in chemotherapy, and there are indications that they are also formed during alcohol metabolism. A balance between pro-oxidant and antioxidant capacities is normally maintained in cells, and a disbalance in favour of pro-oxidants has been called oxidative stress. The most recent reviews on the many aspects in this field from our group are available (Sies 1986, 1987a–c; Sies et al. 1987a, b).

In the present report, some of the aspects recently studied in our group will be discussed.

2 Generation of Photoemissive Species

In previous work we have studied the generation of photoemissive species during quinone redox cycling in subcellular fractions, isolated cells, and the intact organ, using rat liver as starting material. This work has recently been reviewed (Wefers and Sies 1986, 1987). The cytotoxic and carcinogenic effects of quinones as well as the therapeutic effects of quinone anticancer drugs can be related to the formation of semiquinone free radicals which may react with essential biomolecules of the cell. Due to their capability of redox cycling with the formation of superoxide anion radicals, injury to the cell can also occur via reactive oxygen species. The process of formation of reactive species includes the generation of electronically excited species. The photoemissive decay of these molecules per-

Nigam et al. (Eds.), Eicosanoids,
Lipid Peroxidation and Cancer
© Springer-Verlag Berlin Heidelberg 1988

mits the direct monitoring of these species via single-photon counting. Using menadione (2-methyl-1, 4-naphthoquinone) as a model compound, we have shown that the intact haemoglobin-free perfused rat liver responds to infusion of the quinone with an increased photoemission, as do isolated hepatocytes and microsomal fractions. The bulk of the photoemission is in the red spectral region, with about 80% of the intensity being emitted at wavelengths greater than 620 nm. These and other data point to the formation of singlet molecular oxygen during menadione redox cycling in intact cells. The protective function of DT diaphorase against the formation of photoemissive species was further studied using the purified enzyme (Prohaska et al. 1987).

A clinically used agent which contains a quinone moiety is mitomycin C. In recent work with isolated rat liver microsomes we detected photoemissive species during mitomycin C induced redox cycling (Napetschnig and Sies 1987). The increase of photoemission in deuterium oxide as well as greater than 90% of the intensity occuring at wavelengths greater than 610 nm suggest that singlet oxygen is a photoemissive species generated by this system. Glutathione disulfide accumulated during the reaction. We proposed that the superoxide anion radicals formed during redox cycling of mitomycin C react with GSH. The generation of glutathionyl radicals followed by oxygen addition then would lead to the formation of photoemissive species and GSSG.

Using an enzymatic model system comprised of horseradish peroxidase or hemin, we obtained further evidence for this process (Medeiros et al. 1987). Horseradish peroxidase (HRP) catalysed the oxidation of reduced giutathione. This reaction was accompanied by light emission, attributed to the generation of singlet oxygen. The chemiluminescence was directly related to thiyl radical formation, as deduced from the correlation between the time course of HRP-compound II formation and the emission in the presence of different amounts of H_2O_2. Superoxide dismutase had an inhibitory effect on the chemiluminescence without affecting HRP-compound II formation. This indicates the direct involvement of superoxide radicals in the production of photoemissive species. Likewise, the replacement of HRP by hemin was acompanied by chemiluminescence.

In further studies on the raltionships between oxygen uptake, malondialdehyde formation and the generation of photoemissive species as detected by low-level chemiluminescence during microsomal lipid peroxidation (Noll et al. 1987), it was concluded that there are different subsets of peroxidizing compounds. Differences in the time course of the appearance of the different parameters were postulated to be due to the participation of distinct chemical or special subsets of secondary peroxyl radicals in the membrane.

3 DNA Damage by Singlet Molecular Oxygen

Radiation-induced (and much of the chemically induced) DNA damage is attributable to free-radical reactions, notably the hydroxyl radical. However, non-radical reactions of electronically excited species are also important, e.g. in DNA damage by photo-oxidation, and can explain the effects of photosensitizers which generate singlet molecular oxygen. Although singlet molecular oxygen has long been known to react with constituents of nucleic acids there has been uncertainty as to its importance in eliciting DNA damage. Using a physical source of singlet molecular oxygen by employing a microwave discharge system, we examined the transforming activity of the plasmid pBR 322 in *Escherichia coli*. Taking care to exclude 0 atoms and ozone, we showed that singlet molecular oxygen led to a loss of biologically active DNA (Wefers et al. 1987). The increase of the effectiveness of singlet molecular oxygen in deuterium oxide and the effects of some singlet-oxygen quenchers underscore the importance of this electronically excited state of oxygen in biological processes of DNA damage. A similar study using chemically generated singlet oxygen was performed by Lafleur et al. (1987).

4 Ebselen, a Novel Organoselenium Compound

This synthetic organoselenium compound has been found to exhibit antioxidant capacity (Müller et al. 1984, 1985). Using photoemission as an assay of lipid peroxidation in a system of rat liver microsomes, a lag phase preceding the onset of ascorbate/ADP-Fe-induced lipid peroxidation was increased by the addition of ebselen, whereas the sulphur analogue was inactive. This pertained not only to the low-level chemiluminescence, but also to other parameters of lipid peroxidation such as the evolution of ethane and pentane and the production of thiobarbiturate-reactive material. In addition to this antioxidant activity, the compound acts catalytically in the GSH peroxidase reaction. In our recent work (Müller et al. 1988), we studied the biotransformation of ebselen in the isolated perfused rat liver with the demonstration of a novel type of glucuronide. This selenium-glucuronide was identified by high pressure liquid chromatography and mass spectrometry.

5 Control of Levels of Detoxication Enzymes

It seems that the control of antioxidant capacity is regulated by mechanisms of gene expression and further protein processing. In this regard, we have become interested in one aspect relating to the so-called phase II enzymes. One such enzyme is NADPH: quinone oxidoreductase, also called DT diaphorase. In recent work, we have studied the effect of 5-azacytidine on the activity of DT diaphorase and other enzymes (Wagner et al. 1987). The application of 5-azacytidine is known to lead to a DNA hypomethylation, and evidence is accumulating in the literature that DNA hypomethylation can be of importance in regulating gene expression. The pretreatment of mice with 5-azacytidine was shown to lead to an increase in hepatic DT diaphorase activity. Likewise, specific activities of some GSH S-transferase isozymes were increased, for example the activity with DCNB and with ethacrynic acid.

In more recent work, we have demonstrated that the mRNA was substantially increased for the DT diaphorase, as indicated in Northern blots using pDTD 55, kindly supplied by Dr. C. Pickett (Wagner et al. 1988).

6 Spontaneous Mutagenesis and Oxidative Damage to DNA

In recent work on spontaneous mutagenesis (Storz et al. 1988) *Salmonella typhimurium* strains were examined with respect to oxyR, a positive regulator of defences against oxidative stress. It was shown that deletions of oxyR lead to 10- to 25-fold higher frequencies of spontaneous mutagenesis compared to otherwise isogenic oxyR+ control strains. The high spontaneous mutation frequency in oxyR deletion strains was decreased 3-fold when the strains were grown anaerobically. OxyR deletion strains showed an increase in small deletion mutations and at least three of the six possible base substitution mutations (T:A to A:T, C:G to T:A, and C:G to A:T). However, the largest increase in mutation frequency was observed for T:A to A:T transversions (40- to 146-fold), the base substitution mutation most frequently caused by chemical oxidants. The introduction into oxyR deletion strains of multi-copy plasmids carrying the oxyR-regulated genes for catalase (katG) or alkyl hydroperoxide reductase (ahp) resulted in overexpression of the respective enzyme activities and decreased the number of spontaneous mutants to wild type levels. The introduction into oxyR deletions of a plasmid carrying the gene for superoxide dismutase (sodA) decreased the mutation frequency 5-fold in some strain backgrounds. Strains which contain a dominant oxyR mutation and overexpress proteins regulated by oxyR, showed 2-fold lower spontaneous mutation frequencies. these results and further work (Christman et al. 1985; Hartman et al. 1984; Sargentini and Smith 1985; Farr et al. 1986) indicate that there is an important role of oxidative damage to DNA in spontaneous mutagenesis.

7 Summary

Cells physiologically exposed to oxidative challenge normally maintain a delicate prooxidant/antioxidant balance. Oxidative challenge and carcinogenesis are linked in a number of ways. This refers to the generation of reactive metabolites to form ultimate carcinogens as well as to oxygen-derived species that modulate the process.

Chemically, free-radical compounds and electronically excited compounds (singlet oxygen; excited carbonyls) are of interest, in addition to epoxides, hydroperoxides and other structures. For example, plasmid DNA (pBR322) was found to exhibit a loss of transforming activity when exposed to singlet oxygen generated in a microwave discharge system.

Cellular control of the levels of the above mentioned compounds is exerted both enzymatically and non-enzymatically. The latter includes the role of antioxidants such as vitamins E and C and selenium. Regarding enzymatic defence, one sector includes the phase II group of detoxication enzymes, many of which are under the control of DNA methylation. DNA (cytosine) hypomethylation was found to lead to an enhanced expression of some GSH S-transferases and NADPH: quinone oxidoreductase and other antioxidant enzymes, concomitant with diminished expression of cytochrome P-450 forms. This response, studied in mice, resembles the pattern changes observed in hepatic noduli. The experimenal conditions to manipulate the DNA methylation pattern were the use of an antitumour agent, azacytidine, or the (tumorigenic) methyl-deficient diet.

Acknowledgement. Work from the author's laboratory was supported by Deutsche Forschungsgemeinschaft and National Foundation for Cancer Research.

References

Chance B, Sies H, Boveris A (1979) Hydroperoxide metabolism in mammalian organs. Physiol Rev 59:527–605

Christman MF, Morgan RW, Jacobson FS, Ames BN (1985) Positive control of a regulon for defenses against oxidative stress and some heat-shock proteins in Salmonella typhimurium. Cell 41:753–762

Farr S, d'Ari R, Touati D (1986) Oxygen-dependent mutagenesis in Escherichia coli lacking superoxide dismutase. Proc Natl Acad Sci USA 83:8268–8272

Hartman L, Hartman PE, Barnes WM, Tuley E (1984) Spontaneous mutation frequencies in salmonella: Enhancement of G/C to A/T transitions and depression of deletion and frameshift mutation frequencies afforded by anoxic incubation. Environ Mutagen 6:633–650

Lafleur MVM, Nieuwint AWM, Aubry JM, Kortbeek H, Arwert F, Joenje H (1987) DNA Damage by chemically generated singlet oxygen. Free Radic Res Commun 343–350

Medeiros MHG, Wefers H, Sies H (1987) Generation excited species catalyzed by horseradish peroxidase or hemin in the presence of reduced glutathione and H_2O_2. J Free Radic Biol Med 3:107–110

Müller A, Cadenas E, Graf P, Sies H (1984) A novel biologically active selenoorganic compound. I. Glutathione peroxidase-like activity in vitro and anti-oxidant capacity of PZ 51. Biochem Pharmacol 33:3235–3239

Müller A, Gabriel H, Sies H (1985) A novel biologically active selenoorganic compound. IV. Protective glutathione-dependent effect of PZ 51 (ebselen) against ADP-Fe induced lipid peroxidation in isolated hepatocytes. Biochem Pharmacol 34:1185–1189

Müller A, Gabriel H, Sies H, Terlinden R, Fischer H, Römer A (1988) A novel biochemically active selenoorganic compound. VII. Biotransformation of ebselen in perfused rat liver. Biochem Pharacol (in press)

Napetschnig S, Sies H (1987) Generation of photoemissive species by mitomycin C redox cycling in rat liver microsomes. Biochem Pharmacol 18:3037–3042

Noll T, de Groot H, Sies H (1987) Distinct temporal relation among oxygen uptake, malondialdehyde formation, and low-level chemiluminescence during microsomal lipid peroxidation. Arch Biochem Biophys 252:284–291

Prohaska HJ, Talalay P, Sies H (1987) Direct protective effect of NAD(P)H: Quinone reductase against menadione-induced chemiluminescence of postmitochondrial fractions of mouse liver. J Biol Chem 262:1931-1934

Sargentini JJ, Smith KC (1985) Spontaneous mutagenesis: the roles of DNA repair, replication, and recombination. Mutat Res 154:1-27

Sies H (1985) Oxidative stress: Introductory remarks. In: Sies H (ed) Oxidative stress. Academic, London, pp 1-8

Sies H (1986) Biochemistry of oxidative stress. Angew Chem [Int Edn Engl] 25:1058-1071

Sies H (1987a) Die enzymatische Entgiftung reaktiver Sauerstoffspezies. In: Elstner EF (ed) Reaktive Sauerstoffspezies in der Medizin. Springer, Berlin Heidelberg New York, pp 184-190

Sies H (1987b) Antioxidant activity in cells and organs. Am Rev Respir Dis 136:478-480

Sies H (1987c) Lipid peroxidation and its measurement. In: Walden TL Jr, Hughes HN (eds) Prostaglandin and lipid metabolism in radiation injury. Plenum, New York, pp 379-389

Sies H, Akerboom T, Ishikawa T, Cadenas E, Graf P, Gabriel H, Müller A (1987a) Hepatic and cardiac hydroperoxide metabolism. Role of selenium. In: Combs GF Jr, Spallholz JE, Levander OA, Oldfield JE (eds) Selenium in biology and medicine. Van Nostrand Reinhold, New York, pp 104-114

Sies H, Brigelius R, Graf P (1987b) Hormones, glutathione status and protein S-thiolation. Adv Enzyme Regul 26:175-189

Storz G, Christman MF, Sies H, Ames BN (1988) Spontaneous mutagenesis and oxidative damage to DNA in Salmonella typhimurium. Proc Natl Acad Sci USA, (in press)

Wagner G, Balzer F, Swiers C, Sies H (1987) Induction of DT diaphorase and other detoxication enzymes of phase II by 5-azacytidine. Chem Scripta 27A:95-96

Wagner G, Pott U, Bruckschen M, Sies H (1988) Effects of 5-azacytidine and methyl-group deficiency on NAD(P)H: Quinone oxidoreductae and GSH S-transferase in liver. Relationship to DNA methylation and DNA(Cytosine-5) methyltransferase. Biochem J 251, (in press)

Wefers H, Sies H (1986) Generation of photoemissive species during quinone redox cycling. Biochem Pharmacol 35:22-24

Wefers H, Sies H (1987) Formation of photoemissive species during redox cycling of menadione and the menadione glutathione conjugate. Chem Scripta 27A:109-111

Wefers H, Schulte-Frohlinde D, Sies H (1987) Loss of transforming activity of plasmid DNA (pBR 322) in E. coli caused by singlet molecular oxygen. FEBS Lett 211:49-52

Lipid Peroxidation and Irreversible Hepatocyte Damage

G. Poli, E. Albano, E. Chiarpotto, F. Biasi, R. Carini and
M. U. Dianzani

Department of Experimental Medicine and Oncology of the University, Corso Raffaello 30,
10 125, Torino, Italy

1 Introduction

Membrane changes are key events in the cellular injuries caused by a variety of substances and pathological conditions. When these structural alterations occur in parenchymal tissues the consequent functional impairment can be particularly significant. In this respect the hepatocyte has so far been the type of cell most investigated. Moreover, because lipids rich in unsaturated fatty acids are the major components of biological membranes their physico-chemical derangement has been a fundamental subject for experimental research. Damage to membrane lipids is often a free radical-mediated process resulting in the breakdown of polyunsaturated fatty acids via lipid peroxidation (for a review see Slater 1972).

At present the actual role of lipid peroxidation in the pathogenesis of the irreversible damage of liver cells (and also of other kinds of cells) is undecided. These is debate on whether it can be the cause or is merely a consequence of the fundamental damaging event (for a review see Poli et al. 1987).

In order partially to resolve this question the isolated hepatocyte model is useful since one can observe, under conditions in which lipid peroxidation takes place, a nett temporal separation between accumulation of lipid peroxidation products and cell death. Under such conditions where lipid peroxidation precedes the onset of cell degeneration the addition of strong antioxidants to the cell suspension prevents hepatocyte killing.

2 Experimental Procedures

For hepatocyte isolation, male albino rats from a Wistar strain were used (Nossan, Correzana, Milano, Italy), fed on a standard diet containing 40 mg/kg of α-tocopherol (Piccioni, Brescia, Italy). The animals were permitted free access to food and water until sacrifice under penthobarbital anaesthesia.

Nigam et al. (Eds.), Eicosanoids,
Lipid Peroxidation and Cancer
© Springer-Verlag Berlin Heidelberg 1988

The cell isolation procedure was as previously described (Poli et al. 1979, 1981). Hepatocyte viability was routinely checked before each experiment by the Trypan Blue exclusion test (Poli et al. 1979). For a more precise determination of the proportion of total cells which were viable the leakage of intracellular enzymes into the incubation medium was monitored at zero time and then every hour up to the end of each experiment, usually lasting four hours.

Of the several possible parameters for measuring irreversible damage, the leakage of cell enzymes was chosen, since it has proved to give a reproducible estimate of the amount of cells with well preserved plasmamembrane barriers. Recently, some criticism has been levelled at this technique as a measure of cell death, based on the fact that cells with extensive membrane damage morphologically expressed by blebbing can repair and survive. Although this criticism must be carefully considered, blistering, at least in experimental intoxications carried out by us does not seem closely related to leakage of mitochondrial and cytosolic enzymes. In fact, in the case of CCl_4 intoxication of isolated hepatocytes, blisters are detectable by electron microscopy in significant amounts on the cell surface only a few minutes after addition of haloalkane to the cell suspension (Tomasi and Albano, unpublished data), while enzyme leakage starts to increase significantly relative to the control only after two hours treatment. Furthemore, antioxidants can stop lipid peroxidation and prevent enzyme loss without modifying the ultramicroscopic blebbing. The blebs are probably due to a specific solvent effect of CCl_4.

To monitor the propagation of lipid peroxidation stimulated by the different pro-oxidant compounds tested, the measurement of malonaldehyde (MDA) steady-state concentration still appears the easiest and most reliable method. The procedures used to enrich the hepatocyte antioxidant defences were the addition of promethazine directly into the cell incubation medium at the start of the incubation, or the isolation of liver cells from rats preloaded with α-tocopherol by a single injection (100 mg vitamin E/kg body wt.) 15–16 hours before sacrifice (Pfeifer and McCay 1971).

3 Results

Membrane lipid peroxidation can be stimulated in hepatocytes in single cell suspension by the addition of one of the following compounds in micromolar final concentration: CCl_4, 1,2-dibromoethane, $CBrCl_3$, $FeSO_4$, $FeCl_3$, cumene hydroperoxide, or in millimolar final concentration: paracetamol, ethanol.

The relative pro-oxidant effect of these compounds, evaluated in terms of the increase in MDA in the total cell suspension during 60 min incubation is illustrated in Table 1.

Acutely overloading hepatocytes with iron is, in our model, the treatment producing the greatest oxidative stress, followed by treatment with haloalkanes and organic hydroperoxides. the pro-oxidant effect of ethanol is the weakest we observed, moreover its effects were not very reproducible. In any case, while lipid

Table 1. Rise of malonaldehyde (MDA) steady-state levels without increase of lactate dehydrogenase (LDH) in rat hepatocyte suspensions after one hour poisoning with defined pro-oxidant substances

Experimental groups	MDA production (nmol/10^7 cells)	LDH release (% of total cell content)
Control	1–2	2–4
CCl$_4$ 129 μM	4–6	2–5
Ethanol 20 mM	2–3	2–3
Cumene hydroperoxide 200 μM	7–10	2–3
FeCl$_3$ 100 μM	8–10	2–4
FeCl$_3$-ADP 0.1–2.5 mM	14–18	2–4

A representative range of values (time zero subtracted) is reported for both MDA and LDH. MDA was measured in total cell suspensions, i.e. cells plus incubation medium.

peroxidation is stimulated in all listed conditions after 1 hour of liver cell incubation, at the same time no sign of increased enzyme leakage can be found (Table 1). In fact, to see detectable release of lactate dehydrogenase (LDH), or transaminase (AST, ALT) by the treated hepytocytes, the poisoning has to be prolonged for more than two or three hours.

As a general example of the time course of cell damage, AST values obtained during 4 hours of CCl$_4$ poisoning are reported in Fig. 1 Analogous evidence for paracetamol and 1,2-dibromoethane has been published earlier (Albano et al. 1983, 1984).

With the exception of hepatocyte poisoning by ethanol or cumene hydroperoxide (which have not yet been investigated for their susceptibility to antioxi-

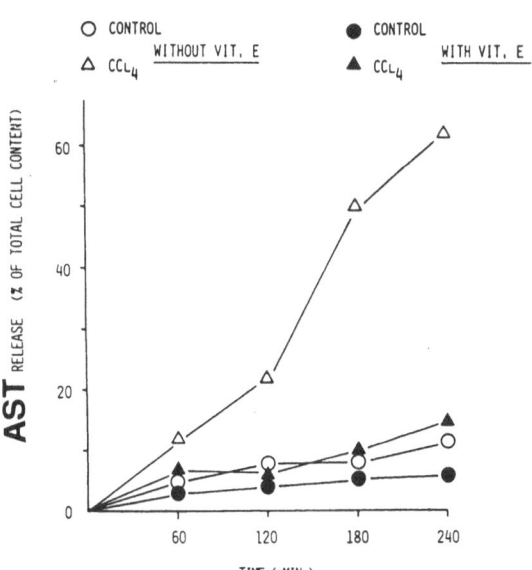

Fig. 1. Time course of AST release after CCl$_4$ poisoning of hepatocytes isolated from normal and vitamin E preloaded rats. Each point represents the mean ± SD of three experiments in duplicate

dants), in all other intoxications the inhibition of the pro-oxidant effect by means of antioxidant treatments led to a significant if not complete inhibition of cell enzyme leakage. This is the case, for instance, for CCl$_4$-treated hepatocytes enriched in their vitamin E content (Fig. 1). Even after 4 hours incubation in the presence of the toxin, not only do they not peroxidise, they do not die.

4 Discussion

Although the supporting data have been obtained primarily using an in vitro experimental model, lipid peroxidation processes seem likely to play a significant role in the hepatotoxicity of defined pro-oxidant compounds. Actually, the list of these substances can be enlarged beyond those we have tested if one reviews the recent literature on this topic, as considered elsewhere in more detail (Poli et al. 1987). For example in acute hepatocyte intoxication with adriamycin (Fariss et al. 1985), ethoxycoumarin (Gerson et al. 1986) and allyl alcohol (Dore et al. 1985, 1986) lipid peroxidation has been shown to be not merely a consequence but the initiating cause of the loss of cell integrity.

Not all toxic compounds possessing pro-oxidant activity have been demonstrated to exert their damaging activity through lipid peroxidation. A typical example of a second group of pro-oxidant toxins is menadione whose liver toxicity has been shown to be independent of its peroxidative action (Orrenius et al. 1984, 1985). Lipid peroxidation is also not involved in the mechanisms leading to liver injury mediated by bipiridyl herbicides, by diethylmaleate, by iodacetamide, by Cd, Hg and Cu (see Poli et al. 1987 for detailed references).

However, even in the cases in which oxidative breakdown of the membrane is not primarily involved in a defined pathogenesis the impairment of cell structure, achieved by other mechanisms of damage, seems to make the hepatocytes more susceptible to the pro-oxidant effect of some toxins (Chiarpotto et al. 1981; Poli et al. 1987). These observations thus support pathogenetic involvement of lipid peroxidation even in conditions not originally due to oxidative stress, but still susceptible of amplification by oxidative stress itself.

Acknowledgements. The authors wish to thank the Ministero della Pubblica Istruzione and the Consiglio Nazionale delle Ricerche, Roma, for supporting this research.

References

Albano E, Poli G, Chiarpotto E, Biasi F, Dianzani MU (1983) Paracetamol-stimulated lipid peroxidation in isolated rat and mouse hepatocytes. Chem Biol Interact 47:249–263
Albano E, Poli G, Tomasi A, Bini A, Vannini V, Dianzani MU (1984) Toxicity of 1,2-dibromoethane in isolated hepatocytes: role of lipid peroxidation. Chem Biol Interact 50:255–265

Chiarpotto E, Olivero J, Albano E, Poli G, Gravela E, Dianzani MU (1981) Studies on lipid peroxidation using whole liver cells: influence of damaged cells on the prooxidant effect ADP-Fe^{3+} and CCl$_4$. Experientia 37:396–397

Dore M, Atzori L, Congiu L (1985) Effect of acrolein on isolated rat hepatocytes. IRCS Med Sci 13:1139–1140

Dore M, Atzori L, Congiu L (1986) Protection by sulphur compounds against acrolein toxicity in isolated rat hepatocytes. IRCS Med Sci 14:595–596

Fariss MW, Pascoe GA, Reed DJ (1985) Vitamin E reversal of the effect of extracellular Calcium on chemically induced toxicity in hepatocytes. Science 227:751–754

Gerson RJ, Serroni A, Gilfor D, Ellen JM, Farber JL (1986) Killing of cultured hepatocytes by the mixed-function oxidation of ethoxycoumarin. Biochem Pharmacol 35:4311–4319

Orrenius S, Thor H, Bellomo G, Moldeus P (1984) Glutathione and tissue toxicity. In: Paton W, Mitchell J, Turner P (eds) Proceedings of the IUPHAR 9th congress of pharmacology. Macmillan, London, pp 57–66

Orrenius S, Rossi L, Eklow-Lastbom L, Thor H (1985) Oxidative stress in intact cells. A comparison of the effects of menadione and diquat in isolated hepatocytes. In: Poli G, Gheeseman KH, Dianzani MU, Slater TF (eds) Free radical in liver injury. IRL, Oxford, pp 99–105

Pfeifer PM, McCay PB (1971) Reduced triphosphopyridine nucleotide oxidase-catalyzed alterations of membrane phospholipids. V. Use of erythrocytes to demonstrate enzyme-dependent production of a component with the properties of a free radical. J Biol Chem 246:6401–6408

Poli G, Gravela E, Albano E, Dianzani MU (1979) Studies on fatty liver with isolated hepatocytes. II. The action of carbon tetrachloride on lipid peroxidation, protein, and triglyceride synthesis and secretion. Exp Mol Pathol 30:116–127

Poli G, Cheeseman KH, Slater TF, Dianzani MU (1981) The role of lipid peroxidation in CCl$_4$-induced damage to liver microsomal enzymes: comparative studies in vitro using microsomes and isolated liver cells. Chem Biol Interact 37:13–24

Poli G, Albano E, Dianzani MU (1987) The role of lipid peroxidation in liver damage. Chem Phys Lipids 45:117–142

Slater TF (1972) Free radical mechanisms in tissue injury. Pion, London

Lipid Peroxidation, Antioxidants and Regenerating Rat Liver

K. H. Cheeseman[1], S. Emery[1], S. Maddix[1], K. Proudfoot[1],
T. F. Slater[1], G. W. Burton[2], A. Webb[2] and K. U. Ingold[2]

1 Introduction

Lipid peroxidation stimulated by ADP-iron or ascorbate-iron is often considerably less active in many liver tumours than in normal liver (Cockerill et al. 1983; Borrello et al. 1985; Cheeseman et al. 1986a, 1988) and a major contributory factor is an increased concentration of lipid-soluble chainbreaking antioxidant material (Cheeseman et al. 1986a, 1988). The decreased rate of lipid peroxidation in the liver tumours may be due to the high number of dividing cells in the tumour sample; on the other hand it may be a feature of malignancy. In order to distinguish between these possibilities it is of interest to study lipid peroxidation and antioxidant content in regenerating liver after partial hepatectomy, where periodic bursts of cell division are known to occur (Hopkins et al. 1973). We report here that dividing normal cells in regenerating liver have a much reduced rate of lipid peroxidation, and an increased content of lipid-soluble chain-breaking antioxidant, analogous to the situation previously observed in liver tumours (Cheeseman et al. 1986a).

2 Materials and Methods

2.1 Chemicals

[Methyl-^{3}H]thymidine was purchased from Amersham International, Amersham, UK. All other chemicals were of the highest quality available and were obtained from BDH (Poole, Dorset), Sigma Chemical Co (Poole, Dorset) or Boehringer Corporation Ltd., (Lewes, Sussex).

---

[1] Dept. of Biology and Biochemistry, Brunel University, Uxbridge, Middlesex UB8 3PH, UK
[2] Division of Chemistry, National Research Council of Canada, Ottawa, KIA OR6, Ontario, Canada

Nigam et al. (Eds.), Eicosanoids,
Lipid Peroxidation and Cancer
© Springer-Verlag Berlin Heidelberg 1988

2.2 Entrainment of Animals and Partial Hepatectomy

Male Wistar rats were obtained from Charles River Ltd. (Margate, Kent) and were entrained to an inverted lighting and feeding schedule for a period of three weeks before partial hepatectomy. The lighting and feeding regimen was that of Hopkins et al. (1973): the rats were housed in a windowless room lit from 2100 hours to 0900 hours and they were allowed access to a standard laboratory diet (Expanded Breeder Diet No. 3, Special Diet Services, Witham, Kent) from 0900 hours to 1700 hours corresponding to the first 8 h of the dark period. One group of animals was maintained on a vitamin E-deficient diet (Diet No. 1253, Special Diet Services) for a period of 8 weeks before the operation. At the time of operation the mean body weight was 270 g. The rats were subject to 65% partial hepatectomy (Higgins and Anderson 1931) under ether anaesthesia. Sham operated animals were subjected to laparotomy.

The rats were killed by cervical dislocation at various intervals after operation and the liver quickly removed. Small necrotic stubs were removed from the partially hepatectomised rats before homogenising the liver. The minor lobes of the livers were homogenised and microsomes prepared as described previously (Slater and Sawyer 1971). Microsomal pellets were rinsed with 0.15 M KCl and stored at $-20°C$ for up to 60 h before resuspension. Post-microsomal supernatant fractions were taken for the assay of thymidine kinase (EC 2.7.1.21) activity; after removal of floating fat, these fractions were stored at $-20°C$.

2.3 Assays of Microsomal Enzyme Activities

Liver microsomes were resuspended in 0.15 M KCl such that 1 ml was equivalent to 1 g wet weight of liver. Cytochrome P_{450} and NADPH:cytochrome c reductase were measured as described previously (Slater and Sawyer 1969) NADPH/ADP-iron-dependent lipid peroxidation was determined by oxygen uptake (Slater 1968). The time between addition of ADP-iron to the microsomal-NADPH suspension and the start of the maximal rate of oxygen-uptake was derived from the recorded traces, and is termed the 'induction period' or lag-time (Burton et al. 1983b).

2.4 Other Procedures

The content of lipid-soluble antioxidant in microsomal suspensions was measured after extraction of the microsomes using a method based on that of Burton et al. (1983a). The extraction was as follows: microsomal suspension (1 ml) was mixed with 2 ml of sodium dodecyl sulphate (25 mM), 3 ml of absolute ethanol and 1 ml of n-heptane. After rotamixing for 1–2 min, phase separation was achieved by a brief centrifugation and the n-heptane phase was taken for analysis. Total microsomal lipid-suluble antioxidant activity was measured in this extract by the inhibited styrene oxidation method of Burton et al. (1983a, b).

Microsomal fatty acids were measured after chloroform-methanol extraction (Esterbauer et al. 1982). Thymidine kinase activity was measured in the post-microsomal supernatant fractions essentially by the method of Ives et al. (1969) with slight modification: non-phosphorylated [^{3}H]thymidine was removed from the ion-exchange filter paper discs (Whatman DE81) by mounting them in a Millipore filtration apparatus and washing each of them with 45 ml of 1 mM ammonium formate and 30 ml of distilled water drawn through by suction. Protein was determined in the various liver fractions by the method of Lowry et al. (1951).

3 Results

Fig. 1 shows the rate of regeneration of the liver mass following partial hepatectomy, expressed as the weight of the regenerating lobes normalised to a body weight of 100 g. For comparison, the weight of the minor lobes of the sham-operated animals is also shown. It can be seen that the increase in liver weight in the hepatectomised animals does not proceed linearly, but with a slight "step-wise" progression, indicative of pulses of cell division.

Fig. 2 gives data obtained for the activity of thymidine kinase during the period of liver regeneration, and corresponding data for sham-operated rats. Marked periodicities are observable at 24-h intervals with maxima at 24, 48 and 72 h post-operation.

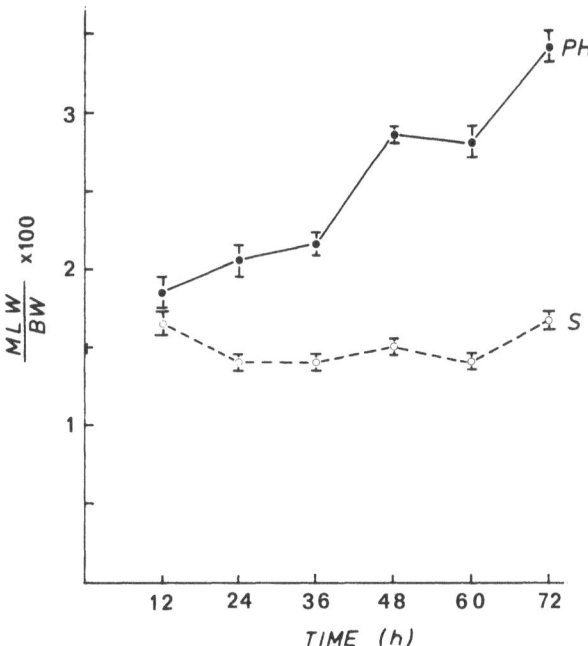

Fig. 1. Regeneration liver mass in partially hepatec-tomised rats (•) over the period 12–72 h post-opera-tion. The values for the minor-lobes if sham-operated rats are also given for comparison (o). Data are expressed as weight of minor liver lobes normalised to 100 g body weight

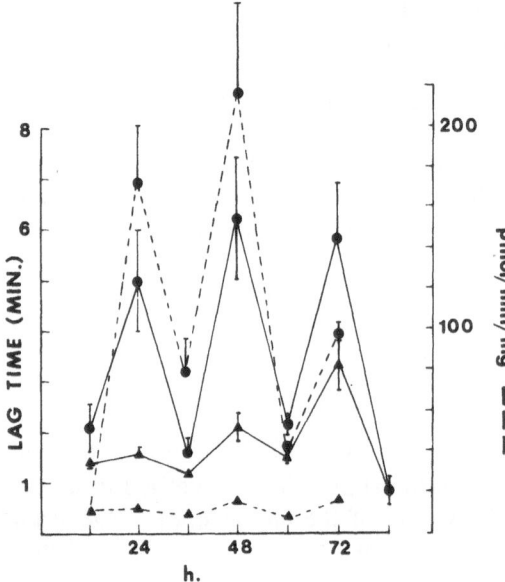

Fig. 2. Changes in thymidine kinase *(broken line)* and in the lag time (induction period) of lipid peroxidation *(solid line)* in liver samples obtained from sham-operated (▲) and partially hepatectomised (●) rats. Mean values are given ± SEM. The numbers of rats used for thymidine kinase assays were 3 at 12 h, 9 at 48 h, and 6 for other time points; for lag times the numbers of rats were 17 at 24 h, 9 at 48 h and 6 for other points

Some typical lipid peroxidation data are illustrated in Fig. 3. All the traces shown are obtained with microsomes from animals at the 48-h time point. It is clear that the microsomes from the sham-operated animal have a short lag-time (induction period) whilst the microsomes from the regenerating liver have an induction period of over 5 min. The lag times ('induction periods', see Sect. 2.3) for NADPH-ADP-iron stimulated lipid peroxidation in microsomal suspensions from hepatectomised animals are included in Fig. 2 and show a similar periodicity to that found for thymidine kinase. When the microsomes were obtained

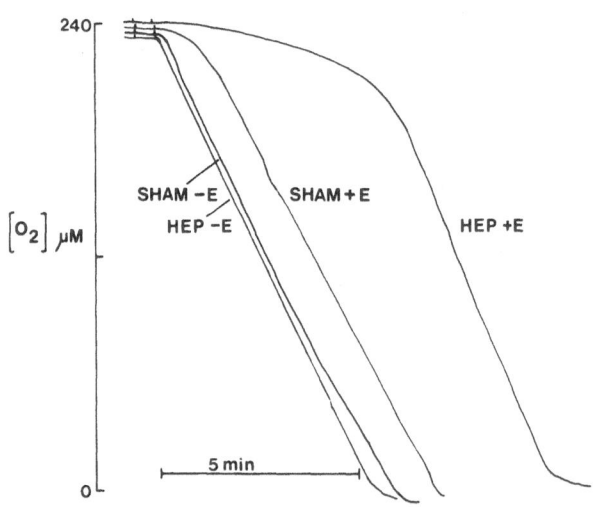

Fig. 3. NADPH/ADP-iron induced lipid peroxidation in rat liver microsomes measured as oxygen uptake. Representative traces obtained with microsome preparations from partially hepatectomised *(HEP)* or sham-operated animals maintained on either a control *(+ E)* or vitamin E-deficient *(− E)* diet. All preparations were made 48 h post-operation

Table 1. Activity of NADPH:cytochrome c reductase and content of arachidonic acid and total lipid-soluble antioxidant in microsomal suspensions prepared from rats at various intervals after either a sham operation or a partial hepatectomy (HEP).

Time (h)	NADPH:cytochrome c reductase		Arachidonic acid		Lipid-soluble antioxidant	
	Sham	HEP	Sham	HEP	Sham	HEP
24	91.4±6.2 (12)	91.1±6.0 (14)	20.7±0.9 (6)	18.4±0.7 (6)	1.49±0.18 (7)	2.35±0.36 (7)[c]
36	85.3±6.1 (6)	86.1±4.5 (6)	24.4±1.0 (6)	20.0±0.7 (6)[b]	1.59±0.11 (7)	1.49±0.07 (7)[e]
48	135.4±20.7 (9)	118.1±15.3 (9)	19.1±1.5 (8)	16.4±2.0 (9)	2.07±0.16 (7)	2.15±0.11 (7)[d]
60	101.3±10.4 (6)	92.4±9.2 (6)	22.9±0.2 (5)	20.4±0.3 (6)[a]	1.10±0.17 (4)	1.17±0.09 (4)[d]

Mean values are given ± SEM; the number of rats is in parenthesis. Reductase activity is given as nmol min^{-1} mg^{-1} protein; arachidonic acid content is given as percentage of total fatty acids and the antioxidant content is given as nmol/mg lipid.
[a] Student's t-test between sham-operated and hepatectomised groups, p < 0.001.
[b] Student's t-test between sham-operated and hepatectomised groups, p < 0.01.
[c] Student's t-test between sham-operated and hepatectomised groups, p < 0.02.
[d] Student's t-test within hepatectomised group, 36-h versus 48-h or 48-h versus 60-h, p < 0.001.
[e] Student's t-test within hepatectomised group, 36-h versus 24-h, p < 0.05.

from an actively regenerating liver 48 h post-operation from animals maintained on a vitamin E-deficient diet the lag time was essentially zero (Fig. 3).

Table 1 gives information on microsomal arachidonic acid, total lipid soluble antioxidant and NADPH:cytochrome c reductase in samples prepared from rats that had either been subject to partial hepatectomy or sham operations. There were no marked changes observable in polyunsaturated fatty acids in the microsomal fractions isolated from the sham-operated or partially-hepatectomised rats. For brevity, only arachidonate values are shown. Although there were statistically significant differences in the amount of arachidonic acid between sham and hepatectomised groups at the 36 h and 60 h time points, these differences are relatively small and unlikely to affect the rate of lipid peroxidation. In fact, these small decreases in arachidonic acid are at time points when the induction periods of lipid peroxidation are at their minima.

The concentration of lipid-soluble antioxidant in liver microsomes from partially hepatectomised rats shows some periodicity in phase with the changes in lipid peroxidation induction periods, but the increased concentration of antioxidant in the hepatectomised groups is only of statistical significance at one time point (24 h, Table 1). Table 1 shows that the activity of NADPH:cytochrome c reductase which catalyses the NADPH/ADP-iron-dependent microsomal lipid peroxidation is not significantly different in the two groups at any time point and shows no periodic cycling.

4 Discussion

An important feature of this series of experiments is the use of the schedule of 12 h dark (including 8 h access to food) and 12 h light. This is of marked benefit in that it induces a pronounced repeated cycle of DNA synthesis and thymidine kinase activity (see Hopkins et al. 1973; Fig. 2) indicating a large degree of synchrony in the cell division. This is very advantageous in studying biochemical events that may be associated with cell division. This partial synchrony in cell division probably explains the slight "step-wise" increase in liver weight during the period of regeneration (Fig. 1).

There are marked periodicities in the induction times of a lipid peroxidation system in liver microsomal suspensions (Fig. 2) and it is interesting to note that this cycling in lipid peroxidation activity is closely synchronised with that of DNA synthesis. Various microsomal components can affect the rate of lipid peroxidation including the content of polyunsaturated fatty acid substrate, the activity of NADPH:cytochrome c reductase (which catalyses NADPH/ADP-iron dependent lipid peroxidation) and the content of antioxidant substances. There were no significant cyclical changes in NADPH:cytochrome c reductase. Microsomal fatty acids did not show any cyclical variations during the period of regeneration when considered as total fatty acid or as polyunsaturated fatty acids. In fact, the only microsomal component that can have produced the effect of increasing the lag times and that was found to have periodicity of behaviour is

lipid-soluble, chain-breaking antioxidant. We have analysed this antioxidant in sham operated and regenerating liver samples both quantitatively and qualitatively and found that it is almost entirely composed of α-tocopherol (Cheeseman et al. 1986b). This is consistent, therefore, with our finding that α-tocopherol is the major lipophilic chain-breaking antioxidant in normal liver and the Novikoff hepatoma (Cheeseman et al. 1986a). There is, in fact, a positive correlation between the microsomal concentration of α-tocopherol and the length of the induction period of lipid peroxidation. However, although there is evidence of cycling, the changes in α-tocopherol and total lipophilic antioxidant do not show the same *marked* periodicity and, in our view, are not sufficient *in themselves* to account for the modulations in induction period. However, if vitamin E deficient animals are used then the induction period is completely absent even in microsomes from regenerating liver. Taken together, these data suggest that α-tocopherol may be acting in concert with some other as yet unidentified factor to decrease lipid peroxidation. We are currently investigating that possibility.

The close temporal relationship between the periodicities of thymidine kinase and peroxidative induction periods are strongly suggestive of connected events and are consistent with the hypothesis (see Slater et al. 1984) that decreased lipid peroxidation has some role in modulating the cell division process. It remains to be proven, however, if this is, in fact, the case or if these are merely coincidental phenomena.

Acknowledgements. K. Cheeseman and S. Emery are supported by the Cancer Research Campaign. Financial assistance was also provided by the Association for International Cancer Research and the National Foundation for Cancer Research.

References

Borrello S, Minotti G, Palombini G, Grattagliano A, Galleotti T (1985) Superoxide-dependent lipid peroxidation and vitamin E content of microsomes from hepatomas with different growth rates. Arch Biochem Biophys 238:588–595

Burton GW, Joyce A, Ingold KU (1983a) Is vitamin E the only lipid-soluble, chain-breaking antioxidant in human blood plasma and erythrocyte membranes? Arch Biochem Biophys 222:281–290

Burton GW, Cheeseman KH, Doba T, Ingold KU, Slater TF (1983b) Vitamin E as an antioxidant in vitro and in vivo. Ciba Found Symp 101:4–18

Cheeseman KH, Collins M, Proudfoot K, Burton GW, Webb AC, Ingold KU, Slater TF (1986a) Studies in lipid peroxidation in normal and tumour tissue. I. The Novikoff rat liver tumour. Biochem J 235:507–514

Cheeseman KH, Collins M, Maddix S, Milia A, Proudfoot K, Slater TF, Burton GW, et al. (1986b) Lipid peroxidation in regenerating rat liver. FEBS Lett 209:191–196

Cheeseman KH, Emery S, Maddix SP, Slater TF, Burton GW, Ingold KU (1988) Studies in lipid peroxidation in normal and tumour tissue. The Yoshida rat liver tumour. Biochem J 250:247–252

Cockerill MJ, Player TF, Horton AA (1983) Studies on lipid peroxidation in regenerating rat liver. Biochim Biophys Acta 750:208–213

Esterbauer H, Cheeseman KH, Dianzani MU, Poli G, Slater TF (1982) Separation and characterization of the aldehydic products of microsomal lipid peroxidation stimulated by ADP-Fe^{2+} in rat liver microsomes. Biochem J 208:129–140

Higgins GM, Anderson RM (1931) Experimental pathology of the liver of the white rat following partial surgical removal. Arch Pathol 12:186–202

Hopkins HA, Campbell HA, Barbiroli B, Potter VR (1973) Food and light as separate entrainment signals for rat liver enzymes. Biochem J 136:955–966

Ives DH, Durham JP, Tucker VS (1969) Rapid determination of nucleoside kinase and nucleotidase activities in tritium labelled substrates. Anal Biochem 28:192–205

Lowry OH, Rosebrough NJ, Farr AL, Randall RJ (1951) Protein measurement with the Folin phenol reagent. J Biol Chem 193:265–275

Slater TF (1968) The inhibiting effects in vitro of phenothiazines and other drugs on lipid peroxidation systems in rat liver microsomes. Biochem J 106:155–160

Slater TF, Sawyer BC (1969) The effects of CCl_4 on rat liver microsomes during the first hour of poisoning in vivo, and the modifying actions of promethazine, Biochem J 111:317–324

Slater TF, Sawyer BC (1971) The stimulatory effects of CCl_4 and other halogeno-alkanes on peroxidative reactions in rat liver fractions in vitro. Biochem J 123:805–814

Slater TF, Benedetto C, Burton GW, Cheeseman KH, Ingold KU, Nodes JT (1984) Lipid peroxidation in animal tumours: a distrubance in the control of cell division? In: Thaler-Dao H, Crastes de Paulet A, Paoletti R (eds) Icosanoids and cancer. Raven, New York pp 21–29

Ugazio G, Gabriel L, Burdinno E (1969) Richere sugli inhibitori della perossidazione lipidica presenti nelle cellule a dell' epatoma ascite di Yoshida. Atti Soc Ital Patol 11:325–341

Effect of Peroxidative Conditions on Human Plasma Low-Density Lipoproteins

H. Esterbauer[1], O. Quehenberger[1] and G. Jürgens[2]

1 Introduction

Human low-density lipoprotein (LDL) is a spherical particle with a diameter of 220 Å and a molecular weight of about 2.5 million. LDL consists of an interior core of esterified cholesterol (about 1500 molecules) and an outer shell of 800 molecules of phospholipids and 500 free cholesterol molecules. Embedded in the shell is the apolipoprotein B-100, a large protein with a molecular weight of 500000, which is recognized by the LDL receptor (Goldstein and Brown 1978) present on the surface of most cells and largely responsible for the controlled catabolism of LDL. It is well established that LDL is the main carrier for cholesterol in the blood-stream and that increased levels of LDL correlate with an increased risk of atherosclerosis. Early atherosclerotic lesions are characterized by deposits of lipid and cholesterol laden foam cells which are derived from monocyte macrophages. Macrophages possess two types of receptor for LDL, one is the LDL receptor which enables the uptake of LDL in a controlled process downregulated by the internalized cholesterol. Uptake by this receptor cannot, therefore, lead to lipid-laden foam cells. In addition to the classical LDL receptor, macrophages posses a scavenger receptor recognizing LDL with modified apo B. Brown and Goldstein (1983) were the first to show that LDL treated with acetic acid anhydride (acetyl-LDL) is taken up through the scavenger receptor, that the uptake is not downregulated by the internalized cholsterol and that macrophages incubated in vitro with acetyl-LDL developed a foam cell like appearance. Since acetylation of LDL does not occur in vivo, the question remained, if, and by what mechanism, LDL is altered in a way that makes it recognizable by the scavenger receptor. Recent studies from several laboratories (Henrikson et al. 1981; Morel et al. 1984; Heinecke et al. 1986; Heinecke 1987) suggest that LDL can become oxidized by the attack of oxygen free radicals. In vitro oxidized LDL is taken up by the scavenger receptor and converts macrophages into lipid-laden foam cells. Moreover, oxidized LDL shows many other features characteristic of the early events in atherogenesis (for review see Jürgens et al. 1987). Thus, it was found that oxidized LDL is chemotactic for circulating

[1] Institute of Biochemistry, University of Graz, Schubertstrasse 1, 8010 Graz, Austria
[2] Institute of Medical Biochemistry, University of Graz, Harrachgasse 21, 8010 Graz, Austria

Nigam et al. (Eds.), Eicosanoids,
Lipid Peroxidation and Cancer
© Springer-Verlag Berlin Heidelberg 1988

monocytes but inhibits the further migration of resident monocyte derived macrophages. Furthermore, oxidized LDL is cytotoxic towards most cells and its accumulation would therefore lead to the death of arterial wall cells.

In vitro oxidation of LDL occurs by incubation with cultured endothelial cells (Henrikson et al. 1981, Morel et al. 1984), smooth muscle cells (Heinecke et al. 1986), monocytes and neutrophils (Cathcart et al. 1985). LDL can also be oxidized in a cell free medium by prolonged exposure to an oxygen saturated buffer, and transition metal ions such as copper or iron can greatly enhance this oxidation process. It is now believed that cell oxidized LDL has structural and functional properties very similar if not identical to LDL oxidized in a cell free medium. This is not surprising, since all the media used for the cell incubation contain Fe and/or Cu in the lower micromolar range. However, it has not yet been elucidated whether the same mechanisms are involved in the oxidation process mediated by cells as that mediated by transition metal ion supplemented buffer only. Strong evidence exists that in both cases free redicals are involved (Jürgens et al. 1987; Esterbauer et al. 1987). In all studies a strong correlation was found between the degree of oxidation, as evidenced by the accumulation of malonaldehyde (MDA), and the uptake by macrophages. The oxidation of LDL is accompanied by a progressive increase of its negative surface charge as measured by its electrophoretic mobility. The increased negative charge is probably due to a neutralization of positive charges of lysine amino groups of the apo B. The apo B contains, in total, 368 lysine residues (Jürgens et al. 1986) and a certain number of them are required for recognition by the classical LDL receptor. The modification of a fraction of critical lysines results in a decreased binding to the apo B/E receptor and an increased recognition by the scavenger receptor (Haberland et al. 1982). Modification of lysines can be achieved by chemical acylation or by treatment with aldehydic lipid peroxidation products such as MDA or 4-hydroxynonenal (HNE). Oxidized LDL also contains fewer lysine residues as compared to native LDL, however, the underlying mechanism leading to the consumption of lysines (or their positive charges) has not been elucidated.

Based on the findings briefly summarized and on own studies (Jürgens et al. 1984, 1986, 1987; Jessup et al. 1986; Koller et al. 1986a, b; Esterbauer et al. 1987) we have proposed the following hypothesis: oxidation of LDL is a free radical process which initiates peroxidation of the polyunsaturated fatty acids (PUFAs) in LDL lipids and their degradation to reacitve aldehydes. The aldehydic lipid peroxidation products react with the ε-amino groups of lysine residues of apo B. The new epitopes thus formed have an increased negative surface charge through the loss of NH_3^+ and are not recognized by the LDL receptor but by the scavenger receptor (Fig. 1).

In the following we review some of our results which favour the above hypothesis.

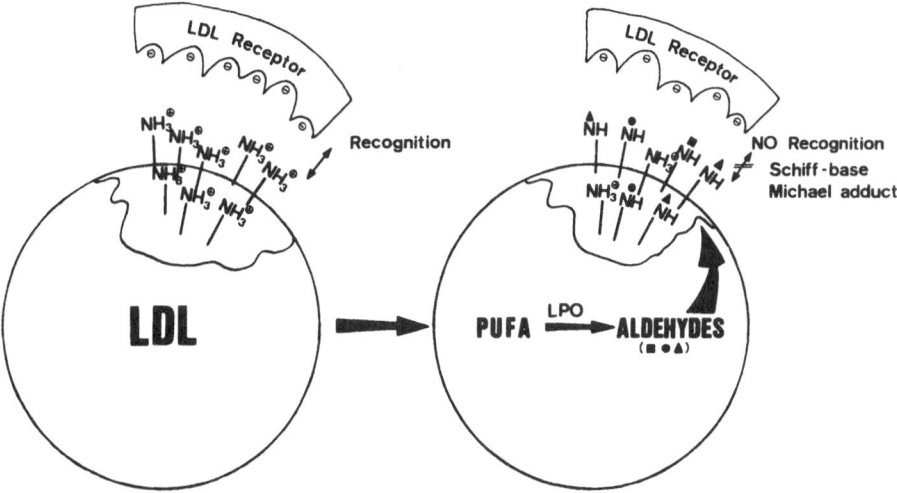

Fig. 1. Scheme showing the principal points of the hypothesis. The LDL receptor recognizes a certain pattern of positive charged NH_3^+ groups of the apo B in native LDL *(Left)*. In oxidized LDL *(right)* aldehydes (■, ●, ▲) formed in situ from PUFAs by lipid peroxidation (LPO) have attacked and thereby neutralized NH_3^+ groups. This LDL with modified apo B is no longer recognized by the LDL receptor but by the scavenger receptor

2 Oxidation of LDL Leads to Loss of Antioxidants and PUFAs

The main PUFAs in LDL are linoleic acid (383 nmol/mg LDL) and arachidonic acid (36 nmol/mg LDL) (Able 1). Other PUFAs, which are only present in trace amounts (in total less than 5 nmol/mg LDL), are 18:3, 20:3, 22:5 and 22:6. The LDL is rich in α-tocopherol which amounts to 2.6 nmol/mg LDL. This means that the ratio of α-tocopherol to PUFAs is 1:160, which is high compared to other systems such as microsomes where the ratio is about 1:600. In addition to

Table 1. PUFAs and antioxidants in native LDL, in LDL oxidized 24 hours in the absence of copper and in LDL oxidized 3 hours in the presence of 10 μM $CuCl_2$

	Native LDL[a]	Oxidized LDL	
		Without Cu, 24 h[b]	Cu, 3 h
Linoleic acid	383.6 ± 78.5	266 ± 63	35 ± 10
Arachidonic acid	35.5 ± 17.1	12 ± 4	3 ± 1
α-Tocopherol	2.60 ± 0.62	0	0
γ-Tocopherol	0.20 ± 0.07	0	0
β-Carotene	0.13 ± 0.07	0	0
Lycopine	0.07 ± 0.04	0	0

All values in nmol/mg LDL.
[a] Mean ± SD from 8 independent LDL preparations.
[b] Mean ± SD from 4 experiments.

α-tocopherol LDL contains small amounts of other lipophilic antioxidants namely γ-tocopherol, β-carotene and lycopine (Table 1).

Incubation of LDL in the absence of transition metal ions leads within 3 hours to a strong decrease in the amount of vitamin E and β-carotene and if these antioxidants are lowered to a threshold level of about 30% of the original values their antioxidant capacity is no longer sufficient to further protect the PUFAs against oxidation. After 24 hours the LDL is completely depleted of the antioxidants and about 30% of the 18:2 and 67% of the 20:4 are consumed. If the incubation system is supplemented with EDTA or BHT no decrease of the antioxidants of PUFAs is observed within 24 hours. On the other hand, supplementation with Cu(II)chloride dramatically enhances the rate and extent of oxidation. Figure 2 shows, as an example, the time course of the consumption of the 18:2 and 20:4 caused by 10 μM CuCl₂ and in the absence of Cu. Already after 1 hour the LDL is completely depleted of all antioxidants (data not shown) and after 3 hours 97% of 20:4 and 79% of 18:2 are degraded.

The susceptibility of LDL to oxidation shows a great variability which appears to depend on the donor. In general LDL preparations with a high vitamin E content (above 2.8 nmol/mg LDL) were more resistant than those with lower values. Moreover, LDL samples with comparable vitamin E and β-carotene content still showed great differences in their oxidizability and we assume that this is due to variable concentrations of an additional antioxidant. According to fluorescence studies this might be a vitamin A derivative. Thus the lipid extract of native LDL shows an emission spectrum identical to retinol with two maxima at 470 and 510 nm. If LDL is incubated with increasing concentrations of copper a

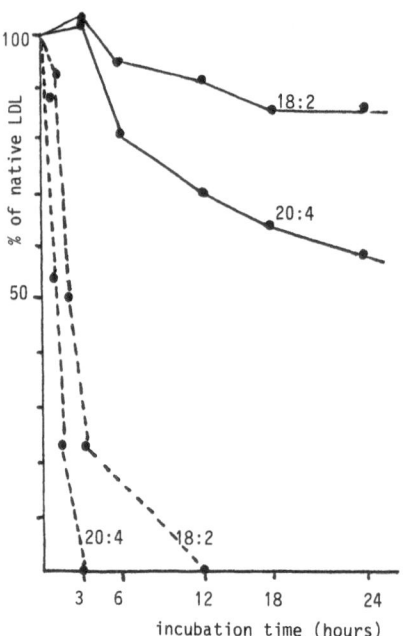

Fig. 2. Time course of the disappearance of linoleic acid and arachidonic acid during oxidation of LDL in absence and presence of copper. LDL (1.5 mg/ml) was incubated in oxygen saturated phosphate buffer at room temperature (—) and in oxygen saturated phosphate buffer supplemented with 10 μM CuCl₂ at 37°C (- - -). The lipids were extracted, treated with BF₃/MeOH and the fatty acid methyl esters were analyzed by capillary GC as described (Esterbauer et al. 1987)

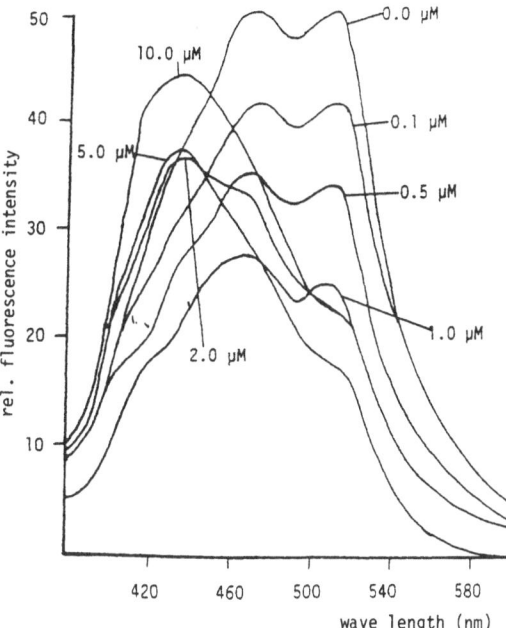

Fig. 3. Effect of oxidation on the fluorescence of the LDL lipids. LDL (1.12 mg/ml) was incubated at 37°C for 3 h with different concentrations of CuCl$_2$ ranging from 0 to 10 µM. The lipids were extracted with CHCl$_3$/MeOH 2:1 and the emission spectra were recorded at an excitation wavelength of 360 nm as described (Koller et al. 1987). The spectra show that the fluorophore with the double maximum at 470 510 nm (likely a retinol derivative) present in native LDL disappears with increasing degree of oxidation, i.e. increasing copper concentration and a new fluorophore with a maqximum at 430 nm is formed

steady decrease in the intensity of this fluorescence occurs (Fig. 3). LDL preparations with a high vitamin E content and a high lipid fluorescence were in all cases most resistant against oxidation.

3 Formation of Aldehydes

It is now well established by studies in many biological systems (for review see Esterbauer 1985) that lipid peroxidation is accompanied by the generation of a great variety of aldehydic lipid degradation products such as alkanals, 2-alkenals, 2,4-alkadienals, malonaldehyde and 4-hydroxyalkenals. The latter class of aldehydes received great attention (Esterbauer 1985) since they can produce many metabolic disturbances and could therefore be, at least in part, responsible for the cytotoxic and genotoxic effects associated with free radical formation in cells and tissues. As described above, exposure of LDL to pro-oxidative condtions leads to a time dependent consumption of PUFAs. That this consumption is indeed due to degradation by a lipid peroxidation process is evidenced by the formation of aldehydes, which occurs more or less concommitently with the decrease of fatty acids. Table 2 lists the aldehydes which have so far been identified in oxidized LDL. The methods for their identification and quantification was in principle the same as used in the studies of rat liver microsomes or hepatocytes (Poli et al. 1985). Briefly, 2,4-dinitrophenylhydrazine was added to the LDL solution, the hydrazone derivatives formed were extracted with dichlorom-

Table 2. Aldehydes in native LDL, in LDL oxidized 24 hours in the absence of copper and in LDL oxidized in the presence of 10 μM CuCl₂ for 60 minutes and 3 hours respectively.

	Native LDL[a]	Oxidized LDL		
		Without Cu, 24 h[a]	Cu, 60 min	Cu, 3 h
4-Hydroxy hexenal	0	0.24 ± 0.04	1.36	1.53
4-Hydroxy octenal	0	0.18 ± 0.16	0.25	1.42
4-Hydroxy nonenal	0.29 ± 0.18	0.59 ± 0.33	0.93	4.97
Propanal	0.24 ± 0.14	0.64 ± 0.47	0.53	1.03
Butanal	0.33 ± 0.17	0.54 ± 0.46	0.72	0.65
Pentanal	0.09 ± 0.08	0.24 ± 0.05	0.36	0.89
Hexanal	0	1.79 ± 1.28	2.03	9.97
2,4-Heptadienal	0	0.29 ± 0.14	0.17	0.88
Malondialdehyde	0.46 ± 0.13	2.6 ± 1.9	5.98	16.51

All values in nmol/mg LDL.
[a] Mean ± SD from 4 different LDL preparations.

ethane, preseparated into different classes by TLC and finally analysed by HPLC. The presence of 4-hydroxynonenal, 4-hydroxyoctenal and 4-hydroxyhexenal was proven additionally by GC/MS. In this case the LDL solution is treated with pentafluorobenzylhydroxylamine hydrochloride, extracted with hexane, the oximes were then silylated and analysed by GC/MS. From the mass spectra of the various peaks it was clearly evident that the samples prepared from oxidized LDL contained 4-hydroxynonenal, 4-hydroxyoctenal and 4-hydroxynonenal with the characteristic fragmentation masses at m/e = 152, 138 and 110, respectively. The LDL oxidized for 24 hours in the absence of cop-

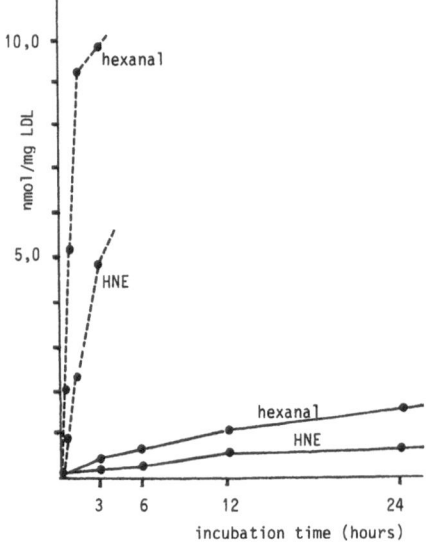

Fig. 4. Time course of the formation of HNE and hexanal during oxidation of LDL in the absence (—) and presence of copper (- - -). Conditions as in Fig. 2. The aldehydes were converted to their dinitrophenyl hydrazones and analyzed by HPLC as described (Poli et al. 1984)

per contained, in addition to malonaldehyde, on average 4.5 nmol of other aldehydes per mg LDL, with hexanal, 4-hydroxynonenal and propanal being the
major products. In the copper oxidized LDL the aldehyde concentration was
considerably higher and the total aldehydes, excluding malonaldehyde,
amounted to 23 nmol/mg LDL, with hexanal, 4-hydroxynonenal and 4-hydroxy-
hexenal as major products. The kinetics of the generation of the major aldehydes in the copper supplemented and copper free oxidation system is shown in
Fig. 4. If these kinetics are compared to the kinetics of the fatty acid decrease
(Fig. 2) it is evident, that both phenomena are time linked and that the aldehydes
only increase when the fatty acids decrease. It is worth noting that several native
LDL preparations already contained aldehydes, whereas others where completely free of these PUFA degradation products. Since antioxidants (BHT,
EDTA) were always present during the isolation period, we can exclude with
certainty the possibility that the aldehydes in native LDL are artefacts due to
lipid peroxidation during the isolation. We assume that these aldehydes are rem-
nants from lipid peroxidation which already occured in vivo.

4 Modification of Native LDL by Aldehydes

In a series of studies Haberland et al. (1982) have shown that treatment of LDL
with MDA results in a modification which is taken up by the macrophage scavenger receptor. It was also shown that MDA reacts preferentially with the ε-
amino group of lysine residues in the apo B and that the MDA modified LDL
has an increased negative surface charge. The binding of MDA most probably
occurs through formation of Schiff bases, which leads to a neutralisation of positive charges from NH^{3+} groups and consequently to an increase of the nett
negative charge on the LDL particle. The concentrations of MDA used in these
studies were around 0.1 M which is extremely high in the biological sense, since
such concentrations could never be produced by a lipid peroxidation process.
Compared to MDA, 4-hydroxy-nonenal exhibits a much higher capacity to modify LDL (Jürgens et al. 1984, 1986; Esterbauer et al. 1987) and we have therefore
proposed that this aldehyde is a more likely candidate for being responsible for
LDL modification under in vivo lipid peroxidation condtions. When LDL (22
mg/ml) was incubated for 3 hours with 5 mM HNE in total 225 mol HNE were
bound to one mol LDL. The amino acid analysis of apo B revealed that 128
amino acid residues were modified. The residues attacked were mainly lysine
and tyrosine and to a lesser extent histidine and serine. The two cysteine residues
present in apo B were also blocked by HNE. In many previous reports (for review see Esterbauer 1985) it was stated that HNE is a reagent highly specific for
thiol groups. In the light of the findings with LDL this assumption should be
revised since the loss of amino acids caused by HNE can only be explained by a
covalent binding of the aldehyde to the amino acid side chains. The mechanism
by which lysine or the other three amino acids (tyr, his, ser) react with HNE is
not yet elucidated. The fact that from the 368 lysines present in apo B, only 45
reacted with HNE strongly suggests that the microenvironment of the lysines

governs the reactivity of the ε-amino group. A plausible explanation could be that that the reactive lysines possess the amino group in the deprotonated NH_2-form, which is a strong nucleophile and could react with HNE by 1,2 or 1,4-addition reactions to Schiff bases or Michael addition products.

That HNE attacks the protein moiety was also demonstrated by SDS-polyacrylamide electrophoresis. The native LDL gives the single apo B-100 Band, whereas the protein prepared from LDL pretreated with 1 mM NHE gives, in addition to the B-100 band, two bands representing two higher molecular weight forms of apo B. A similar effect was observed with LDL pretreated with 200 Mm MDA. Yet another effect produced by treatment of LDL with HNE is a strong increase of the fluorescence in the 420–450 nm range. The newly formed fluorophore is mainly associated with the apo B and has an excitation maximum at 360 nm and an emission maximum at 430 nm. A similar if not identical fluorophore with Ex 360/Em 430 nm is present in the apo B of LDL oxidized in a cell free medium in the absence or presence of copper, and, even more important, in a very high intensity in the apo B from endothelial cell modified LDL. This suggests, but of course does not prove, that the 430 nm fluorophore in cell modified or oxidized LDL is due to in situ generated HNE. Treatment of LDL with MDA also yields a fluorescent apo B, but with an Ex at 400 nm and an Em at 470 nm, which is significantly different to the fluorophore of oxidized or cell modified LDL.

As can be expected from the finding that HNE binds to lysine residues, HNE treated LDL has an increased negative surface charge as compared to the native LDL. The change of the surface charge can be measured by agarose gel electrophoresis as relative electrophoretic mobility (rem), i.e. ratio of migration distance of modified LDL to native LDL. The effect of HNE on the rem is concentration dependent and the rem increased from 1.05 to 1.87 upon treatment with 0.2 and 5 nM HNE respectively. On an equimolar basis (5 nM) HNE and 2,4 -heptadienal change the surface charge most efficiently (rem 1.87, 1.31), whereas hexanal and MDA were less effective and gave only a rem of 1.13 and 1.10. Since all 4 aldehydes together with many others are generated during oxidation of LDL it seems plausible that they all contribute either additively or synergistically to the enhancement of the electrophoretic mobility of oxidized LDL.

That aldehyde treatment alters the cellular uptake of LDL was demonstrated for both, MDA and HNE. Haberland et al. (1982) showed that the recognition of MDA modified LDL by the LDL receptor steadily decreases as the number of blocked lysines increases and that when more than 30 molecules of MDA are bound per LDL particle, which corresponds to a loss of 16% of all lysines the recognition by the scavenger receptor abruptly increases. Jessup et al. (1986) reported that HNE treated LDL shows a reduced recognition by the LDL receptor on fibroblasts. The decrease of the uptake correlated well with the number of HNE molecules incorporated into the LDL and its relative electrophoretic mobility. The strongest modification tested by these authors had 12.5% of the lysines in apo B blocked and, consistent with the report by Haberland et al. (1982) that a minimum of 16% blocked lysines is necessary, this HNE modified LDL was not recognized by the scavenger receptor. Attempts to block by HNE more than 16% of the lysine residues failed in so far as such strong modifications led to the precipitation of LDL. It would be interesting to determine whether the exposure of LDL to a mixture of different aldehydes as for example MDA + HNE + 2,4-heptadienal renders LDL more recognizable for the scavenger receptor than the exposure to only one aldehyde.

5 Do In Situ Generated Aldehydes Attack apo B?

The fact that aldehydes, such as HNE, MDA or others, added externally to native LDL can interact and thereby modify the apo B does not prove that aldehydes generated in situ from the lipids during LDL oxidation are in fact causally involved in the structural and functional alterations of apo B occuring during oxidation of LDL. So far only a few investigations have been made to clarify this problem. As mentioned previously the LDL oxidized by cells or in a cell free medium possesses a strong fluorescent chromophore with Ex 360/Em 430 nm and this chromophore could result from HNE generated in situ during the oxidation process, since a chromophore with exactly the same maxima is formed if native LDL is incubated with HNE. Much stronger evidence for the importance of in situ generated lipid peroxide decomposition products was reported by Steinbrecher (1987). This author oxidized LDL (5 µM CuSO₄, 20 hours) containing phosphatidylcholine with isotopically labelled arachidonic acid and found a high proportion of the radioactivity covalently bound to the apo B. Moreover, in

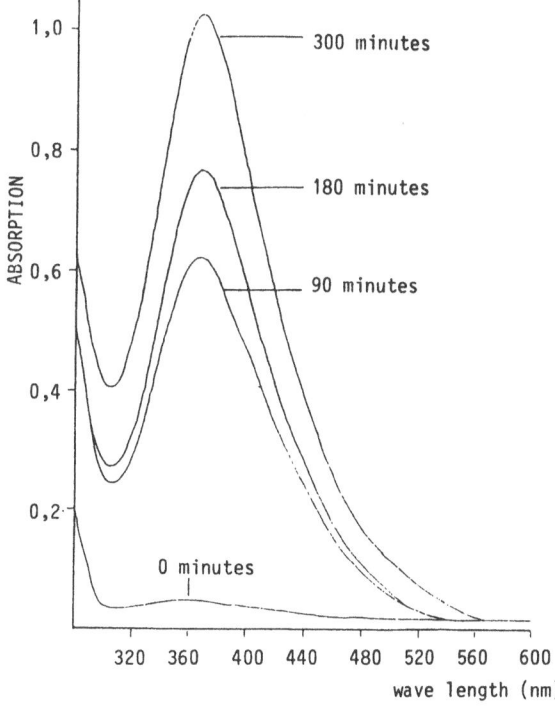

Fig. 5. Protein bound aldehydes in apo B of oxidized LDL measured as DNPH derivatives. LDL (1.5 mg/ml) was oxidized at 37°C in the presence of 10 µM CuCl₂. After 0 (control), 90, 180 and 300 min the LDL solution was reacted with dinitrophenyl hydrazine, the protein was isolated, freed from excess reagent, dissolved in 5% SDS and the vis-spectrum was recorded. Based on the 369 nM maximum the amount of protein bound aldehydes is 3 (control), 42 (90 min), 51 (180 min) and 68 (300 min) nmol/mg LDL

the oxidized LDL about 32% of the lysines were modified, most probably through the binding of aldehydes to the ε-amino groups. Consistent with this is the finding of our laboratory that the apo B of oxidized LDL contains carbonyl groups which are not present in native LDL and the number of which increases with increasing oxidation time i.e. degree of oxidation (Fig. 5).

The most serious evidence against the involvement of in situ generated aldehydes in apo B modification is their concentration. If LDL (1.5 mg/ml) is oxidized in buffer the amount of aldehydes (MDA + all other aldehydes) present after 24 hours is about 7.1 nmol/mg LDL, which corresponds to an overall concentration of about 11 nmol/ml or 11 μM. In the oxidation system with 10 μM $CuCl_2$ the overall concentration reached after 3 hours is about 60 μM. On the other hand, if native LDL is incubated in buffers supplemented with HNE, MDA or any one of the other aldehydic lipid peroxidation products in the concentration range of 10-100 μM no significant alterations of the properties of LDL occur, in particular such aldehyde concentrations do not produce an LDL which is recognized by the scavenger receptor. Does this now mean that aldehydes cannot be responsible for the modification of apo B seen in oxidized LDL? Not necessarily, since the peculiar conditions of in situ generation cannot be reproduced by externally added aldehydes. In situ the aldehydes are formed within the LDL particle and because of their lipophilic nature only a small fraction of them diffuses into the surrounding outer aqueous phase of the incubation medium, whereas the major part remains associated with the LDL particle (Esterbauer et al. 1987) where the total concentration can reach levels in the range of 10 mM (absence of copper) to 50 mM (presence of copper). Such concentrations would certainly be high enough to rapidly attack accessible amino acid residues of the apo B embedded in the lipid domain of the LDL particle. Furthermore, it should be considered that less lipophilic aldehydes emanating from the LDL particle would build up around its immediate surface an aldehyde rich aqueous layer in which a dynamic equilibrium between their delivery from the LDL lipids and their diffusion into the distant medium exists. This is again a situation, which cannot be mimicked by experiments in which native LDL is exposed to low concentrations of aldehydes. It seems important that in the future more emphasis is given to this aspect since it would be of relevance not only for oxidized LDL but for many other studies where aldehydic lipid peroxidation products were assumed to be involved in the damgaging processes. Neither in peroxiding microsomes (Koster et al. 1986) nor in suspensions of isolated hepatocytes exposed to pro-oxidative conditions (Poli et al. 1985), is the overall aldehyde concentration sufficiently high to explain toxic effects, but in all cases the aldehydes are unequally distributed and mainly concentrated within the membranes where they reach mM concentrations and could diffuse laterally and reach critical targets in the membrane. The present knowledge, however, does not allow a definite conclusion that such a mechanism is valid.

Acknowledgments. The authors work was supported by the Association for International Cancer Research, UK and by the Austrian Science Foundation, project no. P6176B.

References

Brown MS, Goldstein JL (1983) Lipoprotein metabolism in the macrophage: implications for cholesterol deposition in atherosclerosis. Annu Rev Biochem 52:223–261

Cathcart MK, Morel DW, Chisolm GM (1985) Monocytes and neutrophils oxidize low density lipoprotein making it cytotoxic. J Leukocyte Biol 38:341–350

Esterbauer H (1985) Lipid peroxidation products: formation, chemical properties and biological activities. In: Poli G, Cheeseman KH, Dianzani MU, Slater TF (eds) Free radicals in liver injury. IRL, Oxford, pp 29–47

Esterbauer H, Jürgens G, Quehenberger O, Koller E (1987) Autoxidation of human low density lipoprotein: loss of polyunsaturated fatty acids and vitamin E and generation of aldehydes. J Lipid Res 28:495–509

Goldstein JL, Brown MS (1978) Low-density lipoprotein pathway and its relation to atherosclerosis. Annu Rev Biochem 46:897–930

Haberland ME, Fogelman AM, Edwards PA (1982) Specificity of receptor-mediated recognition of malondialdehyde-modified low density lipoprotein. Proc Natl Acad Sci USA 79:1712–1716

Heinecke JW (1987) Free radical modification of low-density lipoprotein: mechanisms and biological consequences. Free Radic Biol 3:65–73

Heinecke JW, Baker L, Rosen H, Chait A (1986) Superoxide-mediated modification of low density lipoprotein by arterial smooth muscle cells. J Clin Invest 77:757–761

Henriksen T, Mahoney EM, Steinberg D (1981) Enhanced macrophage degradation of low density lipoprotein previously incubated with cultured endothelial cells: recognition by receptors for acetylated low density lipoproteins. Proc Natl Acad Sci USA 78:6499–6503

Jessup W, Jürgens G, Lang J, Esterbauer H, Dean RT (1986) The interaction of 4-hydroxynonenal-modified low density lipoproteins with the fibroblast apo B/E receptor. Biochem J 234:245–248

Jürgens G, Lang J, Esterbauer H (1986) Modification of human low-density lipoprotein by the lipid peroxidation product 4-hydroxynonenal Biochem Biophys Acta 875:101–114

Jürgens G, Hoff HF, Chisolm GM, Esterbauer H (1987) Modification of human serum low density lipoprotein by oxidation-characterization and pathophysiological implications. Chem Phys Lipids 45:315–336

Koller E, Quehenberger O, Jürgens G, Wolfbeis OS, Esterbauer H (1986a) Investigation of human plasma low density lipoprotein by three-dimensional fluorescence spectroscopy. FEBS Lett 198:229–234

Koller E, Jürgens G, Quehenberger O, Esterbauer H (1986b) Fluorescence properties of native, 4-hydroxynonenal-modified and autoxidized low density lipoprotein. In: Rotilio G (ed) Superoxide and superoxide dismutase in chemistry, biology and medicine. Elsevier, Amsterdam, pp 116–118

Koster JF, Slee RG, Montfoort A, Lang J, Esterbauer H (1986) Comparison of the inactivation of microsomal glucose-6-phosphatase by in situ lipid peroxidation-derived 4-hydroxynonenal and exogenous 4-hydroxynonenal Free Radic Res Commun 1:273–287

Morel DW, DiCorleto PE, Chisolm GM (1984) Endothelial and smooth muscle cells alter low density lipoprotein in vitro by free radical oxidation. Arteriosclerosis 4:357–364

Poli G, Dianzani MU, Cheeseman KH, Slater TF, Lang J, Esterbauer H (1985) Separation and characterization of the aldehydic products of lipid peroxidation stimulated by carbon tetrachloride or ADP-iron in isolated rat heptatocytes and rat liver microsomal suspensions. Biochem J 227:629–638

Steinbrecher UP (1987) Oxidation of human low density lipoprotein results in derivatization of lysine residues of apolipoprotein B by lipid peroxide decomposition products. J Biol Chem 262:3603–3608

Studies on the Oxygen Dependence of Lipid Peroxidation

H. de Groot and T. Noll

Institut für Physiologische Chemie I der Universität Düsseldorf, Moorenstraße 6, 4000 Düsseldorf, FRG

1 Introduction

Molecular oxygen enters lipid peroxidation by addition to a carbon-centered lipid free-radical derived from an unsaturated fatty acid of a membrane phospholipid. During this reaction a lipid peroxyl radical is formed and lipid peroxidation enters its propagation stage, during which the number of free-radical-carried chains increases exponentially (Lundberg 1961). In addition to its fundamental role as a reactant and amplifier in the propagation stage, O_2 may also be involved in the initial generation of free-radicals from non-radical precursors. In iron-induced lipid peroxidation, O_2 contributes to the generation of reactive iron-O_2 intermediates (perferryl ion, $Fe^{2+}O_2$ or $Fe^{3+}O_2^{\cdot-}$; ferryl ion, FeO^{2+} or FeO^{3+}; Aust et al. 1985) which ultimately attack the unsaturated fatty acids within biological membranes. In CCl_4-induced lipid peroxidation, O_2 acts as an inhibitor of the cytochrome P-450-linked generation of the haloalkane free-radicals (de Groot and Noll 1988).

2 Iron-Induced Lipid Peroxidation

In NADPH-supplemented rat liver microsomes iron-induced lipid peroxidation is characterized by a biphasic reaction (Fig. 1). The duration of the lag phase can be related to the occurrence of chain-breaking antioxidants within the microsomal membrane such as vitamin E (Burton et al. 1983; Cadenas et al. 1984) and/ or to the kinetics of the initiation stage. In the first case, it is assumed that the initially formed lipid peroxyl radicals (LOO˙) readily interact with vitamin E (Fig. 2), thereby keeping the steady state concentration of LOO˙ at low levels. Consequently the propagation cycle reactions starting from LOO˙ do not proceed at a significant rate until the chain-breaking antioxidant is consumed. In the second case, it must be taken into account that the formation of lipid free-radicals (L˙) by attack of the reactive iron-O_2 complexes on unsaturated fatty

Nigam et al. (Eds.), Eicosanoids,
Lipid Peroxidation and Cancer
© Springer-Verlag Berlin Heidelberg 1988

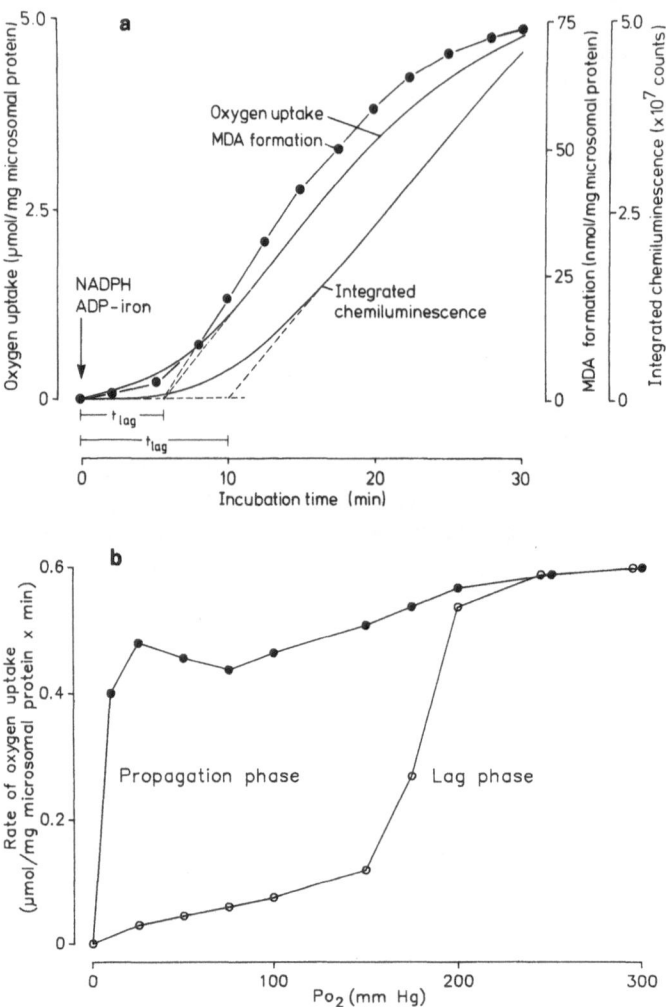

Fig. 1a, b. NADPH/ADP-iron-induced lipid peroxidation in liver microsomes from phenobarbital-pretreated rats. **a** Time course of O_2 uptake, malondialdehyde *(MDA)* formation, and low-level chemiluminescence at a steady state Po_2 of 30 mmHg (from Noll et al. 1987). **b** O_2 dependence of the rate of microsomal O_2 uptake during the lag phase and the propagation phase (From de Groot and Noll 1988)

acids proceeds at a relatively low rate and is the only initiation step during the early phase of iron-induced lipid peroxidation. If, however, accumulating amounts of lipid hydroperoxides (LOOH) are formed, the propagation cycle reactions are accelerated due to so-called secondary initiation reactions, and the rate of the overall process of lipid peroxidation sharply increases. The prolonged lag phase of low-level chemiluminescence, as well as the distinct time displacement of light emission in comparison with O_2 uptake and malondialdehyde formation, can be explained by the quadratic dependence of the light intensity on

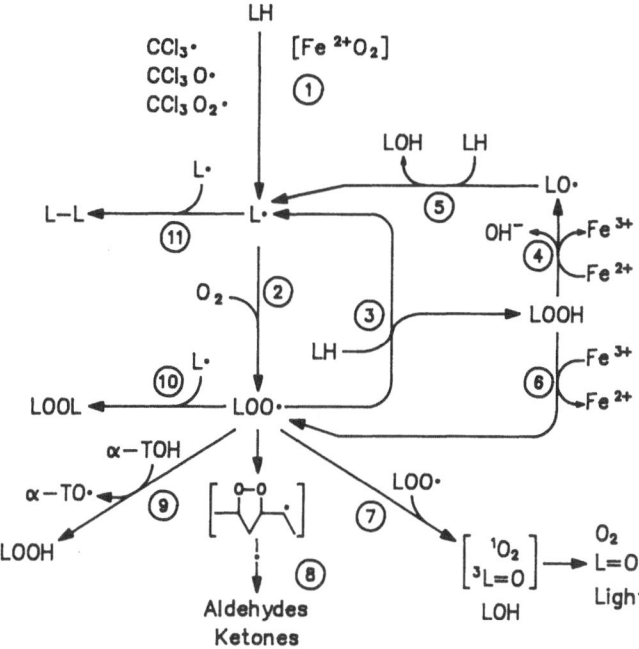

Fig. 2. Relationships between initiation, propagation, and termination reactions in iron and CCl_4-induced lipid peroxidation. The induction step of lipid peroxidation (1) by the iron or the CCl_4 system involves the formation of certain potent oxidants, e.g. $Fe^{2+}O_2$-complex and CCl_4-derived free-radicals ($CCl_3^{\cdot}$, $CCl_3O^{\cdot}$, $CCl_3O_2^{\cdot}$), which attack a polyunsaturated fatty acid of a membrane lipid *(LH)* generating a carbon-centered lipid radical *(L')*. In the presence of O_2, L' is rapidly converted to the respective lipid peroxyl radical *(LOO')* (2) which can be considered as a branching point within the peroxidation process. LOO' may attack another LH leading to a lipid hydroperoxide *(LOOH)* while generating a new L' *(3)*. In the presence of transition metals such as iron the LOOH is decomposed either to a lipid alkoxyl radical *(LO')* (4) or a LOO' (6). Both free-radicals formed can govern the self-propagation cycle as secondary initiators (5, 6). Collision of two secondary LOO' are assumed to generate two different electronically excited species, singlet molecular oxygen $(^1O_2)$ and a compound in the triplet state (possibly a carbonyl compound, $^3L = O$) (7). The LOO' may internally cyclise to form a five-membered endoperoxide which is assumed to decompose to low molecular weight ketones and aldehydes, like malondialdehyde (8). LOO' may interact with vitamin E (α-TOH), the intrinsic chain-breaking antioxidant of the microsomal membrane (9). Termination of chain reactions occur along reactions (10, 11)

the LOO'concentration, and the occurrence of different subsets of secondary LOO' (Noll et al. 1987).

The plots of the respective rates of the lag phase and the propagation phase of microsomal O_2 uptake versus Po_2 (Fig. 1) clearly reveal that the overall rate of iron-induced lipid peroxidation responds differently to the actual Po_2 in each phase. A decrease in the steady state Po_2 below 200 mmHg leads to a significant decrease in the overall rate of lipid peroxidation during the lag phase, while the overall rate of lipid peroxidation during the phase of propagation becomes limited by O_2 only at Po_2 levels below 30 mmHg.

The low O_2 affinity of the lag phase presumably reflects the O_2 dependence of the steady state level of the reactive iron-O_2 complex. Its remains constant over a wide Po_2 range. However, with decreasing Po_2 there is an increase in the steady state concentration of Fe^{2+}. If the latter increases to levels which favour the decomposition of the iron-O_2 complex via an Fe^{2+}-catalyzed autoxidation reaction, the steady state concentration of the iron-O_2 complex and hence the rate of the initiation of lipid peroxidation decreases (de Groot and Noll 1988). The high O_2 affinity of the propagation phase of the peroxidative breakdown of lipids may result from the fact that the O_2 addition reaction to an $L^{\cdot}$ is essentially diffusion limited. For this reason, self-termination reactions, due to increases in the steady state concentration of $L^{\cdot}$, gain significance only at relatively low Po_2.

3 CCl_4-Induced Lipid Peroxidation

CCl_4-mediated lipid peroxidation is initiated by CCl_4-derived free radicals ($CCl_3^{\cdot}$, $CCl_3O_2^{\cdot}$, $CCl_3O^{\cdot}$) which are reductively formed by catalysis of microsomal cytochrome P-450 (Cheeseman et al. 1985; de Groot and Noll 1988) (Fig. 2). In contrast to iron-induced lipid peroxidation, lipid peroxidation induced by CCl_4 occurs at the maximal rate immediately after the addition of the haloalkane to NADPH-supplemented rat liver microsomes and its rate slows down during further incubation (Fig. 3). These kinetics can be explained by the fact that the formation of the CCl_4-derived free radicals, and hence the initiation of lipid peroxidation, proceeds at maximum rate directly after the addition of CCl_4. With further incubation the rate of the formation of the CCl_4-derived free radicals decreases due to increasing inactivation of cytochrome P-450 (de Groot and Haas 1980; Noll and de Groot 1984).

The activation of CCl_4 to CCl_4-derived free radicals by cytochrome P-450 is also the reason for the unusual O_2 dependence of CCl_4-induced lipid peroxidation (Fig. 3). CCl_4-induced lipid peroxidation is maximal at Po_2 around 5 mmHg and markedly decreases at Po_2 below and above this value. At Po_2 above 5 mmHg the rate of the overall process of CCl_4-mediated lipid peroxidation is determined by the rate of initiation by the CCl_4-derived free radicals (Fig. 2). Since the formation of these radicals at the haem moiety of cytochrome P-450 is inhibited by O_2 and thus increases with decreasing Po_2, lipid peroxidation increases with decreasing Po_2 as well. At Po_2 below 5 mmHg the rate of the overall process is presumably determined by the rate of the propagation phase. Due to increasing termination reactions, the rate of propagation, and thus the overall rate of lipid peroxidation, decreases with decreasing Po_2 (de Groot and Noll 1988).

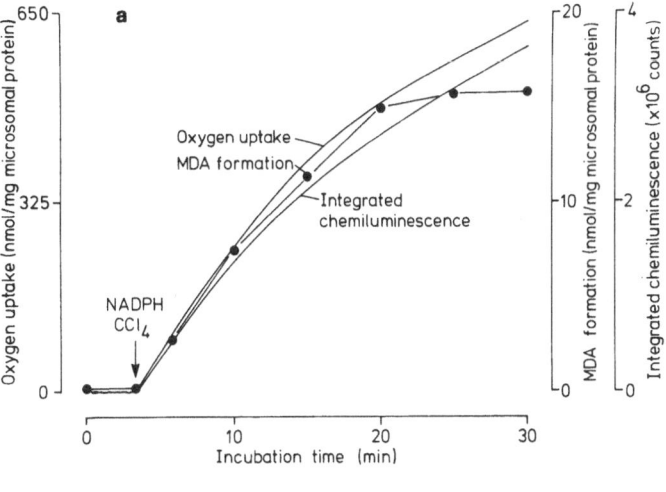

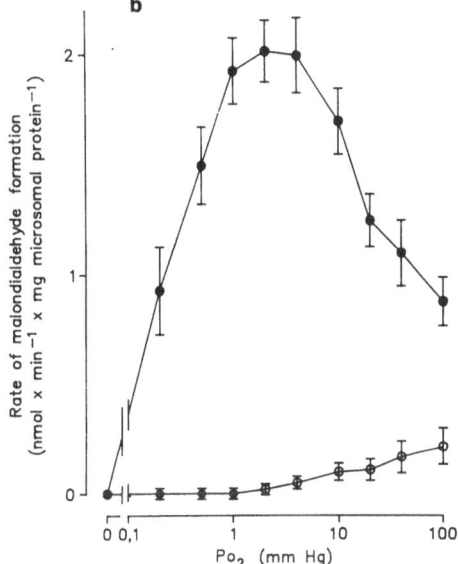

Fig. 3a, b. NADPH-CCl$_4$-induced lipid peroxidation in liver microsomes from phenobarbital-pretreated rats. **a** Time course of O$_2$ uptake, malondialdehyde *(MDA)* formation, and low-level chemiluminescence at a steady state Po$_2$ of 0.5 mmHg (from Noll et al. 1987). **b** O$_2$ dependence of the initial rate of microsomal MDA formation. (From Noll and de Groot 1984)

4 Conclusions

In mammalian tissues the highest Po$_2$, of around 100 mmHg, are found in the lungs and in the arteries. In all other tissues the Po$_2$ are significantly lower. For instance, in liver Po$_2$ between 1 and 60 mmHg have been recorded with a mean around 20 mmHg (Kessler et al. 1984). In solid mammalian tumours relatively large areas of very low Po$_2$ typically exist. These areas of Po$_2$ below 5 mmHg may represent up to 50% of the solid tumour mass (Vaupel 1977).

Both examples given here demonstrate how crucial it can be for the study of lipid peroxidation and for the assessment of its physiological and pathological significance to consider the actual Po_2. Initiation of iron-mediated lipid peroxidation is decisively dependent on O_2 over the whole physiological Po_2 range. In addition propagation of iron-mediated lipid peroxidation becomes dependent on O_2 at Po_2 values close to the mean Po_2 found in tissue. On the other hand, CCl_4-mediated lipid peroxidation, like the lipid peroxidation induced by other polyhalogenated alkanes, only occurs at physiological Po_2, with a distinct maximum at Po_2 located at the lower end of the physiological Po_2 range. The O_2 dependence of lipid peroxidation may also be determined by factors not mentioned here, such as the fatty acid composition of the membrane lipids and the antioxidant status of the cell.

Acknowledgments. This study was supported by the Ministerium für Wissenschaft und Forschung des Landes Nordrhein-Westfalen and by the Deutsche Forschungsgemeinschaft, Schwerpunktprogramm "Mechanismen toxischer Wirkungen von Fremdstoffen".

References

Aust SD, Morehouse LA, Thomas CE (1985) Role of metals in oxygen radical reactions. J Free Radic Biol Med 1:3-25

Burton GW, Cheeseman KH, Doba T, Ingold KU, Slater TF (1983) Ciba Found Symp 101:4-14

Cadenas E, Ginsberg M, Rabe U, Sies H (1984) Evaluation of alpha-tocopherol antioxidant in microsomal lipid peroxidation as detected by low-level chemiluminescence. Biochem J 223:755-759

Cheeseman KH, Albano EF, Tomasi A, Slater TF (1985) Biochemical studies on the metabolic activation of halogenated alkanes. Environ Health Perspect 64:85-101

De Groot H, Haas W (1980) O_2-Independent damage of cytochrome P-450 by CCl_4-metabolites in hepatic microsomes. FEBS Lett 115:253-256

De Groot H, Noll T (1988) The role of physiological oxygen partial pressures in lipid peroxidation. Theoretical considerations and experimental evidence. Chem Phys Lipid 44, (in press)

Kessler M, Höper J, Harrison DK, Skolasinska K, Klövekorn WP, Sebening F, Volksholz HJ, et al. (1984) Tissue oxygen supply under normal and pathological conditions. In: Lübbers DW, Acker H, Leniger-Follert E, Goldstick TK (eds) Oxygen transport to tissue vol 5. Plenum, New York, pp 69-80

Lundberg WO (1961) Autoxidation and antioxidants. Wiley, New York

Noll T, de Groot H (1984) The critical steady-state hypoxic conditions in carbon tetrachloride-induced lipid peroxidation in rat liver microsomes. Biochim Biophys Acta 795:356-362

Noll T, de Groot H, Sies H (1987) Distinct temporal relation among oxygen uptake, malondialdehyde formation, and low-level chemiluminescence during microsomal lipid peroxidation. Arch Biochem Biophys 252:284-291

Vaupel P (1977) Hypoxia in neoplastic tissue. Microvasc Res 13:399-408

Iron-Induced Lipid Peroxidation and Inhibition of Proliferation in Animal Tumour Cells

A. Hammer, H. M. Tillian, E. Kink, M. Sharaf El Din,
R. J. Schaur and E. Schauenstein

Institute of Biochemistry, Karl-Franzens-University of Graz, Schubertstraße 1,
8010 Graz, Austria

While many pathobiochemical consequences of lipid peroxidation like hepato-toxicity, ageing and artherosclerosis have been described in the literature, no non-destructive biochemical function of lipid peroxidation has been demonstrated unequivocally. A negative association between the extent of lipid peroxidation in tumour cells and their proliferative behaviour has been postulated (Bartoli and Galeotti 1979). Experimental evidence is available in the literature for the following factors, which may contribute to this negative association: (a) Decreased content of polyunsaturated fatty acids, (b) Shift of the balance between pro-oxidant (e.g. iron) and antioxidant (e.g. glutathione) substances to the antioxidant side. (c) decreased activity of peroxidative enzymes (e.g. NADPH-cytochrome c reductase), (d) reduced access of oxygen as a consequence of increased tumour membrane rigidity.

Oxidation studies with and without antioxidants demonstrate that lipid peroxides and radicals generated during peroxidation reactions inhibit cell proliferation. The effect of lipid peroxidation on cell growth was summarized by Cornwell and Morisaki (1984).

For our studies we assumed as a working hypothesis that products of lipid peroxidation play a regulatory role with respect to the proliferation of tumour cells. In order to test this hypothesis we developed an experimental model in which we can modulate the extent of lipid peroxidation. Using the rapidly dividing Ehrlich ascites tumour (EAT) cells, suspended in Hanks' solution at 37°C, lipid peroxidation was stimulated by adding increasing amounts of iron. After incubation for up to 1 hour we measured the proliferative activity of the cells and various parameters of lipid peroxidation.

In order to find a suitable chemical form for the administration of iron, various amino acids were tested for their capacity to augment the effect of Fe(II) on lipid peroxidation as measured in terms of the formation of thiobarbituric acid reactive substances (TBArS) (Winkler et al. 1984a). Histidine and its 3-methyl-derivative turned out to be the most effective amino acids, the action of which couid be ascribed to the presence of the imidazole group.

The chelate complex Fe(II)-histidinate (Fe(II)/His) is readily taken up by the tumour cells. The kinetics of the uptake of Fe(II)/His and its binding to cellular proteins were studies using colorimetric methods (Sharaf El Din et al., unpublished data). Total iron was determined by the 1,10-phenanthroline/ascorbate method and ferric iron by the reaction with 5-sulpho-salicylic acid, the differ-

Nigam et al. (Eds.), Eicosanoids,
Lipid Peroxidation and Cancer
© Springer-Verlag Berlin Heidelberg 1988

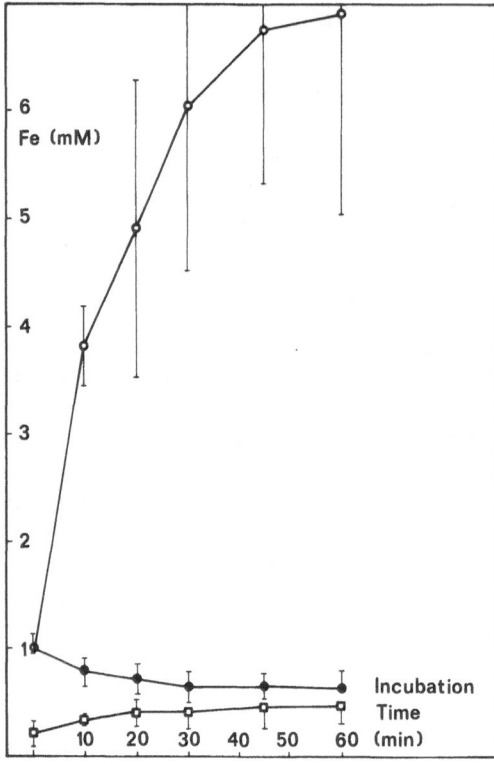

Fig. 1. Uptake of Fe(II)-histidinate: time dependence of the concentration of soluble and total iron within EAT cells and in the suspension medium. 27×10^6 cells were incubated in Hanks's buffer containing Fe(II)-histidinate (1 mM Fe. 10 mM His): after various time intervals soluble and total iron concentration within the cells and total iron concentration in the suspension medium were determined in triplicate: mean $\pm$ SD ($n = 3$) are shown. *Open circles*, total intracellular iron; *open squares*, water-soluble intracellular iron, *closed circles*, extracellular iron

ence being attributable to ferrous iron. Total iron decreased rapidly in the suspension medium and a corresponding increase of total iron was observed within the cells when the cellular proteins were solubilized with urea (Fig. 1). When EAT cells were saturated with a high concentration of iron (1 mM), it was found that more than 90% were bound to water-insoluble proteins and less than 10% were associated with soluble proteins, while no unbound iron was detectable by ultrafiltration. Apparently only a very small portion of the intracellular iron is directly available as redox catalyst for lipid peroxidation.

Although in the absence of EAT cells Fe(II)/His was readily oxidized to Fe(III)/His by oxygen this oxidation was strongly retarded by the tumour cells (Fig. 2). In the presence of EAT cells reduced and oxidized iron disappeared from the medium at almost identical rates.

The proliferative activity of EAT cells was determined by measuring the the growth rate of the cells in vivo after incubation in vitro and reimplantation i.p. Trypan-blue exclusion tests for viability were performed simultaneously. The extent or lipid peroxidation was estimated by the following parameters: (a) thiobarbituric acid reacting substances (TBArS); (b) the formation of lipid hydroperoxides was measured by a modification of the fluorimetric method of Cathcart

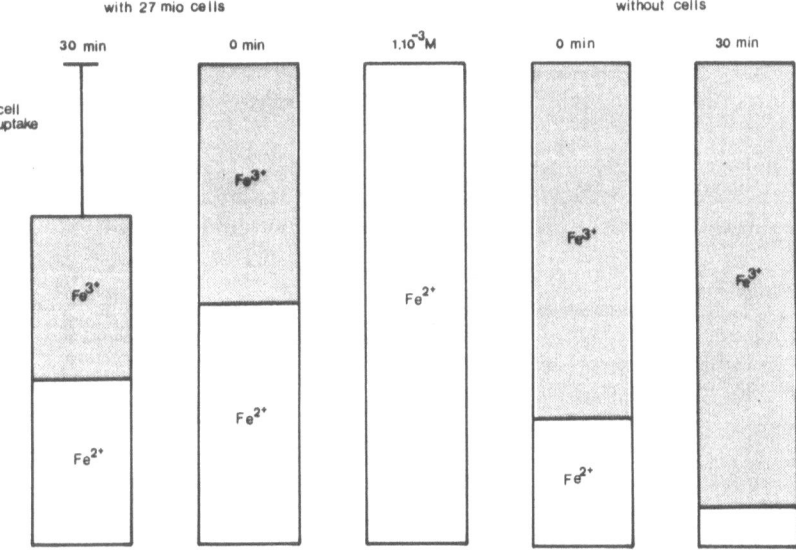

Fig. 2. Extracellular concentration of ferrous and ferric iron in absence and presence of EAT cells. The change of the redox state of iron in the suspension medium with time is shown. A solution of Fe(II)-histidinate (1 mM Fe, 10 mM His; *centre bar*) was prepared in Hanks's buffer. At zero time and after 30 min of incubation the concentration of ferrous and ferric iron was determined in absence and presence of 27×10^6 EAT cells

et al. (1984); (c) As a possible chemical link between lipid peroxidation and proliferation 4-hydroxy-nonenal (HNE) was determined as a hydrazone derivative by reversed phase HPLC and its identification ascertained by on-line uv-spectroscopy (Winkler et al. 1984b).

Five days after reimplantation the total number of tumour cells had decreased with increasing iron concentration (Fig. 3), showing half-maximal inhibition at

Fig. 3. Viability and inhibition of proliferation of EAT cells in vivo after treatment with Fe(II)-histi-dante. Percentages of try-pan-negative cells after reimplantation into mice i.p. ($n = 24$), compared with the control, are shown

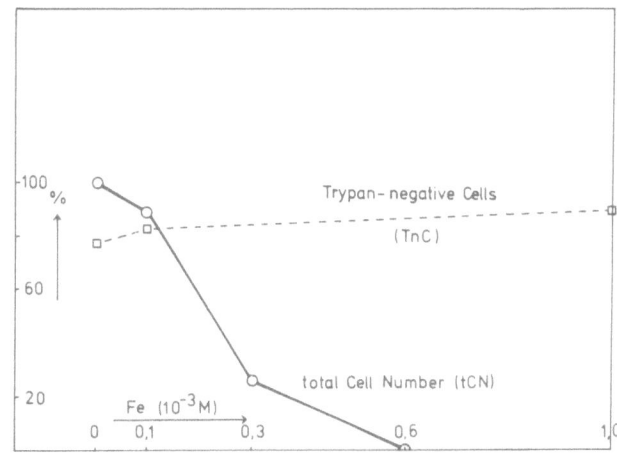

about 0.22 mM. At the same concentration range-from 0.10 mM Fe onwards –
HNE progrssively increased (Fig. 4), while both the concentration of LOOH
(Fig. 5) and of TBArS (Fig. 6) showed a biphasic dependence on the concentration of iron. The viability of the cells was independent of the iron concentration
used.

These results support the hypothesis that HNE plays a role as a chemical messenger between lipid peroxidation and proliferation. It is well known that it can
reversibly block essential thiol groups of proteins and low molecular weight thiol

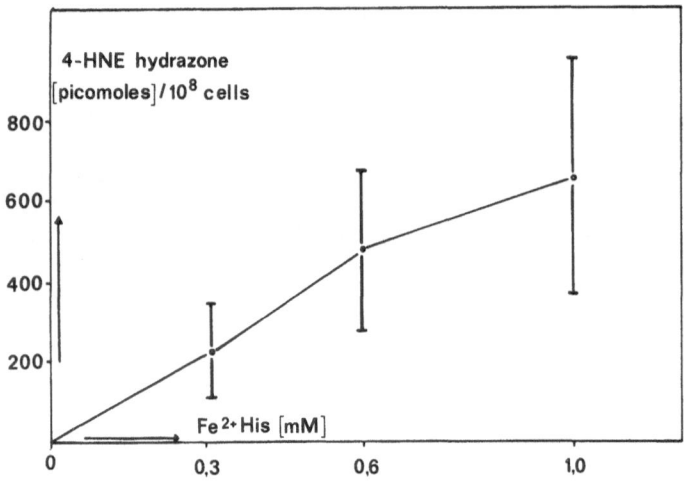

Fig. 4. Dependence of HNE on the concentration of Fe(II)-histidinate. 100×10^6 EAT cells
were incubated in 14 ml Hanks's buffer with increasing concentrations of Fe(II)-histidinate at
37°C for 20 min. Prior to derivatization with 7 ml 2,4-dinitrophenylhydrazine (1.8 mM) in HCl
(1 N) for 24 hours at room remperature in the dark 0.1 ml desferral (1%) were added as iron-
chelating agent. After extraction by dichloromethane the hydrazones were separated by TLC
and HNE quantified by HPLC as described by Winkler et al. (1984b; $n = 3$)

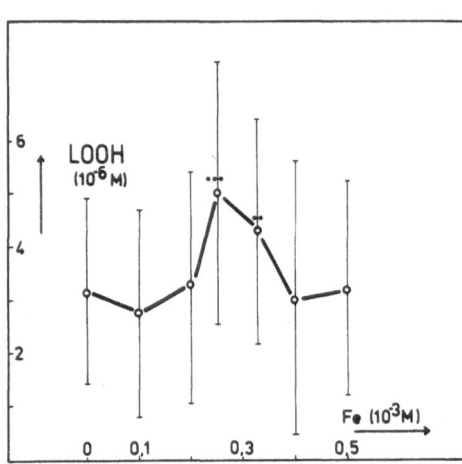

Fig. 5. Dependence of LOOH on the
concentration of Fe(II)-histidinate. After
extraction with dichloromethane LOOH
was determined fluometrically according
to Cathcart et al. (1984) by reaction with
dichlorofluorescin. Hydrogen peroxide
served as a standard: n varied between 6
and 40. Asterisks indicate a significant
difference compared with the control in
the paired double-sided Student's t-test:
** $= 2\,p < 0.01$; *** $= 2\,p > 0.0025$

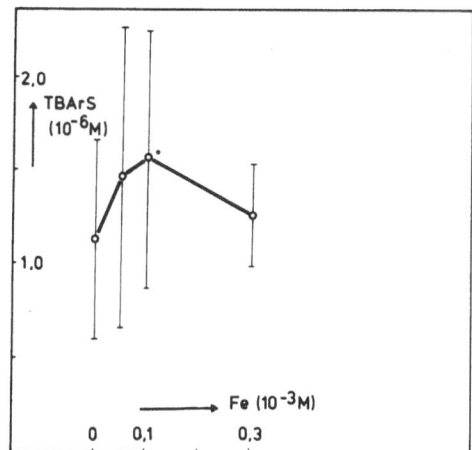

Fig. 6. Dependence of TBArS on the concentration of Fe(II)-histidante. FBArS values are expressed as malondialdehyde equivalents; n = 12. *Asterisk* indicates a significant difference compared with the control: $2f < 0.01$

containing compounds by forming thio-acetals. When applied exogenously it inhibits the biosynthesis of DNA as well as of RNA and proteins and exhibits a higher affinity for protein-thiols in the nucleus of EAT cells than in the cytoplasm (Khoschsorur et al. 1981). A thiol compound which is thought to have a protective function with regard to lipid peroxidation is glutathione (GSH). It was found that reduced as well as total GSH are decreased in the presence of Fe(II)/His (Schaur et al. 1987). This finding, which contradicts the hypothesis that GSH may act as a mere redox mediator of Fe(II)/His induced lipid peroxidation, points to a consumption of GSH by several pathways, one of which might be the reaction with HNE.

In summary the effects of Fe(II)/His on EAT cells may be interpreted as lending new support to the hypothesis that among the many regulating factors for the proliferation ot tumour cells the extent of lipid peroxidation might be significant. Our results are in agreement with the statement of Burton et al. (1983). The decreased rate of peroxidation in some tumours may be associated with a relief from an overall 'coarse' control of DNA synthesis by endogenously produced 4-hydroxy-alkenals.

Acknowledgements. This work was supported by the Association for International Cancer Research, St. Andrews, UK. The excellent technical assistance of Ing. D. Celotto is gratefully acknowledged.

References

Bartoli GM, Galeotti T (1979) Growth-related lipid peroxidation in tumour microsomal membranes and mitochondria. Biochim Biophys Acta 574:537–541
Burton GW, Cheeseman KG, Ingold KU, Blater TF (1983) Lipid antioxidants and products of lipid peroxidation as potential tumour protective agents. Biochem Soc Trans 11:261–262

226   A. Hammer et al.

Cathcart R, Schwiers E, Ames BN (1984) Detection of picomole levels of lipid hydroperoxides using a dichlorofluorescein fluorescent assay. Methods Enzymol 105:352–358

Cornwell DG, Morisaki N (1984) Fatty acid peroxides in the control of cell proliferation: Prostaglandins, lipid feroxides, and cooxidation reactions. In: Pryor WA (ed) Free radicals in biology, vol G. Academic, New York, pp 95–148

Khoschsorur G, Schaur RJ, Schauenstein E, Tillian HM, Reiter M (1981) Intracellular effect of hydroxyalkenals on animal tumors. Z Naturforsch 360:572–578

Schaur RJ, Winkler P, Hayn M, Schauenstein E (1987) Decrease of glutathione and fatty acids during iron-induced lipid peroxidation in Ehrlich ascites tumor cells. Chem Monthly 118:823–830

Winkler P, Schaur RJ, Schauenstein E (1984a) Selective promotion of ferrous ion-dependent lipid peroxidation in Ehrlich ascites tumor cells by histidine as compared with other amino acids. Biochim Biophys Acta 796:226–231

Winkler P, Lindner W, Esterbauer H, Schauenstein E, Schaur RJ, Khoschsorur GA (1984b) Detection of 4-hydroxynonenal as a product of lipid peroxidation in native Ehrlich ascites tumor cells. Biochim Biophys Acta 796-237

Ethane and *n*-Pentane Formation in a Reconstituted Microsomal Lipid Peroxidation System

H. Kappus and J. Kostrucha

Free University of Berlin, FB 3, WE 15, Augustenburger Pl. 1, 1000 Berlin 65, FRG

1 Introduction

Carcinogenesis has been related to lipid peroxidation (McBrien and Slater 1982), because some reaction products formed have DNA damaging (Ueda et al. 1985; Brambilla et al. 1986; Winter et al. 1986), mutagenic (Cajelli et al. 1987) or tumour promoting (Cerutti 1985; Kensler and Taffe 1986) activities. For example, the carcinogenic effects of the non-genotoxic peroxisome proliferators have been related to lipid peroxidation induced by increased peroxisome H_2O_2 formation (Reddy et al. 1982; Goel et al. 1986). Since after treatment with such compounds hydroxyl radicals are formed in peroxisomes in vivo (Elliott et al. 1986) it is suggested that DNA damage occurs indrectly by lipid peroxidation products.

On the other hand, a number of anticancer drugs are able to induce lipid peroxidation which could be the cytotoxic mechanism of these drugs (for review see Kappus 1985).

Therefore, interest in the relationship between lipid peroxidation and cancer is accumulating. However, most methods applied to measure lipid peroxidation in vivo are invasive or unspecific. Therefore, the measurement of alkanes like ethane or *n*-pentane has been recommended as a simple and relevant in vivo technique (for review see Kappus 1985). However, very recent findings indicate that in contrast to malondialdehyde formation alkane formation is inhibited by oxygen when measured in liver microsomes (Kostrucha and Kappus 1986; Reiter and Burk 1987) questioning the reliability of this parameter. We wondered whether the oxygen effect described can also be seen in a simple enzymatic lipid peroxidation system comprised of liposomes, complexed iron ions and NADPH-cytochrome P-450 reductase. Furthermore, we were interested in the underlying mechanism.

We found that depending on the oxygen concentration ethane or *n*-pentane and malondialdehyde formation are inversely related in this reconstituted lipid peroxidation system. Thus alkane measurements have to be evaluated with much caution.

Nigam et al. (Eds.), Eicosanoids,
Lipid Peroxidation and Cancer
© Springer-Verlag Berlin Heidelberg 1988

2 Methods

All chemicals and biochemicals were of the purest grade available and were purchased either from Merck, Darmstadt, FRG, or Boehringer, Mannheim , FRG. The gases used were the same as described previously (Kostrucha and Kappus 1986).

Liposomes comprised of either microsomal or synthetic phospholipids were used. Microsomal phospholipids were extracted from rat liver microsomes by chloroform/methanol (2:1 v/v). Synthetic phospholipids contained a mixture (1:1, w/w) of L-α-phosphatidyl-choline-1-palmitoyl-2-linoleoyl and L-α-phosphatidylcholine-1,2-dipalmitoyl. Preparation of liver microsomes from male Wistar rats, isolation of microsomal NADPH-cytochrome P-450 reductase and preparation of liposomes were performed as described elsewhere (Kostrucha and Kappus 1987).

Lipid peroxidation was performed by incubating liposomes (1 mg lipid/ml) with $FeCl_2$ (various concentrations), ADP (1.7 mM), EDTA (0.11 mM) and isolated NADPH-cytochrome P-450 reductase (2.5 U/ml) in 50 mM Tris-HCl-buffer (pH 7.5) at 37°C as already described (Kostrucha and Kappus 1987). The closed incubation system, which allowed incubations under different atmospheres of oxygen in helium, has been described previously (Kappus and Muliawan 1982).

Ethane and n-pentane formation were determined by gas chromatography (Kappus and Muliawan 1982; Kostrucha and Kappus 1986). Malondialdehyde was estimated in the trichloroacetic acid supernatant using the thiobarbituric acid assay (Kappus et al. 1977).

3 Results

Malondialdehyde formation in liposomes containing microsomal phospholipids increased continuously with increasing oxygen concentrations in the atmosphere above the incubation mixture up to a plateau which was reached between 10%–20% oxygen depending on the concentration of Fe^{2+} (data not shown). Malondialdehyde formation also depended on NADPH-cytochrome P-450 reductase (data not shown). Fe^{2+} alone, without the complexing agents ADP and EDTA, did not result in malondialdehyde formation (data not shown). These results fully agree with previous lipid peroxidation studies carried out in the same enzymatic system (Pederson et al. 1973; Morehouse et al. 1984; Aust et al. 1985).

On the other hand, relatively little ethane and n-pentane were formed under high oxygen concentrations (20%–50%) even if the concentrations of Fe^{2+} were varied (Fig. 1). However, ethane and n-pentane formation increased considerably when the oxygen concentration in the atmosphere above the incubation mixture was lowered, being maximal at 0% O_2 (Fig. 1). But this oxygen concentration is not identical with the oxygen content in the liposomal mixture (Kostrucha and Kappus 1986). Fig. 2 shows the relationships between malondialdehyde for-

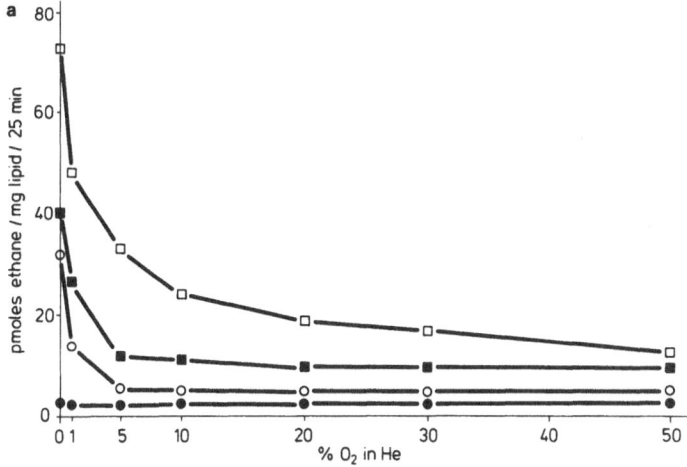

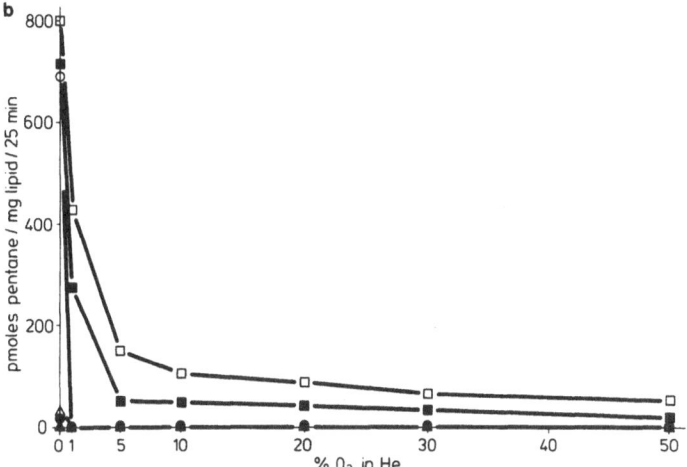

Fig. 1a, b. Effect of different head space concentrations of oxygen in helium on alkane formation in liposomes (microsomal phospholipids) incubated with microsomal NADPH-cytochrome P-450 reductase, NADPH, ADP, EDTA and various $FeCl_2$ concentrations: ●—● 100 µM, ○—○ 150 µM, ■—■ 175 µM, □—□ 200 µM. **a** Ethane. **b** *n*-Pentane

mation on the one hand and ethane or *n*-pentane formation on the other hand. An inverse relationship between malondialdehyde and ethane formation is seen when the oxygen concentration is varied, i.e. high oxygen leads to high malondialdehyde but low ethane or *n*-pentane formation and vice versa. Measuring *n*-pentane this relationship is not as good as with pentane (Fig. 2).

Fig. 3 demonstrates that the inhibitory effect of oxygen on alkane formation also occurs in liposomes comprised of synthetic phospholipids containing linoleic acid as peroxidizable fatty acid. Only *n*-pentane but no ethane was

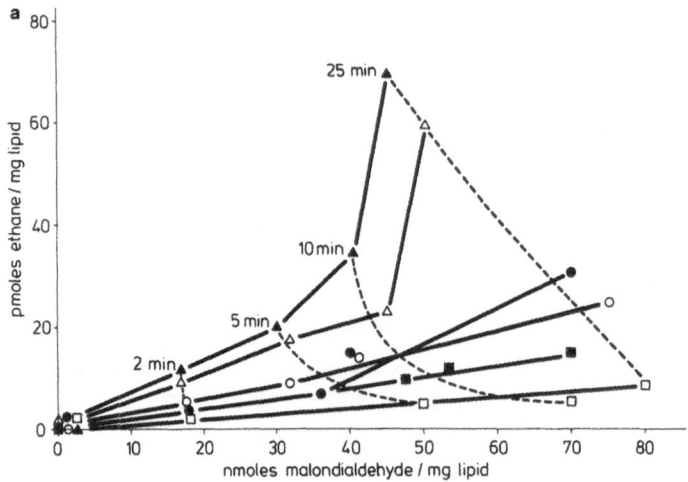

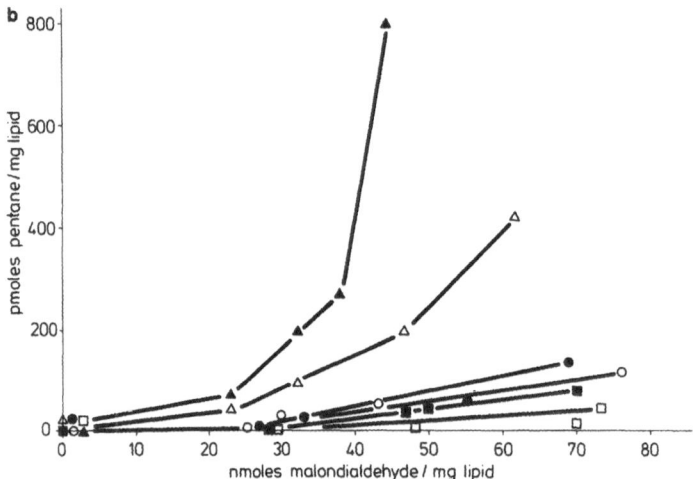

Fig. 2a, b. Correlation of alkane and malondialdehyde formation in liposomes (microsomal phospholipids) indcubated with microsomal NADPH-cytochrome P-450 reductase, NADPH, ADP, EDTA and 200 μM FeCl₂ under various head space concentrations of oxygen in helium: ▲—▲ 0%, △—△ 1%, •—• 5%, ○—○ 10%, ■—■ 20%, □—□ 50%. **a** Ethane. **b** n-Pentane

formed. Malondialdehyde also could not be detected even under higher oxygen concentrations (data not shown).

4 Discussion

Our results on the oxygen dependency of alkane formation in the reconstituted microsomal lipid peroxidation system are in good agreement with the results obtained in whole microsomes (Kostrucha and Kappus 1986). This is particu-

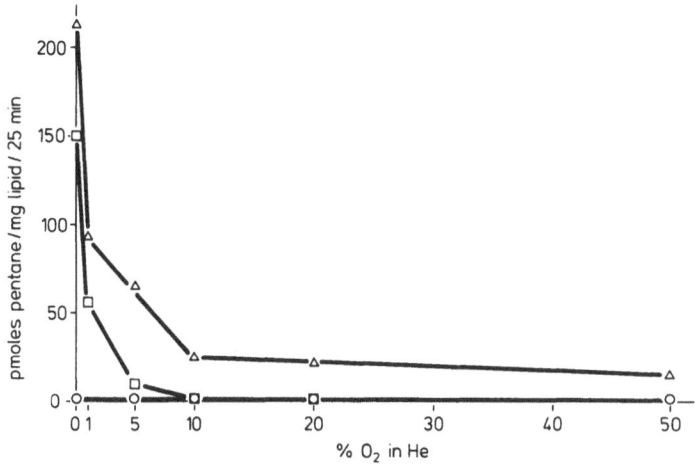

Fig. 3. Effect of different head space concentrations of oxygen in helium on *n*-pentane formation in liposomes (synthetic phospholipids) incubated with microsomal NADPH-cytochrome P-450 reductase, NADPH, aDP, EDTA and various $FeCl_2$ concentrations: o—o 150 μM, □—□ 175 μM, △—△ 200 μM

larly important because, in microsomes, besides NADPH-cytochrome P-450 reductase, cytochrome P-450 is also present which has been shown to be able to participate in the lipid peroxidation process (Ekström and Ingelman-Sundberg 1986). Furthermore, cytochrome P-450 is able to metabolize (hydroxylate) alkanes, especially *n*-pentane (Terelius and Ingelman-Sundberg 1986). Because we obtained similar oxygen dependence curves of alkane formation in microsomes and in the reconstituted system, we conclude that, at least under our conditions (Kostrucha and Kappus 1986), cytochrome P-450 is without influence; this agrees with recent findings (Davis et al. 1987). Other factors like the content of vitamin E present in microsomes are probably also not crucial to this oxygen dependent alkane formation.

However, the present and the previous studies (Kostrucha and Kappus 1986; Reiter and Burk 1987) clearly demonstrate that high oxygen concentrations inhibit alkane formation, whereas malondialdehyde release increases with increasing oxygen concentrations. This inverse relationship was shown in microsomes as well as in the reconstituted system used here, although we were unable to completely remove oxygen from the incubation mixture. This was due to the fact that we did not want to disturb the phospholipid organization of liposomes by gassing the incubation mixture itself. Therefore, it is likely that the actual concentration of oxygen dissolved in the liposomal membrane was considerably higher than indicated in Fig. 1. In a non-enzymatic microsomal lipid peroxidation system Reiter and Burk (1987) found at very low concentrations a continuous increase in alkane formation up to a peak which was followed by a continuous decrease at higher oxygen concentrations. This latter phase is similar to our findings in the enzyme catalyzed microsomal and liposomal systems (Kostrucha and

Kappus 1986; Fig. 1). Therefore, from our results it should not be concluded that the high alkane formation observed under 0% oxygen in the atmosphere above the incubation mixture is due to a non-oxidative reaction. It is, rather, conceivable that during lipid peroxidation low oxygen concentrations favour the formation of hydroperoxides which are the precursors for alkanes, whereas endoperoxide formation which results in malondialdehyde is favoured by higher oxygen levels (Kostrucha and Kappus 1986).

On the other hand, our results with the synthetic phospholipids containing only linoleic acid as peroxidizable fatty acid are not in agreement with such a simple competitive mechanism: As also shown here linoleic acid containing only two double bonds does not release malondialdehyde during lipid peroxidation. But in these experiments the oxygen dependency of pentane formation was almost the same as observed with the microsomal phospholipids, although a competition between hydroperoxide and endoperoxide formation cannot occur with linoleic acid. Thus, the suggestion of Reiter and Burk (1987) is more likely, i.e. under low oxygen concentrations the ethyl or pentyl radicals formed during the lipid peroxidation process react with phospholipids abstracting a hydrogen atom, whereas under higher oxygen concentrations these radicals bind to molecular oxygen.

Irrespective of the underlying mechanism our results demonstrate that depending on the oxygen concentration malondialdehyde and alkane formation differ. Reiter and Burk (1987) found that in a non-enzymatic microsomal system malondialdehyde correlated with the decrease of unsaturated fatty acids. They concluded that malondialdehyde is a better parameter of lipid peroxidation than alkane formation. However, it is well known that malondialdehyde formation is only one part of the whole lipid peroxidation process, that it has to be released from peroxidized fatty acids and that it is rapidly metabolized in the organism (For review see Kappus 1985). Therefore, malondialdehyde is a good indicator for lipid peroxidation only in certain in vitro systems like microsomes. However, based on the present findings of an inhibitory effect of oxygen, alkane formation can similarly not be recommended as the only quantitative measure of lipid peroxidation. In tissues where the oxygen concentration is low, such as the centrilobular area of the liver and in many tumour tissues or under hypoxia, lipid peroxidation would probably be over-estimated when measuring alkane formation. In tissues with a high oxygen conentration such as lung and heart or under hypoxia lipid peroxidation could be under-estimated or even overlooked using this technique. A very complicated situation may occur when the metabolism of a chemical inducing lipid peroxidation also leads to hypoxia as is the case during redox cycling of various chemicals such as the anticancer drugs adriamycin, mitomycin c etc. (Kappus and Sies 1981; Kappus 1986).

Because the model systems used here and in our previous study (Kostrucha and Kappus 1986) are representative in regard to the lipid peroxidation process, the effect of oxygen on alkane formation is of high importance, especially when lipid peroxidation is induced by drugs or when it is stimulated endogenously, e.g. by antioxidant deficiencies. Therefore, when trying to correlate lipid peroxidation and cytotoxicity or carcinogenicity several different methods and not only alkane measurement should be applied. Otherwise misleading results may be obtained.

Acknowledgement. This study has been supported by the Deutsche Forschungs-gemeinschaft, Bonn, FRG.

References

Aust SD, Morehouse LA, Thomas CE (1985) Role of metals in oxygen radical reactions. J Free Radic Biol Med 1:3-25

Brambilla G, Sciabà L, Faggin P, Maura A, Marinari UM, Ferro M, Esterbauer H (1986) Cytotoxicity, DNA fragmentation and sister-chromatid exchange in Chinese hamster ovary cells exposed to the lipid peroxidation product 4-hydroxynonenal and homologous aldehydes. Mutat Res 171:169-176

Cajelli E, Ferraris A, Brambilla G (1987) Mutagenicity of 4-hydroxynoneal in V79 Chinese hamster cells. Mutat Res 190:169-171

Cerutti PA (1985) Prooxidant states and tumor production. Science 227:375-381

Davis HW, Suzuki T, Schenkman JB (1987) Oxidation of esterified arachidonate by rat liver microsomes. Arch Biochem Biophys 252:218-228

Ekström G, Ingelman-Sundberg M (1986) Mechanisms of lipid peroxidation dependent upon cytochrome P-450 LM_2. Eur J Biochem 158:195-201

Elliott BM, Dodd NJF, Elcombe CR (1986) Increased hydroxyl radical production in liver peroxisomal fractions from rats treated with peroxisome proliferators. Carcinogenesis 7:795-799

Goel SK, Lalwani ND, Reddy JK (1986) Peroxisome proliferation and lipid peroxidation in rat liver. Cancer Res 46:1324-1330

Kappus H (1985) Lipid peroxidation: Mechanisms, analysis, enzymology and biological relevance. In: Sies H (ed) Oxidative stress. Academic, London, pp 273-310

Kappus H (1986) Overview of enzyme systems involved in bioreduction of drugs and in redox cycling. Biochem Pharmacol 35:1-6

Kappus H, Muliawan H (1982) Alkane formation during liver microsomal lipid peroxidation. Biochem Pharmacol 31:597-600

Kappus H, Sies H (1981) Toxic drug effects associated with oxygen metabolism: Redox cycling and lipid peroxidation. Experientia 37:1233-1241

Kappus H, Kieczka H, Scheulen M, Remmer H (1977) Molecular aspects of catechol and pyrogallol inhibition of liver microsomal lipid peroxidation stimulated by ferrous ion-ADP-complexes or by carbon tetrachloride. Naunyn Schmiedebergs Arch Pharmacol 300:179-187

Kensler TW, Taffe BG (1986) Free radicals in tumor promotion. Adv Free Radic Biol Med 2:347-387

Kostrucha J, Kappus H (1986) Inverse relationship of ethane or n-pentane and malondialdehyde formed during lipid peroxidation in rat liver microsomes with different oxygen concentrations. Biochim Biophys Acta 879:120-125

Kostrucha J, Kappus H (1987) No evidence for lysophospholipid formation during peroxidation of phospholipids by NADPH-cytochrome P-450 reductase and iron ions. Arch Toxicol 60:170-173

McBrien DCH, Slater TF (eds) (1982) Free radicals, lipid peroxidation and cancer. Academic, London

Morehouse LA, Thomas CE, Aust SD (1984) Superoxide generation by NADPH-cytochrome P-450 reductase: The effect of iron chelators and the role of superoxide in microsomal lipid peroxidation. Arch Biochem Biophys 232:366-377

Pederson TC, Buege JA, Aust SD (1973) Microsomal electron transport. J Biol Chem 248:7134-7141

Reddy JK, Lalwani ND, Reddy MK, Qureshi SA (1982) Excessive accumulation of autofluorescent lipofuscin in the liver during hepatocarcinogenesis by methyl clofenapate and other
  hypolipidemic peroxisome proliferators. Cancer Res 42:259–266
Reiter R, Burk RF (1987) Effect of oxygen tension on the generation of alkanes and malondialdehyde by peroxidizing rat liver microsomes. Biochem Pharmacol 36:925–929
Terlius Y, Ingelman-Sundberg M (1986) Metabolism of n-pentane by ethanol-inducible cytochrome P-450 in liver microsomes and reconstituted membranes. Eur J Biochem 161:303–308
Ueda K, Kobayashi S, Morita J, Komano T (1985) Site-specific DNA damage caused by lipid
  peroxidation products. Biochem Biophys Acta 824:341–348
Winter CK, Segall HJ, Haddon WF (1986) Formation of cyclic adducts of deoxyguanosine with
  the aldehydes trans-4-hydroxy-2-hexenal and trans-4-hydroxy-2-noneal in vitro. Cancer Res
  46:5682–5686

The CCl$_4$-Induced Increase of Free Arachidonate in Hepatocytes

E. Chiarpotto, F. Biasi, G. Marmo, G. Poli and M. U. Dianzani

Department of Experimental Medicine and Oncology of the University, Corso Raffaello 30, 10125, Torino, Italy

1 Introduction

It has recently been postulated that the CCl$_4$-induced stimulation of peroxidative breakdown of membrane lipids leads to hepatocyte death through a phospholipase A$_2$ (PLA$_2$) activation which would produce cytolytic amounts of lysophosphatides (Ungemach 1985).

Direct evidence of PLA$_2$ activation has actually been obtained in isolated hepatocytes treated with relatively high concentrations of CCl$_4$, and the finding has been related to the concomitant deregulation of intracellular Ca^{2+} concentration provoked by the toxin (Glende and Pushpendran 1986). However, proof has still to be provided of a major role for PLA$_2$ activation in CCl$_4$-induced irreversible damage to hepatocytes.

Experiments related to these problems have been then carried out and the results are reported here. The involvement of both lipid peroxidation and haloalkylation in the increase of the free arachidonate content of hepatocytes provoked by CCl$_4$ has been assessed by the addition of well studied antioxidants and metabolic inhibitors. The role of PLA$_2$ activation in CCl$_4$-induced cell death was investigated by monitoring the leakage of cytoplasmic enzymes in the presence or in the absence of phospholipid inhibitors.

2 Experimental Procedures

Reagents were obtained from the following sources: [^{3}H]arachidonic acid from the Radiochemical Centre, Amersham, Bucks, UK; promethazine-HCl from May and Baker, Dagenham, UK; N,N'-Diphenyl-p-phenylenediamine (DPPD) from Merck, Darmstadt, FRG; indomethacin, mepacrine and metyrapone from Sigma Chemical Company, St. Louis, USA; 2-diethyl-aminoethyl-2,2-diphenylvalerate (SKF 525A) from Smith Kline and French Laboratories Ltd., Herts, UK.

Nigam et al. (Eds.), Eicosanoids,
Lipid Peroxidation and Cancer
© Springer-Verlag Berlin Heidelberg 1988

Male rats of the Wistar strain weighing 200–250 g were used. When necessary, rats were injected intraperitoneally with vitamin E (100 mg/kg, dissolved in 1 vol ethanol followed by the addition of 9 vol 16% v/v Tween 80 in 0.9% NaCl) 15 hours before sacrifice.

Hepatocytes, isolated by the collagenase perfusion method (Poli et al. 1979), were suspended in Ham's F-12/horse serum medium to a concentration of 1×10^7 cells/ml.

Cell phospholipids were prelabelled by incubating the suspension with 3 mM [^{3}H]arachidonic acid (specific activity 200 Ci/mmol; 2.5 µCi/ml) for 45 min at 37°C. The labelling was stopped by diluting the cell suspension with a defined balanced salt solution (Poli et al. 1979). Cells were centrifuged at 400 g for 5 min and resuspended in the same medium to 5×10^6 cells/ml. Aliquots of this suspension were treated (or not) with CCl$_4$ in the presence or in the absence of different compounds at 37°C for up to 180 min.

After incubation, the activation of phospholipase A$_2$ was evaluated by determining the intracellular free arachidonate as follows: cell lipids were extracted with chloroform-methanol (2:1 v/v) mixture; the chloroform layer was dried down and the pellet was processed by tlc. on silica gel plates. The spot corresponding to free arachidonate, identified by a standard, was scraped off and processed for radioactivity measurement.

To evaluate [^{3}H]arachidonate incorporation into phospholipids, hepatocytes (5×10^6 cells/ml) were incubated with the above mentioned [^{3}H]arachidonate solution in the presence or in the absence of different concentrations of CCl$_4$ at 37°C up to 120 min. At the end of the incubation, total cell phospholipids were determined using the method of Casini and Farber (1981).

As an index of irreversible damage the release into the cell incubation medioum of α-oxoglutarate-L-aspartate amino transferase (AST) was determined. Lipid peroxidation was monitored in hepatocyte suspension by the TBA method (Poli et al. 1979).

3 Results

3.1 Increase of ^{3}H-Free Arachidonate Level in CCl$_4$-Treated Rat Hepatocytes

A small rise in the level of free arachidonate in the CCl$_4$ treated cells compared with controls is evident as early as 30 min after poisoning. The intracellular content of arachidonic acid after 60, 120 and 180 min shows, in the treated hepatocytes, a relative increase of about 1.5-, 1.4- and 1.8-fold respectively (Fig. 1). The time course of this change is not significantly modified when vitamin E-loaded hepatocytes are employed (Fig. 1).

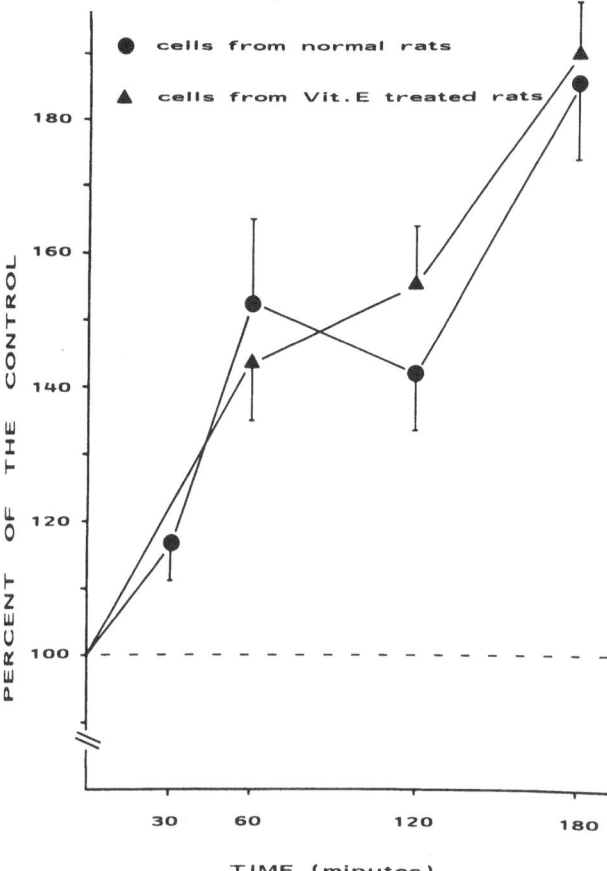

Fig. 1. Intracellular free arachidonate level of rat hepatocytes poisoned with 172 µM CCl₄. Values are means ± SD of four experiments and are expressed as percent increase as to the control taken as 100%. Liver cells isolated from normal rats then treated with CCl₄ (•). Liver cells isolated from vitamin E preloaded animals then treated with CCl₄ (▲)

3.2 Prevention of the CCl₄-Induced Increase of Intracellular Free Arachidonate by PLA₂ Inhibitors

Indomethacin and mepacrine, added to the cell suspension at the start of the poisoning, respectively exert about 50% and 100% protection against the increase in arachidonate concentration (Table 1).

3.3 CCl₄-Induced Delay of [³H]Arachidonate Incorporation into Liver Cell Phospholipids

Both concentrations of the haloalkane tested interfere rapidly with the labelling of lipid membrane. Such an effect attenuates with time until it becomes insignificant 2 hours after treatment (Fig. 2).

Table 1. Effect of PLA_2 inhibitors on intracellular free arachidonate levels in control and CCl_4 treated isolated hepatocytes

Experimental groups	Intracellular free [³H]arachidonate (cpm/10^7 cells)
Control	2761 ± 320
CCl_4 172 μM	4475 ± 146^a (62%)
Indomethacin 50 μM	3444 ± 214
+ CCl_4	4762 ± 247^a (38%)
Mepacrine 50 μM	4097 ± 460
+ CCl_4	4396 ± 552^b (7%)

[³H]-Prelabelled cells were incubated at 37°C for 120 min in the presence or in the absence of the different substances. Values are means ± SD of three experiments in duplicate. Values in parentheses are percent increase with respect to the corresponding control.
[a] Significant compared with the control group $p < 0.001$.
[b] Not significant compared with the corresponding control group.

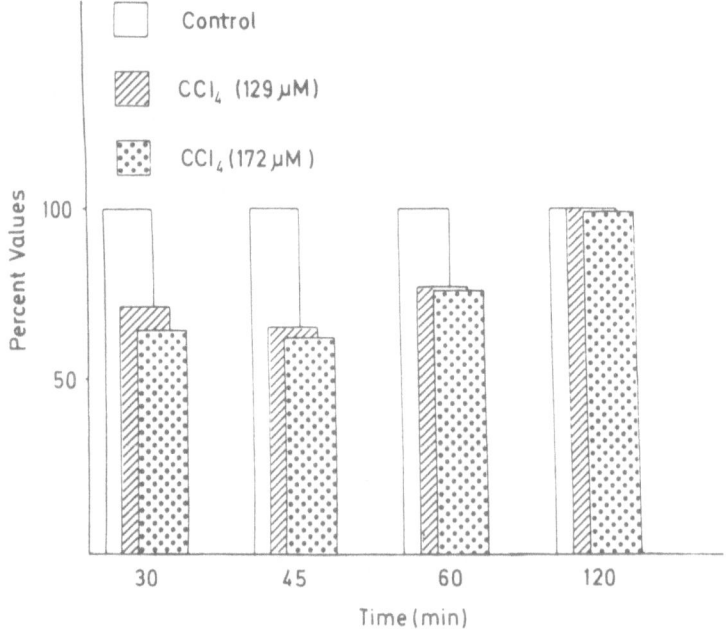

Fig. 2. [³H]arachidonate labelled phospholipids of CCl_4-treated hepatocytes. Values are expressed as percent of [³H]arachidonate incorporated into phospholipids of hepatocytes (5×10^6 cells/ml) treated or not with different concentrations of CCl_4

3.4 Lack of Protection by Antioxidants and Metabolic Inhibitors Against the Perturbation of Free Arachidonate Levels due to CCl$_4$

The enrichment of the antioxidant potential of poisoned cells either by vitamin E loading (see Fig. 1), or by external addition of promethazine and DPPD (diphenyl-phenylenediamine) (Table 2) does not modify the alteration under study, despite a complete inhibition of the CCl$_4$ pro-oxidant effect. Metyrapone and SKF 525A consistently failed to elicit protection, but some role for haloalkylation in PLA$_2$ activation cannot be ruled out since the inhibitors so far tested determined only showed an inconsistent but minimal reduction in CCl$_4$ covalent binding to total protein (data not shown). No drugs at the reported concentrations induced per se hepatocyte leakage of cytoplasmic enzymes or trypan blue stainability.

3.5 Lack of Protection by PLA$_2$ Inhibitors Against CCl$_4$-Induced Irreversible Damage to Hepatocytes

The addition of indomethacin or mepacrine, which show a marked phospholipase inhibition in the intact cell model, to the hepatocyte suspension just before CCl$_4$-poisoning does not alter the leakage of intracellular enzyme provoked by the haloalkane (Table 3).

Table 2. Effect of antioxidant compounds or metabolic inhibitors on intracellular free arachidonate levels in control and CCl$_4$-treated isolated hepatocytes

Experimental groups	Intracellular free [^{3}H]arachidonate (cpm/10^7 cells)
Control	2761 ± 320
CCl$_4$ 172 μM	4475 ± 146^a (62%)
Promethazine 50 μM	2959 ± 448
+CCl$_4$	5282 ± 306^a (78%)
Diphenylphenylenediamine 10 μM	3707 ± 827
+CCl$_4$	5971 ± 698^b (61%)
Metyrapone 100 μM	4346 ± 614
+CCl$_4$	6736 ± 570^c (54%)
SKF 525A 100 μM	4728 ± 502
+CCl$_4$	7862 ± 588^c (66%)

[^{3}H]-Prelabelled cells were incubated at 37°C for 120 min in the presence or in the absence of the different substances. Values are means $\pm$ SD of three experiments in duplicate. Values in parentheses are percent increase with respect to the corresponding control.
[a] Significant compared with the control group $p < 0.001$.
[b] Significant compared with the corresponding control $p < 0.05$.
[c] Significant compared with the coresponding control $p < 0.005$.

Table 3. Effect of phospholipase A_2 inhibitors on CCl_4-induced lipid peroxidation and cell death after four hours treatment

Experimental groups	MDA production (nmol/10^8 cells)	AST release (% of total content)
Control	32±2	0.5±0.7
CCl_4 172 μM	456±27 (14.2)	28.0±3.0
Quinacrine 50 μM	23±2	2.0±1.0
+CCl_4 172 μM	351±30 (15.2)	30.0±4.0
Indomethacin 50 μM	26±3	1.5±1.0
+CCl_4 172 μM	438±17 (16.8)	38.0±8.0

All values are means ± SD of four experiments. Numbers in parentheses are ratio of CCl_4-treated values to the corresponding controls, with or without inhibitors.

4 Discussion

A significant increase in the intracellular pool of free arachidonic acid can be observed in intact rat hepatocytes following incubation with very low concentrations of CCl_4 (0.13–0.17 mM). The amounts of CCl_4 used are in the range of those recovered in the liver after acute poisoning of the rat (Glende and Pushpendran 1986).

Activation of phospholipase A_2 (PLA$_2$) is postulated to be the mechanism mainly responsible for the rise of cytosolic arachidonate, a mechanism indirectly supported by the protective effect of the PLA$_2$ inhibitors indomethacin and mepacrine. In addition, the CCl_4-induced delay in arachidonic acid incorporation into the cell membrane can partly account for its intracellular accumulation during the early phases of the poisoning, whilst eventually the late increase of cytosolic calcium (Long and Moore 1986) could strongly stimulate PLA$_2$ activity.

The lack of any protective effect when antioxidant compounds (promethazine, DPPD) or antioxidant procedures (hepatocyte preloading with vitamin E) were employed demonstrates that lipid peroxidation-derived products are not involved in the stimulation of phospholipase by CCl_4. At the present state of knowledge one cannot exclude the possibility that CCl_4 active metabolites are operating in the activation of PLA$_2$. In any case, even if the experiments actually in progress indicate such an involvement, we expect that other mechanisms are also effective because of the low likelihood of metabolism of the haloalkane at the plasmamembrane level where the highest concentration of PLA$_2$ is located.

In similar fashion to that already described for CCl_4-induced activation of phospholipase C (Lamb and Schwertz 1982; Schwertz and Lamb 1982) phospholipase A_2 stimulation could also be exerted by CCl_4 through a direct physical action.

With regard to the proposed role of the increase in PLA$_2$ activity during CCl_4 acute poisoning as an essential linking phenomenon between the increased breakdown of polyunsaturated fatty acid and cell death, this is not supported by the present data. Further evidence against that hypothesis are the results of similar experiments of Stacey and Klaassen (1982) and of Nicotera et al. (1986).

These authors, in fact, did not achieve any protection by PLA$_2$ inhibitors against diethyl maleate and cystamine-induced hepatocyte killing. Although we cannot yet exclude some role of PLA$_2$ activation in CCl$_4$-induced hepatocyte death, we at present must agree that conclusive evidence in favour of such a role is still missing.

Acknowledgements. The authors wish to thank the Ministero della Pubblica Istruzione and the Consiglio Nazionale delle Ricerche, Roma, for supporting this research.

References

Casini A, Farber J (1981) Dependence of the carbon-tetrachloride-induced death of cultured hepatocytes on the extracellular calcium concentration. Am J Pathol 105:138–148

Glende EA Jr, Pushpendran CK (1986) Activation of phospholipase A$_2$ by carbon tetrachloride in isolated rat hepatocytes. Biochem Pharmacol 35:3301–3307

Lamb RG, Schwertz DW (1982) The effects of bromobenzene and carbon tetrachloride exposure in vitro on the phospholipase C activity of rat liver cells. Toxicol Appl Pharmacol 63:216–229

Long RM, Moore L (1986) Inhibition of liver endoplasmic reticulum calcium pump by CCl$_4$ and release of a sequestered calcium pool. Biochem Pharmacol 35:4131–4137

Nicotera P, Hartzell P, Baldi C, Svensson SA, Bellomo G, Orrenius S (1986) Cystamine induces toxicity in hepatocyte through the elevation of cytosolic Ca^{2+} and the stimulation of a non-lysosomal proteolytic system. J Biol Chem 261:14628–14635

Poli G, Gravela E, Albano E, Dianzani MU (1979) Studies on fatty liver with isolated hepatocytes. II. The action of carbon tetrachloride on lipid peroxidation, protein, and triglyceride synthesis and secretion. Exp Mol Pathol 30:116–127

Schwertz DW, Lamb RG (1982) The influence of carbon tetrachloride metabolism on the carbon tetrachloride-induced activation of rat liver cell phospholipase C acticity. Toxicol Appl Pharmacol 65:402–412

Stacey NH, Klaassen CD (1982) Effects of phospholipase A$_2$ inhibitors on diethylmaleate-induced lipid peroxidation and cellular injury in isolated rat hepatocytes. J Toxicol Environ Health 9:439–450

Ungemach FR (1985) Plasma membrane damage of hepatocytes following lipid peroxidation: involvement of phospholipase A$_2$. In: Poli G, Cheeseman KH, Dianzani MU, Slater TF (eds) Free radicals in liver injury. IRL, Oxford, pp 127–134

Lipid Peroxidation Products and Carcinogenesis: Preliminary Evidence of *n*-Alkanal Genotoxicity

G. Brambilla[1], A. Martelli[1], E. Cajelli[1], R. Canonero[2] and U. M. Marinari[2]

1 Introduction

A number of experimental findings, mainly accumulated in the last few years, have been interpreted as indicating that the radical-induced peroxidative breakdown of biomembrane lipids might play a role in the initiation and/or in the promotion phase of the carcinogenetic process. As a matter of fact, the observations listed in Table 1, even if representing an incomplete review of the pertinent findings, are sufficient to show that lipid peroxidation products have been found to be present during tumour induction, to form adducts on DNA, to cause lesions of DNA, and to give positive responses in a variety of short-term screening assays for the detection of chemical carcinogens. Until now the majority of genotoxicity studies have been performed on the best known lipid peroxidation product, malondialdehyde, and comparatively little attention has been devoted to assessing the mutagenic and clastogenic potential of the other aldehydic products generated by the oxidative breakdown of polyunsaturated fatty acids. Recently we have demonstrated that 4-hydroxynonenal, a major, and probably the most active, product of the peroxidation of liver microsomal lipids, shares with some homologous aldehydes the capability of inducing in mammalian cells, at micromolar concentrations, DNA fragmentation, point mutations and sister chromatid exchange (Brambilla et al. 1986; Cajelli et al. 1987).

In this report we present some preliminary results of a study undertaken to evaluate the cytotoxic, DNA-damaging and mutagenic activities of some *n*-alkanals produced in the course of lipid peroxidation. The interest in *n*-alkanal genotoxicity has been prompted by the experimental evidence indicating that they represent the major class among the non-polar carbonyl compounds obtained from liver microsomes peroxidized with ADP-iron (Esterbauer et al. 1982). In fact the proportion of *n*-alkanals – propanal, hexanal, and in minor amounts butanal and pentanal – contained in the microsomal supernatant fraction and in the microsomal sediment were approximately 30% and 50% of the total carbonyl compounds present. A similar yield of *n*-alkanals was observed by Poli et al. (1985) in both isolated rat hepatocytes and liver microsomal suspensions when lipid peroxidation was stimulated by carbon tetrachloride or ADP-iron.

Institute of Pharmacology[1], Viale Benedetto XV, 2, and Institute of General Pathology[2], Via L. B. Alberti, 2; University of Genoa, 16132 Genoa, Italy.

Nigam et al. (Eds.), Eicosanoids,
Lipid Peroxidation and Cancer
© Springer-Verlag Berlin Heidelberg 1988

Table 1. Experimental evidence supporting the hypothesis of an initiating and/or promoting carcinogenic effect of lipid peroxidation products

Experimental observations	References
Tumour induction	
– Initiating activity of malondialdehyde as a carcinogen for mouse skin	Shamberger et al. (1974)
– Lipofuscin accumulation in the liver during peroxisome proliferators-induced hepatocarcinogenesis	Reddy et al. (1982)
– Occurrence of membrane lipid peroxidation in liver of rats treated with carcinogenic peroxisome proliferators	Goel et al. (1986)
– Lipid peroxidation in the liver nuclear fraction of rats fed with a choline and methionine-deficient diet inducing the development of hepatocellular carcinoma	Rushmore et al. (1984)
– Increased levels of conjugated dienes in the liver microsomal lipids of rats fed with a cholinedeficient diet promoting the development of hepatomas	Perera et al. (1984)
– Increased peroxidability of liver microsomes and malondialdehyde production during the phenobarbital-induced promotion of hepatocarcinogenesis in rats	Vo et al. (1986)
– Lipid peroxidation of both mitochondrial and microsomal membranes in rats fed with the hepatocarcinogen methapyrilene	Perera et al. (1985)
– Inhibitory effect of dietary antioxidants on epidermal lipid peroxidation and UV-carcinogenesis	Black and Lenger (1984)
– Toxic effects of adriamycin in rat liver nuclei coupled with peroxidation of the nuclear membrane phospholipids	Mimnaugh et al. (1985)
Interactions with DNA	
– Formation of fluorescent products by DNA-malondialdehyde reaction	Reiss et al. (1972)
– Formation of pyrimido-purine nucleosides from the reaction of malondialdehyde with nucleic acids	Seto et al. (1983)
– Formation of cyclic propanodeoxyguanosine adducts in DNA upon reaction with acrolein	Chung et al. (1984)
Genotoxic effects	
– Occurrence of DNA damage in mammalian cells after exposure to propanal, hexanal, 4-hydroxynonenal and malondialdehyde	Marinari et al. (1984)
– DNA-fragmentation and sister chromatid exchange in mammalian cells exposed to 4-hydroxynonenal and homologous aldehydes	Brambilla et al. (1986)
– Induction of oxidative DNA damage by lipid oxidation products released from activated macrophages	Lewis et al. (1986)
– Increase of frameshift mutant frequency induced by malondialdehyde in S. typhimurium	Mukai and Goldstein (1976)
– Lethal and mutagenic effects of malondialdehyde on E. coli with different DNA-repair capacities	Yonei and Furui (1981)
– Mutagenic activity of malondialdehyde in the Salmonella tester strain TA102	Lewin et al. (1982)
– Unequivocal demonstration that highly purified malondialdehyde is a weak mutagen for S. typhimurium	Basu and Marnett (1983)

Table 1. Continued

– Weak mutagenic activity of malondialdehyde as detected by a screening for genetic mosaics and for sex-linked recessive lethal mutations in *Drosophila megalogaster*	Szabad et al. (1983)
– Induction of mutations in the *S. typhimurium* strain TA104 by hexenal, hexadienal and 4-hydroxypentenal	Marnett et al. (1985)
– Mutagenic activity in bacteria of the reaction mixture of NADPH-dependent microsomal lipid peroxidation	Akasaka and Yonei (1985)
– Dose-dependent increase of mutation frequency in murine L5178Y lymphoma cells treated with malondialdehyde	Yau (1979)
– Dose-dependent production of micronuclei and chromosomal aberrations in primary cultures of rat skin fibroblasts exposed to malondialdehyde	Bird et al. (1982)
– Dose-dependent increase of 6-thioguanine-resistant mutants in V79 Chinese hamster cells after exposure to 4-hydroxynonenal	Cajelli et al. (1987)

2 Materials and Methods

2.1 Chemicals

Propanal, butanal, hexanal and *N*-nitrosodimethylamine (DMN) were purchased from E. Merck (Darmstadt, FRG); ethyl methane-sulfonate (EMS) from ICN Pharmaceutical Inc. (New York, New York); 6-thioguanine from Sigma Chemical Co., (St. Louis, Missouri) [methyl-^{3}H]thymidine (specific activity 23–25 Ci/mmol) from the Radiochemical Centre (Amersham, UK). All other chemicals, reagent grade, were obtained from E. Merck (Darmstadt, FRG).

2.2 Cell Culture and Treatment

V79 cells (N.I.H., Coriell Institute for Medical Research, Camden, New Jersey), an established fibroblast-like line derived from Chinese hamster lungs, were cultured in α-MEM (Flow Laboratories, Milan, Italy) supplemented with 5% foetal calf serum. For the determination of alkanal cytotoxicity, 6×10^5 cells were inoculated into 25 cm^2 plastic flasks and, after 24 h, were exposed for 60 min in serum-free medium at the indicated concentrations of the test compound. Cell viability was measured immediately after exposure by cloning efficiency. Cells were trypsinized and seeded in 60 mm platic dishes (200 cells/dish); after incubation for 7 days, colonies were fixed, stained and counted.

Hepatocytes were isolated from the liver of male Sprague-Dawley rats (200–300 g) by collagenase perfusion (Williams 1976), that yielded cells over 80% viable. Isolated hepatocytes were suspended in Williams medium E (WME, Flow Laboratories, Milan, Italy) containing 10% fetal calf serum, and aliquots of this suspension (2×10^6 cells) were inoculated in 60 mm plastic dishes. After an attachment interval of 90 min the cultures were exposed to the test compound in serum-free medium for 20 h. Cell viability was measured immediately after treatment by trypan blue dye exclusion; at least 1000 cells/sample were examined.

2.3 Determination of Mutagenic Activity

The quantitative assay for induced frequency of 6-thioguanine-resistant (TG^r) cells was a modification (Cajelli et al. 1987) of that described by Myhr and di Paolo (1978). The duration of exposure to the test chemical was 60 min. At the end of the 6-day period of mutant expression, the fraction of TG^r cells in the population was determined from 10 dishes (100 mm diameter) by counting mutant colonies 6–7 days after seeding 3.5×10^5 cells per dish. For selection of mutants, 6-thioguanine was added 2 h after seeding to give a final concentration of 10 µg/ml.

2.4 DNA Repair Test

The evaluation of DNA repair synthesis in rat hepatocyte primary cultures was performed essentially according to Williams (1977) by simultaneous exposure to the test compound and [methyl-^{3}H]thymidine (10 µCi/ml). After 20 h incubation, the cultures were washed twice with 0.9% NaCl and treated with 1% sodium citrate solution for 10 min. The cells were then fixed in acetic acid:ethanol (1:3) for three 10-min changes and a section of the plastic air-dried culture dish bottom was cut out, glued to a microscope slide, dipped in Kodak NTB-2 emulsion and exposed for 7 days at 4°C. Autoradiographs were developed and stained in May-Grünwald-Giemsa. Autoradiographic nuclear grains were counted in 50 consecutive cells of each slide. The highest cytoplasmic background count for each cell was substracted from the nuclear count to obtain the nett nuclear grains due to DNA repair synthesis. The data are expressed as the mean of 100 nett nuclear counts obtained from 2 autoradiographs ($\pm$SD). When the increase in the number of nett nuclear grains exceeded 5 as compared with the corresponding control DNA repair induction was estimated to be positive.

3 Results

3.1 Cytotoxicity

Table 2 shows the relative survival of V79 cells exposed for 60 min in serum-free medium to log-spaced concentrations of the three n-alkanals tested as measured by their cloning efficiency. The reduction of cell viability was dose-dependent

Table 2. Relative survival of V79 cells and of rat hepatocytes after exposure to *n*-alkanals

Aldehyde	Concentration (mM)	Relative survival (mean ± SD)	
		V79 cells	Rat hepatocytes
Propanal	1	0.83 ± 0.11	–
	3	0.82 ± 0.06	–
	10	0.62 ± 0.18	0.98 ± 0.01
	30	0.55 ± 0.22	0.94 ± 0.04
	60	0.49 ± 0.10	–
	90	0.27 ± 0.16	0.81 ± 0.04
	270	<0.01	0.60 ± 0.06
Butanal	1	0.91 ± 0.01	–
	3	0.88 ± 0.03	0.96 ± 0.02
	10	0.79 ± 0.08	0.91 ± 0.03
	30	0.63 ± 0.17	0.84 ± 0.03
	60	0.12 ± 0.04	–
	90	<0.01	0.37 ± 0.08
Hexanal	1	0.88 ± 0.03	–
	3	0.81 ± 0.06	0.98 ± 0.02
	10	0.71 ± 0.15	0.90 ± 0.02
	30	0.48 ± 0.08	0.77 ± 0.05
	60	<0.01	–
	90	–	0.02 ± 0.01

Data represent the mean of values obtained from at least 3 independent experiments. The length of the exposure to the three alkanals was 60 min for V79 cells and 20 h for rat hepatocytes. The relative survival was measured by cloning efficiency in V79 cells and by trypan blue exclusion in rat hepatocytes.

for concentrations ranging from 1 to 90 mM. In terms of their capability of drastically reducing plating efficiency (fraction of cloning cells <0.01), the potency was found to increase in the following order: propanal < butanal < hexanal; i.e. in proportion to the chain length. The same trend was observed with the trypan blue exclusion method in rat hepatocytes after 20-h exposure (Table 2). At a first glance they appear to be more resistant than V79 cells to *n*-alkanal cytotoxicity. However, a quantitative comparison is precluded by the differences in both the length of treatment and the method used to estimate the fraction of surviving cells.

3.2 Mutagenic Activity

Data listed in Table 3 indicate that 60 min exposure to the three alkanals tested, in the absence of a metabolic system, produced in V79 cells a dose-dependent increase in the number of TG^r clones. The mutation frequency reached the level of 5.15 × baseline in cells exposed to 90 mM propanal, and of 4.95 × with 30 mM hexanal. Therefore, both these alkanals meet the criterion of mutagenicity adopted by Bradley et al. (1981) who classified a compound as positive in this test if it induced a mutation frequency at least 3 times higher than the sponta-

Table 3. Induction of mutation to 6-thioguanine resistance in V79 cells by propanal, butanal, hexanal, and ethylmethanesulphonate employed as positive control

Treatment conditions	Number of experiments	Relative survival (%)		TG^r mutant/10^6 survivors	
		Mean	Range	Mean	Range
Propanal					
Control	5	100	–	9.7	5.5–16.1
1 mM	1	82	–	7.6	–
3 mM	2	81	74–87	13.3	10.1–16.5
10 mM	3	82	81–84	19.2	16.5–21.9
30 mM	4	63	45–72	25.9	18.4–38.7
60 mM	1	59	–	42.7	–
90 mM	1	36	–	50.0	–
Butanal					
Control	2	100	–	19.2	17.4–21.0
1 mM	1	91	–	29.6	–
3 mM	2	88	86–91	30.7	28.9–32.5
10 mM	2	81	73–84	38.8	24.7–51.2
30 mM	1	77	–	42.4	–
Hexanal					
Control	3	100	–	6.4	5.9– 6.7
3 mM	2	78	72–83	7.3	7.1– 7.4
10 mM	2	61	52–70	14.2	12.8–15.6
30 mM	3	46	37–51	31.7	28.9–36.9
Ethyl-methanesulphonate					
2.5 mM	2	88	–	83.6	52.8–114.5
10.0 mM	2	55	–	204.6	120.3–288.9

neous frequency. With 30 mM butanal, the highest dose tested so far, the frequency of TG clones increased from a control level of 19.2/10^6 survivors to 42.4/10^6 (at a relative survival of 77%). EMS, a potent direct-acting mutagen used as positive control, yielded a markedly higher incidence of TG^r mutants at approximately equitoxic concentrations.

3.3 Induction of DNA Repair

The evaluation of the alkanals ability to induce DNA repair synthesis in primary cultures of rat hepatocytes has so far been limited to propanal. When liver cells were treated for 20 h with concentrations ranging from 1 to 33 mM clear-cut evidence of a positive response was obtained only at the highest dose (Table 4). However, a modest amount of unscheduled DNA synthesis was also observed at lower doses, as indicated by the increase over controls of both the average number of nett nuclear grains and the number of cells with nuclear nett counts ≥ 5. It is worth noting that the occurrence of DNA repair in the hepatocytes exposed to

Table 4. DNA repair synthesis in rat hepatocyte primary cultures after 20 h exposure to propanal

Treatment	Concentration (mM)	Number of nett nuclear grains (Mean $\pm$ SD)
Control		$-1.3 \pm$ 5.3 (12)
Propanal	1 mM	$0.8 \pm$ 4.3 (15)
	3.3 mM	$1.2 \pm$ 5.6 (26)
	10 mM	$1.5 \pm$ 4.7 (28)
	33 mM	11.3 ± 11.6 (78)
DMN	10 mM	76.4 ± 28.8 (100)

Mean of 100 nett nuclear grain counts obtained from 2 autoradiographs. Grain counts include cells with no nuclear labelling encountered in the 50 cells counted for each slide. Numbers in parentheses are the percentages of cells with ≥ 5 net nuclear grains.

the procarcinogen DMN, used as positive control, demonstrates their metabolic competence.

4 Discussion

The first results of the present study, which is still in progress, indicate that millimolar concentrations of propanal, butanal and hexanal produce cytotoxic and mutagenic effects in mammalian cells in the absence of a metabolic system. Moreover, DNA repair synthesis has been detected in metabolically competent hepatocytes exposed to propanal, and it has been previously observed that a short treatment with subtoxic doses of both propanal and hexanal results in the fragmentation of DNA in CHO cells (Marinari et al. 1984). Damage to DNA, its erroneous repair and the consequent occurence of somatic mutations are known to represent the initial event of most carcinogenetic processes (Grover 1979; Ames 1979; Brookes 1980). Therefore, our findings may be considered a further proof of the carcinogenetic potential of lipid peroxidation products.

Propanal, butanal and hexanal have been shown to constitute a significant fraction of the non-polar carbonyl compounds formed from the peroxidative breakdown of rat hepatocyte biomembrane lipids (Esterbauer et al. 1982). According to Poli et al. (1985) the total amounts of these three alkanals which are produced by rat hepatocytes after exposure to carbon tetrachloride and to ADP-iron correspond to 130 and 80 nmoles/10^8cells, respectively. Even if these data give no information about the actual concentration at the DNA level they suggest that the amount of the three alkanals generated inside the cell is not markedly lower than the amount which can be absorbed into a cell exposed to doses found to induce genotoxic effects.

In comparison with 4-hydroxynonenal, which has been shown to induce DNA fragmentation, point mutations and sister chromatid exchange in concentration ranging from 20 to 170 μM (Brambilla et al. 1986; Cajelli et al. 1987), the genotoxic potencies of propanal, butanal and hexanal are about 200-fold lower. However, they have been found to be produced by CCl_4 and ADP-iron-stimulated hepatocytes in amounts between 50- and 20-fold higher than 4-hydroxynonenal (Poli et al. 1985). Consequently, since the level of the effect can be considered as directly proportional to the product [concentration × potency], the contribution of the three n-alkanals studied to possible lipid peroxidation genotoxicity might not be negligible.

Finally, the present and previous results (Marinari et al. 1984; Brambilla et al. 1986; Cajelli et al. 1987) suggest that assays in bacteria (Marnett et al. 1985), which failed to demonstrate the mutagenic activity of hexanal and 4-hydroxynonenal, are less qualified than those in mammalian cells for the assessment of the genotoxic activity of lipid peroxidation products.

Acknowledgements. This work was supported by the Association for International Cancer Research, UK, by the CNR special project "Oncology" (contracts no. 86.00326.44 and 86.00470.44), Italy, and by the Ministero della Pubblica Istruzione, Italy.

References

Akasaka S, Yonei S (1985) Mutation induction in *Escherichia coli* incubated in the reaction mixture of NADPH-dependent lipid peroxidation of rat-liver microsomes. Mutat Res 149:321–326

Ames BN (1979) Identifying environmental chemicals causing mutations and cancer. Science 204:587–593

Basu AK, Marnett LJ (1983) Unequivocal demonstration that malondialdehyde is a mutagen. Carcinogenesis 4:331–333

Bird RP, Draper HH, Basrur PK (1982) Effect of malonaldehyde and acetaldehyde on cultured mammalian cells. Production of micronuclei and chromosomal aberrations. Mutat Res 101:237–246

Black HS, Lenger W (1984) Inhibition of epidermal lipid peroxidation by dietarily administered antioxidants. Proc Am Assoc Cancer Res 25:132

Bradley MO, Bhuyan B, Francis MC, Langenbach R, Peterson A, Huberman E (1981) Mutagenesis by chemical agents in V79 Chinese hamster cells: A review and analysis of the literature. A report of the Gene-Tox Program. Mutat Res 87:81–142

Brambilla G, Sciabà L, Faggin P, Maura A, Marinari UM, Ferro M, Esterbauer H (1986) Cytotoxicity, DNA fragmentation and sister-chromatid exchange in Chinese hamster ovary cells exposed to the lipid peroxidation product 4-hydroxynonenal and homologous aldehydes. Mutat Res 171:169–176

Brookes P (1980) Chemical carcinogenesis. Br Med Bull 36:1–100

Cajelli E, Ferraris A, Brambilla G (1987) Mutagenicity of 4-hydroxynonenal in V79 Chinese hamster cells. Mutat Res 190:169–171

Chung FL, Young R, Hecht SS (1984) Formation of cyclic 1,N^2-Propanodeoxyguanosine adducts in DNA upon reaction with acrolein or crotonaldehyde. Cancer Res 44:990–995

Esterbauer H, Cheeseman KH, Dianzani MU, Poli G, Slater TF (1982) Separation and characterization of the aldehydic products of lipid peroxidation stimulated by ADP-Fe^{2+} in rat liver microsomes. Biochem J 208:129–140

Goel SK, Lalwani ND, Reddy JK (1986) Peroxisome proliferation and lipid peroxidation in rat liver. Cancer Res 46:1324–1330

Grover PL (ed) (1979) Chemical carcinogenesis and DNA. CRC, Boca Raton

Lewin DE, Hollstein M, Christman MF, Schwiers E, Ames BN (1982) A new *Salmonella* tester strain (TA102) with A-T base pairs at the site of mutation detects oxidative mutagens. Proc Natl Acad Sci USA 79:7445–7449

Lewis JG, Hamilton T, Adams DO (1986) The effect of macrophage development on the release of reactive oxygen intermediates and lipid oxidation products, and their ability to induce oxidative DNA damage in mammalian cells. Carcinogenesis 7:813–818

Marinari UM, Ferro M, Sciabà L, Finollo R, Bassi AM, Brambilla G (1984) DNA-damaging activity of biotic and xenobiotic aldehydes in Chinese hamster ovary cells. Cell Biochem Funct 2:243–248

Marnett LJ, Hurd HK, Hollstein MC, Levin DE, Esterbauer H, Ames BN (1985) Naturally occuring carbonyl compounds are mutagens in *Salmonella* tester strain TA104. Mutat Res 148:25–34

Mimnaugh EG, Kennedy KA, Trush MA, Sinha BK (1985) Adriamycin-enhanced membrane lipid peroxidation in isolated rat nuclei. Cancer Res 45:3296–3304

Mukai FH, Goldstein BD (1976) Mutagenicity of malonaldehyde, a decomposition product of peroxidized polyunsatured fatty acids. Science 191:868–869

Myhr BC, di Paolo JA (1978) Mutagenisis by N-acetoxy-2-acetylaminofluorene of Chinese hamster V79 cells is unaffected by caffeine. Chem Biol Interact 21:1–18

Perera MIR, Demetris AJ, Katyal SL, Shinozuca H (1984) Lipid peroxidation as a possible underlying mechanism of liver tumor promotion by a choline-deficient diet. Proc Am Assoc Cancer Res 25:141

Perera MIR, Katyal SL, Shinozuca H (1985) Methapyrilene induces membrane lipid peroxidation of rat liver cells. Carcinogenesis 6:925–927

Poli G, Dianzani MU, Cheeseman KH, Slater TF, Lang J, Esterbauer H (1985) Separation and characterization of the aldehydic products of lipid peroxidation stimulated by carbon tetrachloride or ADP-iron in isolated rat hepatocytes and rat liver microsomal suspensions. Biochem J 227:629–638

Reddy JK, Lalwani ND, Reddy HK, Qureshi SA (1982) Excessive accumulation of autofluorescent lipofuscin in the liver during hepatocarcinogenesis by methyl clofenapate and other hypolipidemic peroxisome proliferators. Cancer Res 42:259–266

Reiss U, Tappel AL, Chio KS (1972) DNA-malondialdehyde reactions: formation of fluorescent products. Biochem Biophys Res Commun 48:921–926

Rushmore TH, Lim YP, Farber E, Ghoshal AK (1984) Rapid lipid peroxidation in the nuclear fractions of rat liver induced by a diet deficient in choline and methionine. Cancer Lett 24:251–255

Seto H, Okuda T, Takesue T, Ikemura T (1983) Reactions of malonaldehyde with nucleic acid. I. Formation of fluorescent pyrimido[1,2-α]purin-10(3H)-one nucleosides. Bull Chem Soc Jpn 56:1799–1802

Shamberger RJ, Andreone TL, Wills CE (1974) Antioxidants and cancer. IV. Initiating activity of malonaldehyde as a carcinogen. JNCI 53:1771–1773

Szabad J, Soós I, Polgár G, Héjja G (1983) Testing the mutagenicity of malondialdehyde and formaldehyde by the *Drosophila* mosaic and the sex-linked recessive lethal tests. Mutat Res 113:117–133

Vo TK-O, Buc-Calderon P, Somer M-P, Taper H, Roberfroid M (1986) Modulation of hepatocarcinogenesis and changes in lipid peroxidation and chemiluminescence (Abstr). Cancer Lett [Suppl 4] 30

Williams GM (1976) Primary and long-term culture of adult rat liver epithelial cells. Methods Cell Biol 14:357–364

Williams GM (1977) Detection of chemical carcinogens by unscheduled DNA synthesis in rat liver primary cell cultures. Cancer Res 37:1845–1851

Yau TM (1979) Mutagenicity and cytotoxicity of malonaldehyde in mammalian cells. Mech Ageing Dev 11:137–144

Yonei S, Furui H (1981) Lethal and mutagenic effects of malondialdehyde, a decomposition product of peroxidized lipids, on *Escherichia coli* with different DNA-repair capacities. Mutat Res 88:23–32

Detection of 4-Hydroxynonenal and Other Lipid Peroxidation Products in the Liver of Allyl Alcohol-Intoxicated Mice

A. Pompella, A. Romani, R. Fulceri, E. Maellaro, A. Benedetti
and M. Comporti

Istituto di Patologia Generale, Università di Siena, Via del Laterino 8, 53100 Siena, Italy

1 Introduction

Allyl alcohol has long been known to produce periportal necrosis of the liver in
rats and mice (Miessner 1891; Piazza 1915). Also it is known that allyl alcohol is
metabolized by the cytosolic enzyme alcohol dehydrogenase to acrolein (Rees
and Tarlow 1967; Serafini-Cessi 1972). The latter is considered as one of the
most important toxic metabolites responsible for the damage induced by allyl
alcohol in liver and other tissues. Acrolein is in fact the most toxic member of
the class of 2-alkenals (Beauchamp et al. 1985; Schauenstein et al. 1977), α-β
unsaturated aldehydes which also include crotonaldehyde, pentenal, hexenal
and so on. Acrolein is a powerful electrophile which reacts even spontaneously
with nucleophiles such as sulphydryl groups (Esterbauer et al. 1975). The reaction is markedly accelerated by the activity of GSH-transferases. Cellular GSH is
primarily involved in the reaction and the result is a dramatic loss of GSH stores
(Hanson and Anders 1978; Zitting and Heinonen 1980; Dawson et al. 1984;
Ohno et al. 1985; Jaeschke et al. 1987). The covalent binding of allyl alcohol
metabolites to liver cells has been demonstrated with various techniques (Reid
1972).

As in the case of other well known GSH depleting agents, such as bromobenzene and acetaminophen, it has been assumed for a long time that allyl alcohol-
induced liver injury is mediated by the covalent binding of reactive metabolites,
acrolein and others, to cellular macromolecules (Reid 1972). Some years ago,
however, it was shown that the addition of acrolein to freshly isolated hepatocytes, causes, together with cell death and release of cellular enzymes (Zitting
and Heinonen 1980; Dawson et al. 1984), the development of lipid peroxidation
(Zitting and Heinonen 1980), as measured by the formation of thiobarbituric
acid reactive products. Lipid peroxidation, as measured by ethane and pentane
exhalation, was found to be markedly increased in mice acutely intoxicated with
allyl alcohol (Jaeschke et al. 1987).

Previous studies from our laboratory (Casini et al. 1985, 1988; Benedetti et al.
1986) demonstrated that lipid peroxidation develops in the livers of bromobenzene-poisoned mice and suggested that lipid peroxidation plays an important
role in bromobenzene hepatotoxicity. In particular it was demonstrated (Benedetti et al. 1986) that 4-hydroxynonenal and other aldehydes derived from lipid
peroxidation can be detected in vivo in the liver after bromobenzene intoxica-

Nigam et al. (Eds.), Eicosanoids,
Lipid Peroxidation and Cancer
© Springer-Verlag Berlin Heidelberg 1988

tion. This demonstration was of particular relevance, since a line of previous studies (Benedetti et al. 1980, 1984a; Esterbauer et al. 1982) has shown the formation of 4-hydroxynonenal and other 4-hydroxyalkenals in in vitro systems, i.e. in liver microsomes, peroxidized in the presence of NADPH-Fe. It was also shown that these aldehydes, in particular 4-hydroxynonenal, are provided with high cytopathological activities (Schauenstein et al. 1977; Benedetti et al. 1981, 1984b; Dianzani 1982; Esterbauer 1982) and it was suggested that these lipid peroxidation products could be the mediators of at least some aspects of the liver injury induced by pro-oxidant agents, such as CCl_4 and others. In the present study we report the in vivo formation of 4-hydroxynonenal and other aldehydes derived from lipid peroxidation in the liver of mice intoxicated with allyl alcohol. These and other results reported in this communication further support the idea that lipid peroxidation plays an important role in allyl alcohol-induced cellular damage.

2 Materials and Methods

Male NMRI albino mice (Charles River) weighing 20–30 g and maintained on a pellet diet (Altromin-Rieper, Bolzano, Italy) were used. The animals were starved 24 h (starvation decreases the hepatic GSH stores and renders the animals more susceptible to the toxic effects of GSH depleting agents), before receiving allyl alcohol (1.5 mmoles/kg body wt.) in saline i.p. or an equivalent volume of saline.

Liver necrosis was assessed by measuring the levels of serum glutamate-pyruvate transaminase (SGPT) activity (optimized U.V. enzymatic method, C. Erba, Milan, Italy). Lipid peroxidation was assessed by measuring the tissue content of malonic dialdehyde as previously reported (Pompella et al. 1987). In previous studies from our laboratory (Pompella et al. 1987) it has been reported that this assay correlates with other more sophisticated analyses to detect lipid peroxidation in vivo. Since, besides malonic dialdehyde, free acrolein reacts with thiobarbituric acid to give an additional absorbance at 530 nm (Witz et al. 1986), the total absorbance was corrected by the absorbance obtained by adding acrolein (2 mM) to appropriate blank samples. It must also be considered that, as it will be shown in the subsequent HPLC analyses, the amount of free acrolein found in the liver of allyl alcohol intoxicated animals was very low, if any, and in any case minimal with respect to that which is necessary to give significant absorption with thiobarbituric acid.

Hepatic GSH was measured according to Sedlak and Lindsay (1986). Protein was determined according to Lowry et al. (1951). Liver mitochondria were obtained as reported by Bielawski and Lehninger (1966).

The assay for 4-hydroxynonenal and other aldehydes was performed according to a procedure similar to that reported previously (Benedetti et al. 1986; see also Esterbauer et al. 1982). All the aldehydes were detected as dinitrophenylhydrazone derivatives. Briefly, liver homogenates were allowed to react for 2 h at room temperature in the dark with an equal volume of the 2,4-dinitrophenylhydrazine reagent. The mixture was extracted with dichloromethane. Phospholipids were removed by silicic acid column chromatography. The hydrazones, neutral lipids and unreacted 2,4-dinitrophenylhydrazine were eluted with chloroform and the eluted material was applied to a TLC plate. After TLC with this developer, the bulk of neutral lipids could be discarded. The hydrazones, which remained near the origin, were scraped off, eluted with methanol and applied again to a TLC plate. TLC was performed using dichloromethane as first developer and benzene as second developer. The bands corresponding to unreacted 2,4-dinitrophenylhydrazine and dinitrophenylhydrazones of acetone, formaldehyde and acetaldehyde, present as contaminants and identified with the use of known standards, were removed. The plate was then divided into three areas. The first area, from the origin up to but not including the dinitrophenylhydrazone of standard 4-hydroxynonenal was referred to as "polar carbonyls". The second area, from the dinitrophenylhydrazone of 4-hydroxynonenal to unreacted 2,4-dinitrophenylhydrazine, was referred to as "medium polar carbonyls". The third area, from unreacted 2,4-dinitrophenylhydrazine to the end, was referred to as "non-polar carbonyls". The products present in each area were eluted with methanol. The dinitrophenylhydrazones of polar carbonyls were redissolved in chloroform and the U.V.-visible spectrum was recorded. The dinitrophenylhydrazones of "medium polar carbonyls" were further purified by TLC, redissolved in chloroform and the U.V.-visible spectrum was recorded. An aliquot of the chloroform solution was used to prepare the dinitrophenylhydrazone of 4-hydroxynonenal by TLC with dichloromethane as developer and the dinitrophenylhydrazone of standard 4-hydroxynonenal applied at both sides. The area corresponding to 4-hydroxynonenal was scraped off, eluted and analyzed by spectroscopy and HPLC.

The dinitrophenylhydrazones of "non-polar carbonyls" were further fractionated by TLC using chloroform/n-hexane (7:3, v/v) as developing mixture. The area corresponding to alkanals and alk-2-enals was identified with the use of standard hydrazones, and the products were eluted and analyzed by spectroscopy and HPLC. The area corresponding to osazones was localized by the characteristic change of the colour (orange-red to blue) after spraying one part of the plate with methanolic solution of KOH (Esterbauer et al. 1982). The U.V.-visible spectra of osazones as well as of the other carbonyls present in the TLC area of non-polar carbonyls were recorded.

3 Results and Discussion

Table 1 shows the time-course of hepatic GSH depletion, lipid peroxidation and liver necrosis in mice intoxicated with allyl alcohol. The GSH depletion is already maximal at 15 min. Lipid peroxidation is increased at 30 and 45 min when no or minor liver necrosis (as measured by the serum transaminase levels) occurs. Lipid peroxidation therefore precedes cell death.

Table 1. Time-course of hepatic glutathione (GSH) depletion, lipid peroxidation (hepatic content of malonic dialdehyde) and liver necrosis (SGPT) following allyl alcohol intoxication

Time after intoxication	GSH (nmol/mg protein)	Malonic dialdehyde (pmol/mg protein)	SGPT U/l
0 min (7)	26.7 ± 2.8	67.4 ± 9.5	91 ± 11
5 min (6)	1.8 ± 0.5	68.2 ± 6.8	97 ± 9
15 min (4)	1.0 ± 0.1	33.4 ± 5.8	88 ± 12
30 min (8)	1.1 ± 0.1	108.7 ± 13.2[a]	82 ± 10
45 min (7)	1.4 ± 0.2	190.6 ± 23.1[b]	134 ± 15[a]
1 h (8)	1.7 ± 0.4	203.5 ± 55.1	731 ± 93
2 h (3)	1.1 ± 0.4	323.7 ± 59.1	1430 ± 717
3–4 h (5)	3.3 ± 0.4	478.6 ± 148.0	4436 ± 2004

Male NMRI albino mice, starved for 24 h, were intoxicated with allyl alcohol (1.5 mmol/kg body wt.). Results are expressed as means ± SEM. The number of animals is reported in parentheses.
[a] $p < 0.05$.
[b] $p < 0.001$.

Table 2. Time-course of mitochondrial glutathione (GSH) depletion following allyl alcohol and bromobenzene intoxications

Intoxication	Time after intoxication	Mitochondrial GSH (nmol/mg protein)	% Decrease	Total hepatic GSH (nmol/mg protein)	% Decrease
Allyl alcohol	0 min (7)	7.4 ± 0.6	–	31.5 ± 1.2	–
	5 min (3)	3.0 ± 0.1	− 59.5	1.3 ± 0.1	− 95.8
	30 min (4)	2.1 ± 0.1	− 71.6	1.0 ± 0.2	− 96.8
	45 min (3)	1.9 ± 0.1	− 74.3	1.5 ± 0.2	− 95.1
	60 min (4)	2.4 ± 0.1	− 67.6	2.4 ± 0.6	− 92.2
Bromobenzene	0 h (13)	7.6 ± 0.3	–	25.9 ± 1.3	–
	1 h (6)	5.9 ± 0.1	− 23.0	–	–
	3 h (6)	5.1 ± 0.4	− 32.4	3.8 ± 0.2	− 85.4
	6 h (5)	3.6 ± 0.4	− 53.2	1.1 ± 0.3	− 95.7
	9 h (6)	3.6 ± 0.2	− 52.8	1.6 ± 0.1	− 93.8

Male NMRI albino mice, starved for 24 h, were intoxicated with either allyl alcohol (1.5 mmol/kg body wt.) or bromobenzene (13 mmol/kg body wt.). Results are given as means ± SEM. The number of animals is reported in parentheses.

The GSH depletion seen following allyl alcohol intoxication is much more rapid and severe than that seen after the intoxications with other GSH depleting agents such as bromobenzene, iodobenzene and diethylmaleate (Casini et al. 1985). A great deal of experimental work (Vignais and Vignais 1973; Jocelyn and Kamminga 1974; Wahlländer et al. 1979; Meredith and Reed 1982) has shown the presence of two pools of GSH in the liver cell. The largest pool (about 85% of total GSH) is located in the cytoplasm and is characterized by a rapid turnover (t½ of about 2 h), while the other pool (10%–15%) is located in the mitochondria and is characterized by a much slower turnover (t½ of about 30 h).

According to several studies (Meredith and Reed 1982, 1983) the depletion of the supply of GSH sequestered within the mitochondria could be a crucial step in the sequence of events leading to cellular death. The mitochondrial GSH level was therefore measured in allyl alcohol-poisoned animals (Table 2) and found to be already decreased by about 60% at 5 min after the intoxication. As can be seen (Table 2), the depletion of the mitochondrial GSH, which was maximal at 30 and 45 min, occurs much earlier in allyl alcohol intoxication compared to that seen with other GSH depleting agents, such as bromobenzene. The rapidity of mitochondrial GSH depletion seems to parallel the rapidity of induction of both liver damage and lipid peroxidation. In bromobenzene intoxication, in fact, lipid peroxidation and liver necrosis do not occur prior to 9 h after the administration of the toxin (Casini et al. 1985, 1988).

Since a large individual variation was observed in the sensitivity of the animals to allyl alcohol (as shown by the dispersion of malonic dialdehyde and SGPT values in Table 1) we decided to analyze for 4-hydroxynonenal and other aldehydes derived from lipid peroxidation in the livers of animals showing relatively high malonic dialdehyde values (higher than 170 pmoles/mg protein). In these animals (sacrificed 1–3 h after the intoxication) the hepatic GSH content was decreased to 2.9 ± 0.3 nmoles/mg protein (that is about 10% of the control values); the SGPT level was 3390 ± 1330 U/l and the malonic dialdehyde content of the liver was 431.3 ± 135.6 pmoles/mg protein.

Figure 1 shows the HPLC analysis of the TLC area with an Rf value equal to that of the dinitrophenylhydrazone of standard 4-hydroxynonenal. As can be seen, a peak with the same retention time of 4-hydroxynonenal is evident in the chromatograms of the intoxicated animals. When the dinitrophenylhydrazone of standard 4-hydroxynonenal was added to the sample, an increase in the height of the peak was seen. Two unidentified peaks with shorter retention times were

Table 3. Carbonyls detected in the liver of allyl alcohol-poisoned mice

	pmol/mg protein	
	Controls	Allyl alcohol
Malonic dialdehyde	56.7 ± 13.9 (3)	431.3 ± 136.5 (6)
Total carbonyls, except malonic dialdehyde	362.7 ± 119.2 (3)	1256.2 ± 159.1 (5)
4-Hydroxynonenal[a]	0 (3)	3.2 ± 1.1 (6)
Polar carbonyls[b]	93.8 ± 54.6 (3)	437.4 ± 86.0 (5)
Medium polar including 4-hydroxynonenal[b]	84.0 ± 40.2 (3)	300.0 ± 27.0 (5)
Non-polar carbonyls[b]		
Osazones	19.2 ± 7.5 (3)	54.6 ± 8.7 (5)
Alkanals + alk-2-enals	70.7 ± 11.7 (3)	191.7 ± 24.8 (5)
Other carbonyls	95.2 ± 12.0 (3)	272.3 ± 31.4 (5)

Results are given as means $\pm$ SEM. Reported in parentheses are the numbers of pools of livers which were analyzed; each pool corresponded to 2–3 livers.
[a] Calculated from the high pressure liquid chromatography (HPLC) analysis of the 2,4-dinitrophenylhydrazone derivative.
[b] The amounts of carbonyls were calculated from the absorbance maximum (in the range of 360–370 nm) and an average molar extinction coefficient (ε) of 25 000.

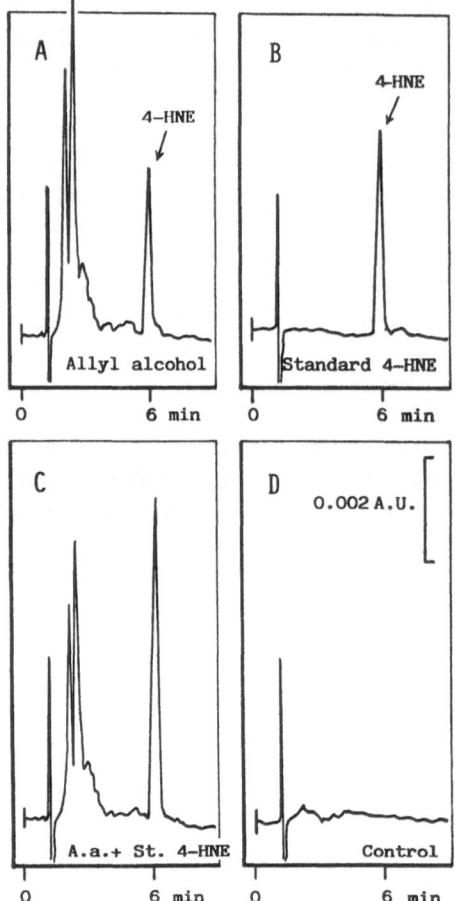

Fig. 1A–D. Detection by high-pressure liquid chromatography of 4-hydroxynonenal (4-HNE) as 2,4-dinitrophenylhydrazone derivative in the livers of allyl alcohol-poisoned mice. **A** Allyl alcohol-treated. **B** 2,4-Dinitrophenylhydrazone derivative of standard 4-HNE. **C** Allyl alcohol-treated plus standard 4-HNE. **D** Control. Separation conditions: reversed-phase column, LiChrospher 100 CH 18 Merck (4 mm × 25 cm); column temperature, 36°C; mobile phase, acetonitrile/water (7:3, v/v); flow rate, 1.5 ml/min; detector wavelength, 350 nm. A sample derived from the liver corresponding to 15 mg of protein was injected

also seen. Almost no peaks were seen in the chromatograms for the control animals and for the blank reagents.

The U.V.-visible spectra of "polar carbonyls", "medium polar carbonyls", alkanals and alkenals, osazones and other carbonyls present in the TLC area of "non-polar carbonyls" are shown in Figs. 2, 3. The amounts of carbonyls present in each different fraction were calculated from the absorbance maximum (in the range of 365–370 nm) by means of the molar extinction coefficient $\varepsilon = 25\,000$ $M^{-1} \cdot cm^{-1}$ reported by Esterbauer et al. (1982).

Figure 4 shows the HPLC analysis of the hydrazones eluted from the area corresponding to alkanals and alk-2-enals. Several aldehydes were present in appreciable amounts in the chromatograms of the intoxicated animals. The aldehydes that could be identified with the use of known standards were butanal, pentanal and hexanal. Acrolein could not be clearly demonstrated. Based on the retention time of the dinitrophenylhydrazone of standard acrolein the peak of acrolein would be hardly distinguishable from that of butanal. Much smaller peaks than those present in the samples of the intoxicated animals were seen in

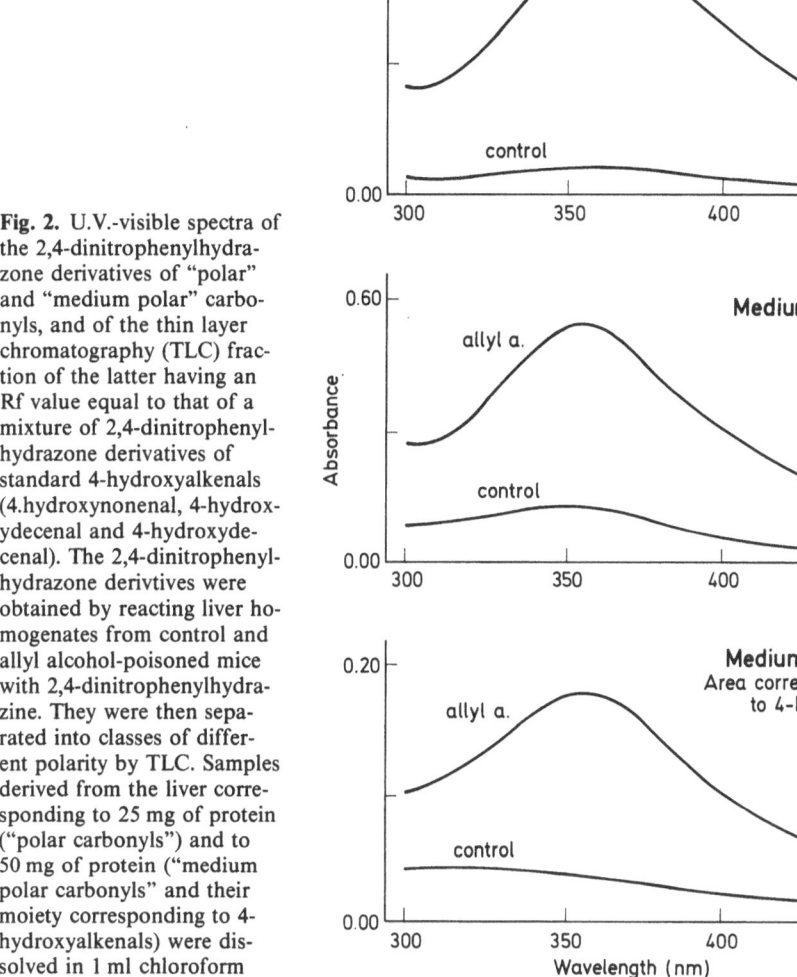

Fig. 2. U.V.-visible spectra of the 2,4-dinitrophenylhydrazone derivatives of "polar" and "medium polar" carbonyls, and of the thin layer chromatography (TLC) fraction of the latter having an Rf value equal to that of a mixture of 2,4-dinitrophenylhydrazone derivatives of standard 4-hydroxyalkenals (4.hydroxynonenal, 4-hydroxydecenal and 4-hydroxydecenal). The 2,4-dinitrophenylhydrazone derivtives were obtained by reacting liver homogenates from control and allyl alcohol-poisoned mice with 2,4-dinitrophenylhydrazine. They were then separated into classes of different polarity by TLC. Samples derived from the liver corresponding to 25 mg of protein ("polar carbonyls") and to 50 mg of protein ("medium polar carbonyls" and their moiety corresponding to 4-hydroxyalkenals) were dissolved in 1 ml chloroform

the samples of the controls and those of blank reagents (small amounts of butanal, pentanal and hexanal present as contaminants).

The quantitative determination of free aldehydes in the liver of allyl alcohol intoxicated animals is reported in Table 3. As in the case of bromobenzene-induced liver injury a relatively large amount of carbonyls can be detected in this model of experimental liver injury. On the basis of the protein content of the liver, the concentration of free carbonyls (except malonic dialdehyde) which can be calculated from the data shown above is 0.25 mM in the liver of allyl alcohol-poisoned mice, which is about 3 fold lower than that found (Benedetti et al. 1986) in the liver of bromobenzene-poisoned animals. This concentration is how-

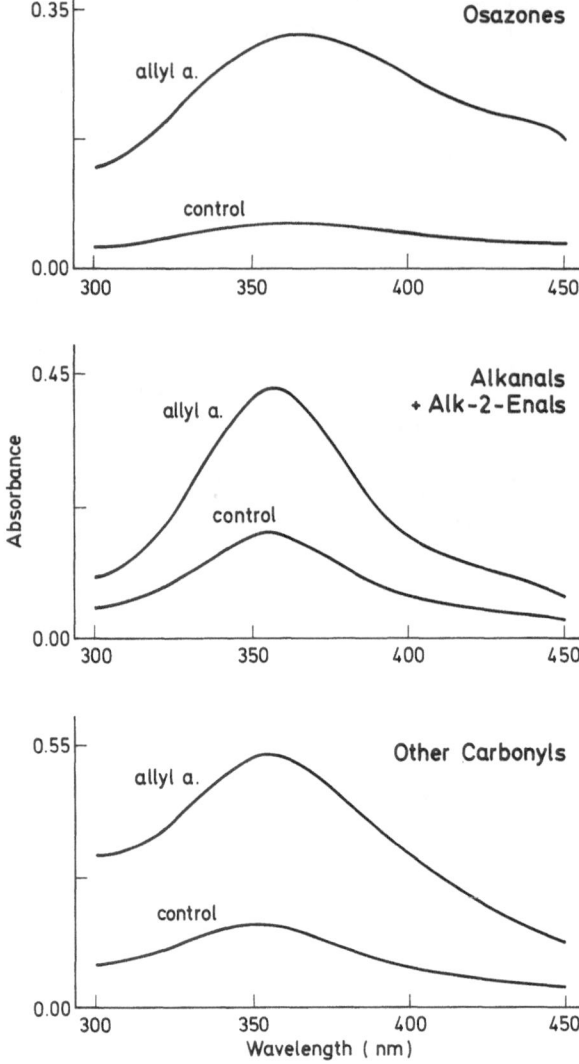

Fig. 3. U.V.-visible spectra of the 2,4-dinitrophenylhydrazone derivatives of osazones, alkanals plus alk-2-enals, and other carbonyls, obtained by thin layer chromatography (TLC) fractionation of the 2,4-dinitrophenylhydrazone derivatives of "non-polar carbonyls". The 2,4-dinitrophenylhydrazone derivatives were obtained by reacting liver homogenates from control and allyl alcohol-poisoned mice with 2,4-dinitrophenylhydrazine (see text for details)

ever sufficient to produce many cytopathological effects in in vitro systems (Esterbauer 1985).

As previously mentioned carbonyl products of lipid peroxidation have been implicated as possible mediators of the liver injury induced by pro-oxidant agents. The relevance of the present as well of the previous study (Benedetti et al. 1986) is therefore the demonstration that 4-hydroxynonenal and other aldehydes having cytopathological activity are in fact produced in the liver in vivo in experimental situations in which the occurrence of lipid peroxidation has been unequivocally demonstrated.

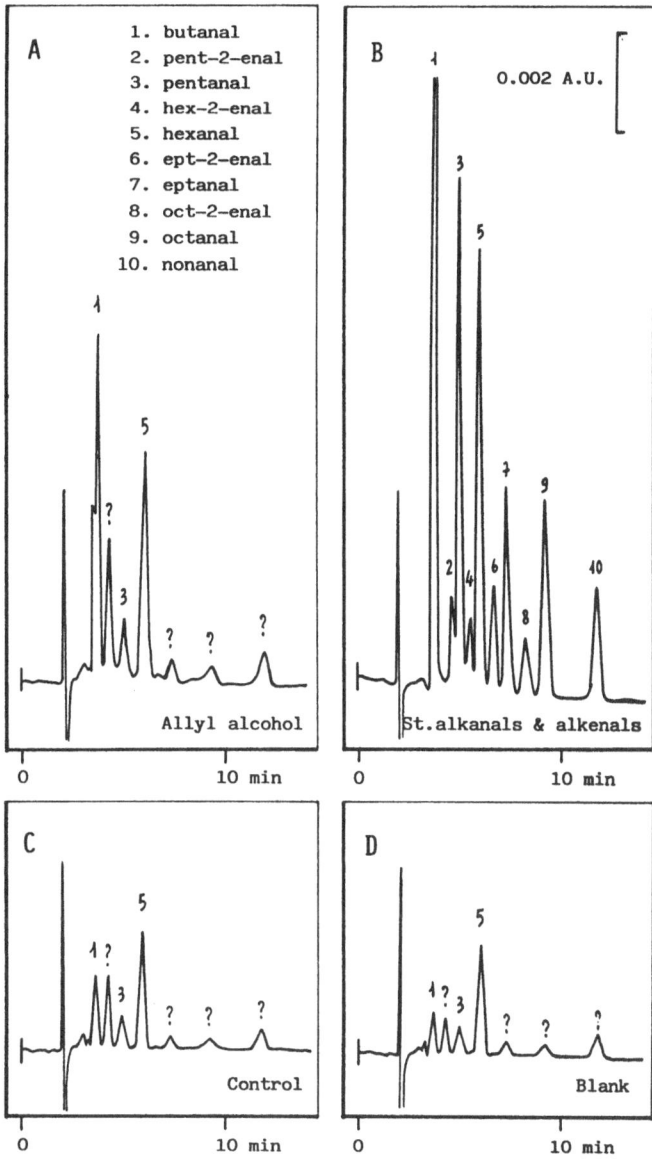

Fig. 4A–D. High-pressure liquid chromatography analysis of the thin layer chromatography (TLC) fraction containing the 2,4-dinitrophenylhydrazone derivatives of alkanals and alk-2-enals, obtained from the livers of allyl alcohol-poisoned mice. **A** Allyl alcohol-treated. **B** 2,4-Dinitrophenylhydrazone derivatives of standard alkanals and alk-2-enals. **C** Control mice. **D** Blank reagents. Peaks are numbered as reported in **A**. The fraction was prepared by TLC separation of the 2,4-dinitrophenylhydrazones obtained by reacting liver homogenates from control and allyl alcohol-poisoned mice with 2,4-dinitrophenylhydrazine. Separation conditions: reversed-phase column, LiChrospher 100 CH 18 Merck (4 mm × 25 cm); column temperature, 50°C; mobile phase, acetonitrile/water (8:2, v/v); flow rate, 1 ml/min; detector wavelength, 350 nm. A sample derived from the liver corresponding to 1 mg of protein was injected

Acknowledgements. This work was supported by a grant from the Association for International Cancer Research (Great Britain). Additional funds were derived by a grant (no. 86.00370.44) from CNR (Italy) Special Project "Oncology", and from CNR Group of Gastroenterology.

References

Beauchamp RO Jr, Andjelkovich DA, Kligerman AD, Morgan KT, d'A. Heck H (1985) A critical review of the literature on acrolein toxicity. CRC Crit Rev Toxicol 14:309–380

Benedetti A, Comporti M, Esterbauer H (1980) Indentification of 4-hydroxynonenal as a cytotoxic product originating from the peroxidation of liver microsomal lipids. Biochim Biophys Acta 620:281–296

Benedetti A, Casini AF, Ferrali M, Fulceri R, Comporti M (1981) Cytotoxic effects of carbonyl compounds (4-hydroxyalkenals) originating from the peroxidation of microsomal lipids. In: Slater TF, Garner A (eds) Recent advances in lipid peroxidation and tissue injury. Brunel, Uxbridge, pp 56–85

Benedetti A, Comporti M, Fulceri R, Esterbauer H (1984a) Cytotoxic aldehydes originating from the peroxidation of liver microsomal lipids. Identification of 4,5-dihydroxydecenal. Biochim Biophys Acta 792:172–181

Benedetti A, Fulceri R, Comporti M (1984b) Inhibition of calcium sequestration activity of liver microsomes by 4-hydroxynonenal originating from the peroxidation of liver microsomal lipids. Biochim Biophys Acta 793:489–493

Benedetti A, Pompella A, Fulceri R, Romani A, Comporti M (1986) Detection of 4-hydroxynonenal and other lipid peroxidation products in the liver of bromobenzene-poisoned mice. Biochim Biophys Acta 876:658–666

Bielawski J, Lehninger AL (1966) Stoichiometric relationships in mitochondrial accumulation of calcium and phosphate supported by hydrolysis of adenosine triphosphate. J Biol Chem 241:4316–4322

Casini AF, Pompella A, Comporti M (1985) Liver glutathione depletion induced by bromobenzene, iodobenzene, and diethylmaleate poisoning and its relation to lipid peroxidation and necrosis. Am J Pathol 118:225–237

Casini AF, Maellaro E, Pompella A, Ferrali M, Comporti M (1988) Lipid peroxidation, protein thiols and calcium homeostasis in bromobenzene-induced liver damage. Biochem Pharmacol, (in press)

Dawson JR, Norbeck K, Anundi I, Moldéus P (1984) The effectiveness of N-acetylcysteine in isolated hepatocytes, against the toxicity of paracetamol, acrolein, and paraquat. Arch Toxicol 55:11–15

Dianzani MU (1982) Biochemical effects of saturated and unsaturated aldehydes. In: McBrien DCH, Slater TF (eds) Free radicals, lipid peroxidation and cancer. Academic, London, pp 129–158

Esterbauer H (1982) Aldehydic products of lipid peroxidation. In: McBrien DCH, Slater TF (eds) Free radicals, lipid peroxidation and cancer. Academic, London, pp 101–128

Esterbauer H (1985) Lipid peroxidation products: formation, chemical properties and biological activities. In: Poli G, Cheeseman KH, Dianzani MU, Slater TF (eds) Free radicals in liver injury. IRL, Oxford, pp 29–47

Esterbauer H, Zollner H, Scholz N (1975) Reaction of glutathione with conjugated carbonyls. Z Naturforsch [c] 30:466–473

Esterbauer H, Cheeseman KH, Dianzani MU, Poli G, Slater TF (1982) Separation and characterization of the aldehydic products of lipid peroxidation stimulated by ADP-Fe^{2+} in rat liver microsomes. Biochem J 208:129–140

Hanson SK, Anders MW (1978) Effect of diethyl maleate treatment, fasting, and time of administration on allyl alcohol hepatotoxicity. Toxicol Lett 1:301–305

Jaeschke H, Kleinwaechter C, Wendel A (1987) The role of acrolein in allyl alcohol-induced lipid peroxidation and liver cell damage in mice. Biochem Pharmacol 36:51–58

Jocelyn PC, Kamminga A (1974) The non-protein thiol of rat liver mitochondria. Biochim Biophys Acta 343:356-362

Lowry OH, Rosebrough NJ, Farr AL, Randall RJ (1951) Protein measurement with the Folin phenol reagent. J Biol Chem 193:265-275

Meredith MJ, Reed DJ (1982) Status of the mitochondrial pool of glutathione in the isolated hepatocyte. J Biol Chem 257:3747-3753

Meredith NJ, Reed DJ (1983) Depletion in vitro of mitochondrial glutathione in rat hepatocytes and enhancement of lipid peroxidation by adriamycin and 1.3-bis(2-chloroethyl)-1-nitrosourea (BCNU). Biochem Pharmacol 32:1383-1388

Miessner H (1891) Berlin Klin Wochenschr 28:819-822

Ohno Y, Ormstad K, Ross D, Orrenius S (1985) Mechanism of allyl alcohol toxicity and protective effects of low-molecular-weight thiols studied with isolated rat hepatocytes. Toxicol Appl Pharmacol 78:169-179

Piazza JG (1915) Zur Kenntnis der Wirkung der Allylverbindungen. Z Exp Pathol Ther 17:318-341

Pompella A, Maellaro E, Casini AF, Ferrali M, Ciccoli L, Comporti M (1987) Measurement of lipid peroxidation in vivo: a comparison of different procedures. Lipids 22:206-211

Rees KR, Tarlow MJ (1967) The hepatotoxic action of allyl formate. Biochem J 104:757-761

Reid WD (1972) Mechanism of allyl alcohol-induced hepatic necrosis. Experientia 28:1058-1061

Schauenstein E, Esterbauer H, Zollner H (eds) (1977) Aldehydes in biological systems. Pion, London, pp 25-102

Sedlak J, Lindsay RH (1968) Estimation of total, protein-bound, and nonprotein sulfhydryl groups in tissue with Ellman's reagent. Anal Biochem 25:192-205

Serafini-Cessi F (1972) Conversion of allyl alcohol into acrolein by rat liver. Biochem J 128:1103-1107

Vignais PM, Vignais PV (1973) Fuscin, an inhibitor of mitochondrial SH-dependent transport-linked functions. Biochim Biophys Acta 325:357-374

Wahlländer A, Soboll S, Sies H (1979) Hepatic mitochondrial and cytosolic glutathione content and the subcellular distribution of GSH-S-transferases. FEBS Lett 97:138-140

Witz G, Lawrie NJ, Zaccaria A, Ferran HE Jr, Goldstein BD (1986) The reaction of 2-thiobarbituric acid with biologically active alpha, beta-unsaturated aldehydes. J Free Radic Biol Med 2:33-39

Zitting A, Heinonen T (1980) Decrease of reduced glutathione in isolated rat hepatocytes caused by acrolein, acrylonitrile, and the thermal degradation products of styrene copolymers. Toxicology 17:333-341

Liver Cell Heterogeneity in Biotransformation of Carcinogenic Polycyclic Hydrocarbons

A. Schlenker, S. A. E. Finch, W. Kühnle, J. Sidhu, A. Stier

Max Planck-Institut for Biophysical Chemistry, Am Faßberg, 3400 Göttingen, FRG

1 Introdruction

Regioselectivity and local appearance of toxic effects in liver tissue are due to intralobular microheterogeneity of distribution of enzymes, biotransforming drugs and carcinogens (Gooding et al. 1978; Wolf et al. 1984; Ullrich et al. 1984).

The mode of metabolic channelling in the sequence of reactions in which reactive intermediates transiently appear is decisive for the biological outcome and depends mainly on a coupling between phase I enzymes (the cytochrome P-450 system of oxygenases) and the phase II enzymes (epoxide hydrolase, UDP-glucuronyltransferases, sulfotransferase). In contrast to most experimental models used to investigate liver cell heterogeneity (Jungermann and Katz 1982) the model of isolated liver cells perifused with substrates to yield fluorescent products permitting measurement of these products by microfluorescence spectroscopy (Stier et al. 1980), 1988) does feature this coupling.

Fluroranthene was shown to be a highly suitable substrate for non-invasive investigation of metabolite pathways of polycyclic hydrocarbons in individual liver cells by fluorescence microscopic techniques. It meets a number of criteria which are important to the practicability of such measurements: (a) the pattern of metabolites is straightforward, (b) their fluorescence properties and (c) the special kinetics of perifusion of cells together with the use of metabolic inhibitors permit distinction of different pathways, particularly the activities of phase I (cytochrome P-450 dependent) and phase II (epoxide hydration, conjugation) reactions of biotransformation and their coupling. Metabolic functions can be measured comparatively in individual cells and metabolic heterogeneity of liver cell populations can be investigated.

Due to the structural symmetry of fluoranthene two main pathways of biotransformation exist (Fig. 1). They differ in the molecular site and perhaps the mechanism of oxygenation, the form of cytochrome P-450 involved and, therefore, in their inducibility, and in the prevailing phase II-reactions. From microsomal incubations (Stier et al. 1980; Rastetter et al. 1982; Babson et al. 1986) three main metabolites, 3-hydroxyfluoranthene, 8-hydroxyfluoranthene and 2,3-dihydro-2,3-dihydroxyfluoranthene arising from both pathways of fluoranthene metabolism are known. the fluorescence characteristics of the two phenols facilitate monitoring of their formation without interference from the fluorescence of

Nigam et al. (Eds.), Eicosanoids,
Lipid Peroxidation and Cancer
© Springer-Verlag Berlin Heidelberg 1988

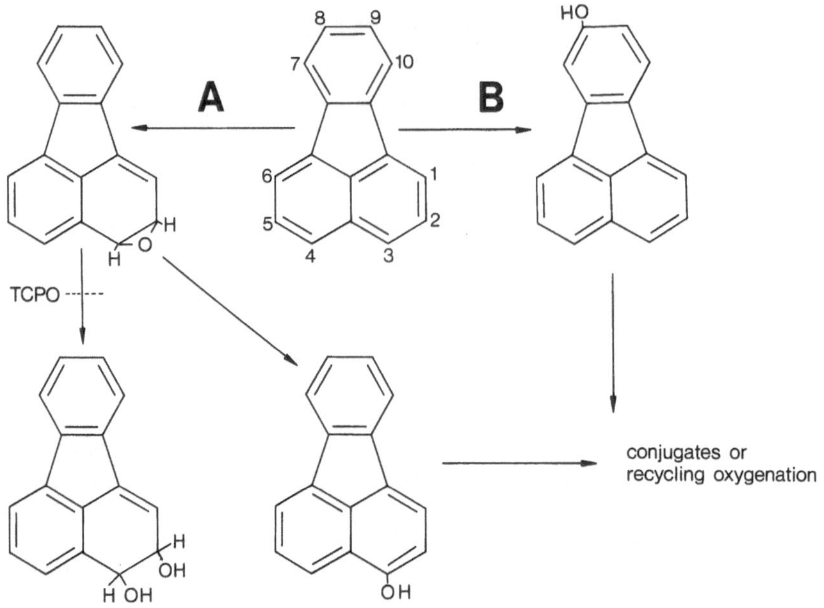

Fig. 1. Oxidative pathways of fluoranthene metabolism in rat liver. (From Babson et al. 1986)

fluoranthene, 2,3-dihydro-2,3-dihydroxyfluoranthene or products of conjugation (glucuronidation and sulfation). It is difficult to differentiate spectroscopically between the two phenols, but under certain experimental conditions either 3- or 8-hydroxyfluoranthene accumulate predominantly. Formation of 3-hydroxyfluoranthene in uninduced liver cells is favoured by inhibition of the epoxide hydrolase by 1,1,1-trichloro-2-propeneoxide, while considerably more 8-hydroxyfluoranthene is formed relative to 3-hydroxyfluoranthene in β-naphthoflavone-induced liver cells, the case to be considered here. It is not yet settled whether this phenol is formed by direct oxygenation of the 8-position of fluoranthene or via the 7,8-epoxyfluoranthene (this path is not shown in Fig. 1) which is either a poor substrate of epoxide hydrolase or may rearrange rapidly to the 8-hydroxyfluoranthene. On inhibition of the epoxide hydrolase in these cells, nearly equal amounts of the two phenols are formed.

The constant perifusion with substrates simplifies the kinetics of the enzyme reactions due to the formation of steady states and diminution of product inhibition of the enzymes, and is more relevant to the in vivo situation than incubations of liver cell suspensions. Thus we can discriminate the cytochrome P-450 activities forming the two phenolic compounds. Furthermore, it is possible to determine the degree of coupling of cytochrome P-450 to phase II enzymes, e. g. to conjugating enzymes in β-naphthoflavone induced cells in the case of the 8-hydroxyfluoranthene path and to epoxide hydratase in the case of the 2,3-epoxide path. Here we show, that the two pathways are separated in two populations of liver cells of about equal size. In addition, we found a small number of cells with a metabolic behaviour quite different from the distribution profiles of these two populations.

2 Materials and Methods

2.1 Chemicals

Fluoranthene from the "Gesellschaft für Teerverarbeitung", Duisburg-Meiderich, FRG, checked for purity by mass spectrometry and TLC, or Fluoranthen reinst from Serva, Heidelberg, FRG, was used. β-Naphthoflavone and 3,3,3-trichloropropene-2,3-oxide (trichloropropene oxide) were obtained from EGA-Chemie, Steinheim, FRG. All other chemicals, biochemicals and solvents for HPLC, were purchased from Boehringer, Mannheim; Merck, Darmstadt; Serva, Heidelberg; or Sigma, München (all FRG) and were of the highest quality available.

2.2 Perifusion of Cells on the Microscopic Stage

Liver cells from male Sprague Dawley rats (200–300 g body weight) were isolated by collagenase (Type IV, Sigma) perfusion (Seglen 1973). In some cases the animals were pretreated with intraperitoneal injections of 80 mg β-naphthoflavone dissolved in corn oil/kg body weight 36 h before cell preparation.

Liver cells were sedimented onto polylysine-coated cover glasses and perifused in a Dvorak-Stotler chamber purchased from Zeiss, Göttingen, FRG. Perifusion was started 45 min after the beginning of their preparation. The buffer used for perifusion was: HEPES 50 mM, pH 7.4 containing NaCl 60 mM, KCl 40 mM. $CaCl_2$ 5 mM, $MgSO_4$ 2 mM, Na_2HPO_4 1 mM, glucose 5 mM, methionine 1 mM. The solution contains 10^{-7} M bovine serum albumin (Sigma, crystalline, lyophilized), fluoranthene (final concentration 10^{-6} M) was added from a stock solution which was 100 times this concentration in fluoranthene and albumin. Cells were perifused at room temperature with a rate of 0.2 ml/min.

2.3 Microfluorescence Spectroscopic Detection of Metabolites

The apparatus used is shown schematically in Fig. 2. Cells were perifused in a Dvorak-Stotler chamber mounted on the stage of a Zeiss Universal microscope. Fluorescence was excited by dark field illumination with the 407.7 nm line of a Hg high-pressure lamp (HBP 100 W/2 Osram, Hamburg, FRG), selected with a 397 nm interference filter, bandwidth 20 nm (Anders, Nabburg, FRG), KG 01,

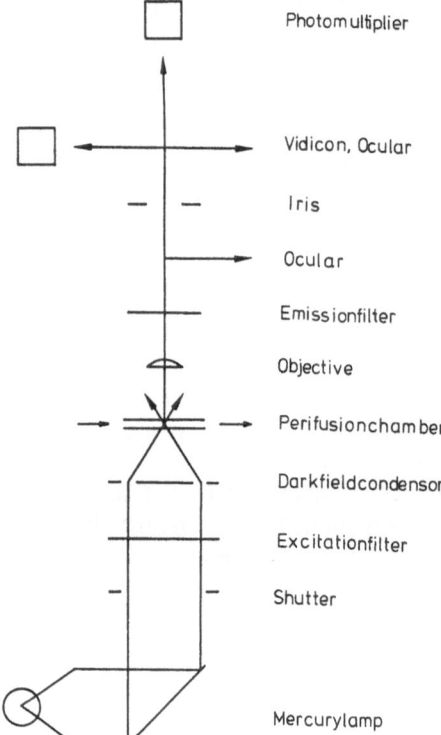

Photomultiplier

Vidicon, Ocular

Iris

Ocular

Emissionfilter

Objective

Perifusionchamber

Darkfieldcondensor

Excitationfilter

Shutter

Mercurylamp

Fig. 2. Fluorescence microscopic instrumentation

BG 38 and KV 370 glasses (Schott, Mainz, FRG). Light emitted beyond the cutoff by KV 450 and KV 470 filters (Schott, Mainz) was collected intermittently, usually for 1 sec every 30 sec, to keep photochemical damage as low as possible. These optical settings allowed us to discriminate between the fluorescence from the two phenolic metabolites and that of fluoranthene and 2,3-dihydro-2,3-dihydroxyfluoranthene (Stier et al. 1988). They favour detection of 3-hydroxyfluoranthene by a factor of 1.2 compared to 8-hydroxyfluoranthene.

Emitted light was detected either with a cooled ($-15\,^\circ$C) type 9862 B photon counting photomultiplier (EMI, Hayes, UK) and detection electronics from Ortec, München, FRG, or a SIT vidicon camera system (OSA 500, B & M Spektronik GmbH, Puchheim, FRG). Digitized signals from photon counting were stored in a multichannel analyzer (Canberra Model 8100, Meriden, USA). The information from the target of the SIT vidicon camera (cooled to $-20\,^\circ$C) was transferred to a two dimensional multichannel analyzer. By comparison with a microphotographic picture the light emitted from individual cells was integrated and corrected for the size of the rounded cells, Only cells which appeared morphologically normal have been used for evaluation.

3 Results and Discussion

3.1 Kinetics in Perifused Cells

The phenols are produced and eliminated with straightforward kinetics comprising 3 typical periods (Stier et al. 1988): (a) linear accumulation, (b) steady state with a stationary phenol concentration and (c) linear elimination on discontinuation of substrate supply. Formal assumption of two sequential zero-order reactions is consistent with this observation. This implies that the cytochomes P-450 producing the phenols, and the UDP-glucuronyltransferases and/or sulfotransferases conjugating them, are substrate saturated. However, in the case of the oxygenation we cannot exclude a stationary, non-saturating concentration of fluoranthene supplied continuously by perifusion of the cells as the cause of the apparent zero-order reactions as we know neither the concentration of fluoranthene in the perifused cells nor the exact Michaelis Menten constants of cytochromes P-450 for this substrate. The transition to the steady state may be due to a partial product inhibition of cytochromes P-450 and/or an activation of conjugating enzymes. At the plateau the processes of formation and elimination are at equilibrium, maintaining a stationary concentration of the phenols. the conjugating enzymes appear to be saturated as evidenced by a linear elimination characteristic on discontinuation of substrate supply which is followed at lower, apparently non-saturating, cellular phenol levels, by an exponential elimination phase. This straightforward kinetic description of the time-dependent phenol concentration as the result of two sequential processes of formation and elimination is supported by the observation that the rate of phenol accumulation and the stationary concentration (plateau) are correlated in individual cells (see Fig. 3): the greater the rate, the higher the plateau. The plateau in β-naphthoflavone-induced cells was on average 1.7 times higher in the presence of trichloropropene oxide

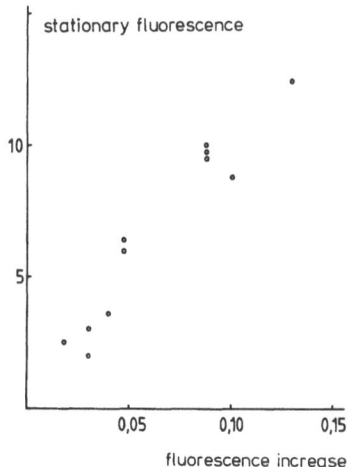

Fig. 3. Correlation of the rate of the initial rise (arbitrary units) and the height of the stationary phase (arbitrary units) of fluorescence sampled with a SIT vidicon camera from metabolites produced in individual rat liver cells during perifusion with a fluoranthene (10^{-6} M) containing buffer

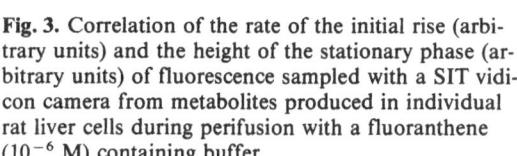

and 2.1 times higher in the presence of trichloropropene oxide plux salicylamide than in unhibited controls (Stier et al. 1988). The height of the plateau was also strongly influenced by induction, being ten times higher in β-naphthoflavone-induced cells than in uninduced cells under comparable conditions.

3.2 Intercellular Heterogeneity of Fluoranthene Biotransformation

The correlation between the two kinetic phases, rate of accumulation of the phenols and maintenance of their stationary concentrations supports the view that both parameters may be regarded as a measure of the coupling of phase II to phase I reactions.

The plateau was, therefore, used as the experimental parameter in an experiment designed to compare the coupling of the two pathways of fluoranthene oxygenation to conjugation in a larger number of cells from β-naphthoflavone-induced rat liver. From experiments with cell suspensions (Stier et al. 1988) we knew that in those cells in which the epoxide hydrolase was inhibited by trichloropropene oxide similar amounts of the two phenols acumulated whereas in non-inhibited cells the product was mainly 8-hydroxyfluoranthene. the distribution of the hieght of the plateau in 52 individual cells, perifused with fluoranthene and observed in different microscopic fields of view with a SIT vidicon camera (see Fig. 4), is broad and somewhat asymmetric, presumably due to the presence of a small subpopulation of lower accumulating capacity. The distribution profile contrasts with the high capacity of a few cells far exceeding the

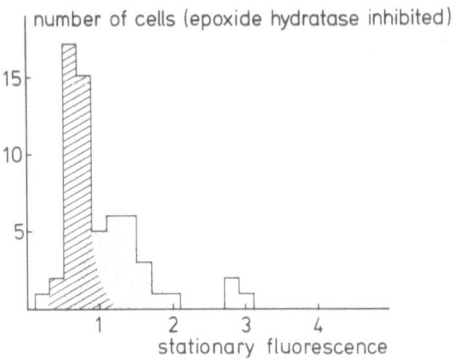

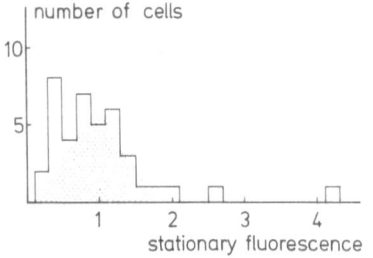

Fig. 4. Distribution of the stationary fluorescence intensity in 60 trichloropropene oxide inhibited and 52 uninhibited β-naphthoflavone induced rat liver cells. The average fluorescence intensity was set to 1 in each distribution

mean; one cell accumulated 4.5 times the average and 45 times the lowest values observed. When, by addition of trichloropropene oxide, 3-hydroxyfluoranthene was accumulated together with 8-hydroxyfluoranthene, a different profile was observed. Apparently a separate population appears with a narrower distribution superimposed on the broader distribution of the remainder.

Fulfilment of the criteria mentioned above constitutes the basis of interpretation of our results relating to the distribution of stationary concentration levels in populations of β-naphthoflavone-induced cells with and without inhibition of the epoxide hydrolase. We assume that two sub-populations α and β of about equal size exist. Subpopulation β accumulates mainly 8-hydroxyfluoranthene via path B, subpopulation α mainly the metabolites of path A (see Fig. 1) namely 2,3-dihydro-2,3-dihydroxyfluoranthene, which does not show up under the spectroscopic conditions used, and a small amount of 3-hydroxyfluoranthene. Presumably different cytochromes P-450 are involved in the two pathways. On inhibition of epoxide hydrolase all metabolites of path A become visible in subpopulation α which has a distinctly narrower distribution profile and lower average value of the stationary phenol concentration than subpopulation β. These results may be explained by a disproportionate expression of the enzymes concerned in different liver cells. β-Naphthoflavone is known to induce cytochrome P-450 predominantly in periportal cells (Foster et al. 1986; Wolf et al. 1984). However, other forms of cytochromes P-450 induced by β-naphthoflavone (Strobel and Lau 1982) may be distributed differently within the liver lobule. A similar intralobular microheterogeneity applies to the phase II enzymes: some forms of UDP-glucuronyltransferase are concentrated periportally, others in the centrilobular region; sulfotransferase is localized predominantly periportally (Ullrich et al. 1984; Conway et al. 1984; El Mouelhi and Kauffman 1986). Therefore, it is not possible to decide from the present data which of the two pathways of fluoranthene metabolism should be ascribed to periportal or centrilobular cells.

Our results demonstrate that the experimental model of perifused liver cells and the use of fluoranthene are useful tools in further definition of phase I to phase II coupling heterogeneity in liver cells. This is of toxicological interest since tight coupling of cytochromes to conjugating enzymes precludes recycling oxygenation with subsequent formation of toxic semiquinone radicals. In this respect the finding that some cells accumulate phenols to a much greater extent than the bulk of the population is remarkable. Perhaps these cells do not express epoxide hydrolase and/or conjugating enzymes and should be considered outstandingly sensitive to tumour induction by some chemical carcinogens as well as to toxic damage. Another aspect is that the degree of coupling with epoxide hydrolase determines the amount of epoxides, potential reactants with proteins and DNA, and therefore ultimate carcinogens; though it should not be overlooked that in the case of fluoranthene as with other polycyclic hydrocarbons, hydratation of epoxides is a precondition for the formation of highly mutagenic vicinal dihydrodiol epoxides (LaVoie et al. 1982; Babson et al. 1986).

References

Babson JR, Russo-Rodriguez SE, Wattley RV, Bergstein PL, Rastetter WH, Liber HL, Andon BM, et al. (1986) Microsomal activation of fluoranthene to mutagenic metabolites. Toxicol Appl Pharmacol 85:355–366

Conway JG, Kauffman FC, Tsukada T, Thurman RG (1984) Glucuronidation of 7-hydroxycoumarin in periportal and pericentral regions of the liver lobule. Mol Pharmacol 25:487–493

El Mouelhi M, Kauffman FC (1986) Sublobular distribution of transferases and hydrolases associated with glucuronide, sulfate and glutathione conjugation in human liver. Hepatology 6:450–456

Foster JR, Elcombe CR, Boobis AR, Davies DS, Sesardic D, McQuade J, Robson RT, et al. (1986) Immunocytochemical localization of cytochrome P-450 in hepatic and extra-hepatic tissues of the rat with a monoclonal antibody against cytochrome P-450. Biochem Pharmacol 35:4543–4554

Gooding PE, Chayen J, Sawyer B, Slater TF (1978) Cytochrome P-450 distribution in rat liver and the effect of sodium phenobarbitone administration. Chem Biol Interact 20:299–310

Jungermann K, Katz N (1982) Metabolic heterogeneity of liver parenchyma. In: Sies H (ed) Metabolic compartmentation. Academic, New York, pp 411–435

LaVoie EJ, Hecht SS, Bedenko V, Hoffmann D (1982) Identification of the mutagenic metabolites of fluoranthene, 2-methylfluoranthene and 3-methylfluoranthene. Carcinogenesis 3:841–846

Rastetter WH, Nchbar RB, Russo-Rodriguez S, Wattley RV (1982) Fluoranthene: synthesis and mutagenicity of four diol epoxides. J Org Chem 47:4873–4878

Seglen PO (1973) Preparation of rat liver cells. Exp Cell Res 76:25–30

Stier A, Nolte KH, Schlenker A, Schumann W, Zuretti FM (1980) Toxicological studies of liver cells by microspectrofluorometry. Arch Toxicol 44:45–54

Stier A, Schlenker A, Sidhu J, Kühnle W, Finch SAE (1988) Heterogeneity of biotransformation of fluoranthene in perifused liver cells. J Cancer Res Clin Oncol, (in press)

Strobel HW, Lau PP (1982) Five forms of cytochromes P-450 from β-naphthoflavone pretreated rats. In: Sato R, Kato R (eds) Microsomes, drug oxidations, and drug toxicity. Wiley, New York, pp 91–92

Ullrich D, Fischer G, Katz N, Bock KW (1984) Intralobular distibution of UDP-glucuronosyltransferase in livers from untreated, 3-methylcholanthrene- and phenobarbital-treated rats. Chem Biol Interact 48:181–190

Wolf CR, Buchmann A, Friedberg T, Moll E, Kuhlmann WD, Kunz HW, Oesch F (1984) Dynamics of the localization of drug metabolizing enzymes in tissues and cells. Biochem Soc Trans 12:60–62

Free Radical Intermediates of Carcinogenic Hydrazines

E. Albano[1], A. Tomasi[2], A. Iannone[2], L. Goria-Gatti[1], V. Vannini[2] and M. U. Dianzani[1]

Substituted hydrazine derivatives include a large number of compounds characterized by high chemical reactivity and a wide spectrum of biological activities, which make them a potential hazard from both a medical and ecological point of view.

Among these compounds methyl-substituted hydrazines have received attention because of their presence in many natural products, including tobacco (Toth 1975), and their extensive use as high energy fuels in the aerospace industry (Back and Thomas 1970).

Exposure to methylhydrazines causes methaemoglobinemia, disturbances to the central nervous system and affects kidney and liver functions (Back and Thomas 1970; Zimmerman 1978). More important, all these conpounds are well-known carcinogens (International Agency for Research on Cancer 1974; Toth 1975, 1977).

In fact, 1,2 dimethylhydrazine induces an high incidence of tumours in the colon and rectum of rodents (Toth 1977) and, when administered in low doses it also causes kidney carcinomas, vascular tumours and histiocytomas in many organs (IARC 1974). Recently 1,2 dimethylhydrazine has been demonstrated to act as a carcinogen in the liver when associated with a promoting treatment with orotic acid (Laurier et al. 1984). Methyl- and 1,1 dimethyl-hydrazines are similarly carcinogenic in the colon and rectum, but, in addition, they produce lung tumours in mice (IARC 1974). In contrast another methyl-substituted hydrazine, N-isopropyl-(2 methyl-hydrazino)-p-tolbutamide (procarbazine), has found applications in the treatment of lymphomas, lymphogranulomas and brain tumours (Goodman and Gilman 1985).

Although many hydrazines can be oxidized by atmospheric oxygen, a number of investigations have produced evidence that both mono- and dialkyl-substituted hydrazines require metabolic activation to produce reactive intermediates. The following enzymatic systems have been implicated in the process (Moloney and Prough 1983; Prough and Moloney 1985): (a) microsomal cytochrome P_{450}-dependent monoxygenase system, (b) microsomal FAD-containing monoxygen-

[1] Dipartimento di Medicina ed Oncologia Sperimentale, Università di Torino, Corso Raffaello 30, 10125 Torino, Italy
[2] Istituto di Patologia Generale, Università di Modena, Via G. Campi 287, 41100 Modena, Italy

Nigam et al. (Eds.), Eicosanoids,
Lipid Peroxidation and Cancer
© Springer-Verlag Berlin Heidelberg 1988

ase system, (c) mitochondrial monoamine oxidase systen, and (d) prostaglandin synthetase system.

The possibility that free radicals might be generated during the oxidation of monoalkyl-hydrazines was first suggested by Prough et al. (1979) to explain the formation of hydrocarbons during the metabolism of these compounds by liver microsomes. This hypothesis was supported by the fact that in chemical systems free radicals can be generated by either one or two electron oxidation of hydrazines (Nelson 1978). In the case of mono- and di-substituted compounds the primary radical cations are very unstable and decompose to produce alkyl radicals from the substituent groups and nitrogen.

By applying ESR spectroscopy free radical species have been observed during the enzymatic oxidation of tetramethylhydrazine by horse-radish peroxidase and prostaglandin synthetase (Kalyanaraman et al. 1983). These enzymes are also able to activate other hydrazine derivatives, including isonicotinic acid 2-isopropylhydrazide (iproniazid), 1-hydrazinophtalazine (hydralazine), procarbazine and methylhydrazine (Sinha 1983, 1984; Augusto et al. 1985; Kalyanaraman and Sinha 1985). In the above studies spin trapping procedures have been used to detect highly unstable free radical species. In recent years the spin trapping technique has also been successfully applied to the identification of free radicals formed during the metabolism of ethyl-, phenylethyl-, isopropyl- and acetyl-hydrazines in either liver microsomes or isolated hepatocytes (Augusto et al. 1981; Ortiz de Montellano et al. 1983; Albano et al. 1985; Albano and Tomasi 1987).

We have observed that either monomethyl-, 1,1 dimethyl- or 1,2 dimethyl-hydrazines give rise to nitroxide adducts when incubated with isolated hepatocytes or liver microsomes in the presence of the spin trap 4-pyridyl-1-oxide-t-butyl nitrone (4-POBN) (Fig. 1). The hyperfine splitting constant of the ESR spectra produced by the three compounds are remarkably similar (aN = 14.97–

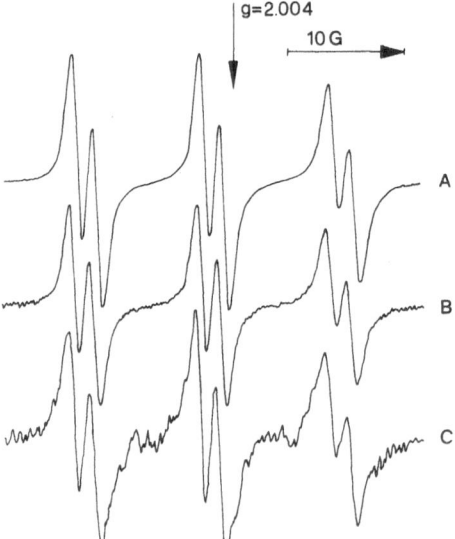

g=2.004

10 G

A

B

C

Fig. 1. ESR spectra of 4-POBN free radical adducts produced in isolated hepatocytes incubated with methylhydrazines. Isolated liver cells (7.5×10^6 cells/ml) were incubated for 30 min at 37°C with 25 nM 4-POBN and 2 nM, respectively, monomnethylhydrazine *(trace A)*, 1,1 dimethylhydrazine *(trace B)* and 1,2 dimethylhydrazine *(trace C)*. Spin adducts were processed as elsewhere reported (Albano et al. 1982)

14.94; aH = 2.33–2.34) and are identical to the spectral features of the 4-POBN adducts prepared by chemical oxidation of the same hydrazines. On this basis they are ascribed to the trapping of methyl free radicals in accordance with previous studies by Augusto et al. (1985).

The various hydrazines display different susceptibilities to free radical activation since the ESR signals produced by monomethylhydrazine are about ten- and fivefold more intense than those formed by equimolar concentrations of 1,1 dimethyl- and 1,2 dimethyl-hydrazines, respectively (Table 1).

In microsomal fractions free radical formation requires the presence of an NADPH-regenerating system and is inhibited by incubation under anaerobic conditions or by heat denaturation of microsomes (Table 2). However, even in

Table 1. Effect of pretreating rats with phenobarbital or cobalt chloride on the intensities of the ESR signals produced by various hydrazine compounds in liver microsomes

	ESR signal intensity (arbitrary units)		
	Controls	Phenobarbital	CoCl$_2$
Monoethyl-hydrazine 2 mM	105	315 (+303%)	55 (−48%)
1,1 dimethyl-hydrazine 2 mM	22	95 (+430%)	10 (−55%)
1,2 dimethyl-hydrazine 2 mM	9	41 (+450%)	2 (−78%)

Animal pretreatments were performed as described in the text. Liver microsomes were incubated 30 min at 37°C using 4-POBN as spin trap except for the acetyl-hydrazine experiments where PBN was employed. The results are expressed in arbitrary units and values in parentheses represent the percentage changes with respect to controls.

Table 2. Effects of treatments which affect the activity of microsomal monoxygenase systems on the free radical activation of monomethyl-hydrazine (MMH), 1,1 dimethyl-hydrazine (1,1 DMH) and 1,2 dimethyl-hydrazine (1,2 DMH) by rat liver microsomes

Treatments	Intensities of the ESR signals (% of the control values)		
	MMH	1,1 DMH	1,2 DMH
Anaerobic incubation	10	3	3
Incubation without NADP$^+$	14	7	6
Heat-inactivated microsomes	15	4	2
SKF 525A 1 mM	72	68	70
Metyrapone 0.5 mM	60	61	56
Carbon monoxide	74	40	54
1-Nitro-isothiocyanate 0.1 mM	27	32	35
Methimazole 1 mM	96	14	92

Phenobarbital-induced liver microsomes were incubated 30 min at 37°C with, respectively, 1 mM monomethyl-hydrazine or 2 mM of either 1,1 dimethyl- and 1,2 dimethyl-hydrazines and 25 mM 4-POBN. Anaerobic conditions were obtained by flushing the incubation flasks with nitrogen for 10 min. Heat inactivated microsomes were boiled for 5 min before being added to the incubation mixture. The values are expressed as percent of the respective controls incubated without inhibitors. The results were calculated as the means of three differents experiments.

these conditions monomethylhydrazine gives rise to a small but detectable ESR signal, probably as a result of spontaneous decomposition catalyzed by traces of transition metals present in the incubation buffer. The induction of the cytochrome P_{450} system by phenobarbital increases by two to four fold free radical production from all the hydrazines tested (Table 1). Conversely, cobalt chloride administration to the rats, which lowers by approx. 60% the cytochrome P_{450} content in the microsomes, causes a decrease in the ESR signal intensities ranging from 48% up to 78% (Table 1). Moreover, the addition of inhibitors of cytochrome P_{450} such as SKF 525A, metyrapone and NITC decreases by 30%, 40% and 70%, respectively the intensities of the ESR signals derived from all three methyl hydrazines (Table 2). Carbon monoxide displays a variable activity since it inhibits by about 56%–60% the activation of the two dimethyl-hydrazines but decreases by only 26% the free radical metabolism of monomethylhydrazine (Table 2). These results are consistent with the effects exerted by cobalt chloride pretreatment or by inhibitors of cytochrome P_{450} on the metabolic activation of phenylethyl-, acetyl- and isopropyl-hydrazines (Ortiz de Montellano et al. 1983; Albano and Tomasi 1987) and indicate that the microsomal monoxygenase system plays a mayor role in the free radical activation of methylhydrazines.

One electron oxidation, as catalyzed by horse radish peroxidase and prostaglandin synthetase, has been shown to produce free radicals from several alkylhydrazines (Sinha 1983, 1984; Augusto et al. 1985). However, the reaction mediated by cytochrome P_{450} appears to involve a two electron transfer process, leading to a postulated diazene intermediate (Prough and Moloney 1985). It has been shown, in fact, that the destruction of cytochrome P_{450} occuring concomitantly with the metabolism of subsituted-hydrazine can be the consequence of the formation of ferrous cytochrome-diazene complexes (Battioni et al. 1983; Moloney et al. 1984).

Prough et al. (1981) have reported that 1,1 dimethylhydrazine is a better substrate for FAD-containing monoxygenase rather than for the cytochrome P_{450} system. Furthermore, 1,1 dialkyldiazenium ions the principal oxidation products of 1,1 disubstituted hydrazines are known to form complexes with cytochrome P_{450} which may terminate its function (Hines and Prough 1980). In microsomes incubated with 1,1 dimethyl-hydrazine methimazole, a competitive inhibitor of FAD-containing monoxygenase enzymes, reduces free radical formation by 86% whereas it has no effect on the activation of monomethyl- and 1,2 dimethyl-derivatives (Table 2). FAD-catalyzed oxidation of 1,1 dimethyl-hydrazine produces stoichiometric amounts of formaldehyde and monomethyl-hydrazine (Prough et al. 1981). It is, therefore, possible that this latter compound might be further metabolized to methyl free radical by cytochrome P_{450} and thus explain the effect of both FAD-dependent and cytochrome P_{450}-dependent monooxygenase inhibitors on the radical activation of 1,1 dimethylhydrazine (Fig. 2).

The formation of alkylating metabolites from 1,2 dimethyl-hydrazine is postulated to proceed through a rather complicated pathway involving azomethane, azoxymethane and methylazoxymethanol as intermediates (Fiala 1977). However, in our hands the incubation of azoxymethane with liver microsomes in the presence of NADPH and 4-POBN does not result in any detectable ESR spectrum, suggesting that the azoxy-derivative is not a precursor of the free radical

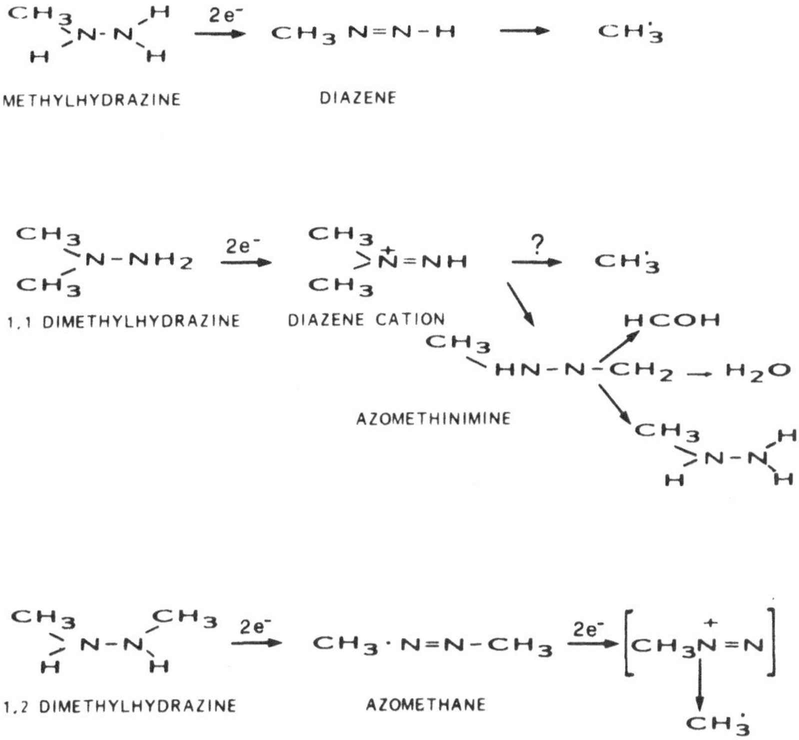

Fig. 2. Scheme of the metabolic pathways which are postulated to be responsible for the free radical activation of methylhydrazines

intermediate. Studies using procarbazine have recently consistently shown that azoprocarbazine, but not azoxyprocarbazine, is metabolized by liver microsomes to a reactive species, possibly the methyl free radical (Prough et al. 1985). Thus it is probable that the metabolism of 1,2 dimethylhydrazine by cytochrome P_{450} or monoamine oxidase would result in the formation of azomethane (Prough and Moloney 1985) which, in its turn, can be further oxidized to methyl free radicals (Fig. 2).

So far the relevance of methyl free radical in relation to the toxic and carcinogenic effects of methyl-substituted hydrazines has to be established. Prough et al. (1985) have shown that the the addition of glutathione to liver microsomes incubated with procarbazine increases the production of methane at the expense of covalent binding of radiolabelled methyl groups to proteins. In our hands glutathione similarly decreases the spin trapping of methyl free radicals, suggesting the possibility that this species could act as an alkylating agent toward macromolecules. Moreover, when incubated with isolated hepatocytes monomethylhydrazine causes a time-dependent depletion of intracellular glutathione and stimulates lipid peroxidation (Fig. 3). These effects are associated with the release of lactate dehydrogenase (LDH) from the cells which argues for a free radical-mediated injury of hepatocytes.

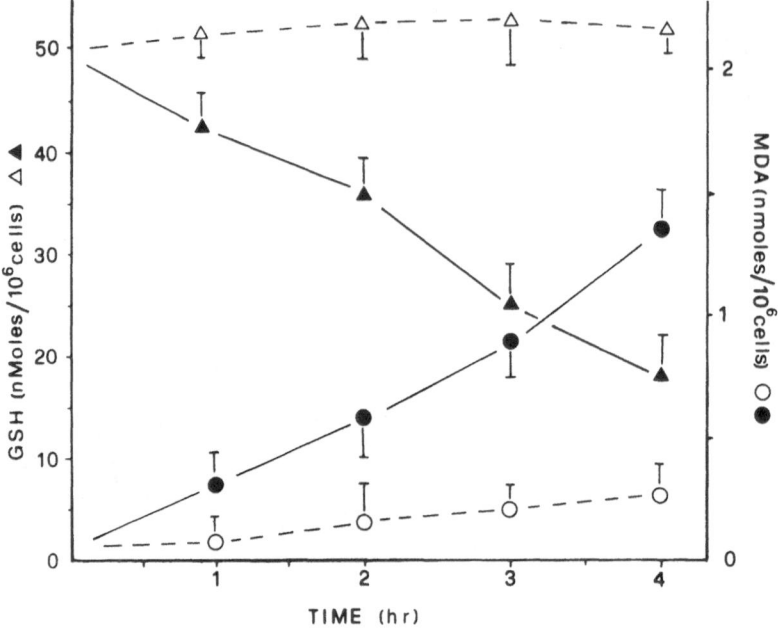

Fig. 3. Time-course of the glutathione depletion and the stimulation of lipid peroxidation induced by monomethylhydrazine in isolated hepatocytes. Liver cells (5×10^6 cell/ml) were incubated at 37 °C in the absence *(open circles)* or in the presence *(filled circles)* of 1 mM monomethylhydrazine. Lipid peroxidation, as measured by malondialdehyde (MDA) production and cellular glutathione were measured at each time-point as previously described (from Albano et al. 1984)

In conclusion, the results obtained indicate that free radical activation of mono and dimethyl hydrazines is taking place in liver cells. the metabolic pathways involved are similar to those observed with other alkyl hydrazines (Augusto et al. 1981; Ortiz de Montellano et al. 1983; Kalyanaraman and Sinha 1985; Albano and Tomasi 1987), although disubstituted derivatives seem to require a preliminary oxidation passage. The detection of methyl free radicals may be important for understanding the mechanism of DNA akylation by methylating hydrazines Bodell et al. (1982) in relation to both the carcinogenic and antitumour activities of these compounds.

Acknowledgements. This work has been supported by the Ministero della Pubblica Istruzione, Project "Patologia da Radicali Liberi" and by the Consiglio Nazionale della Ricerche, Project "Rischio Tossicologico, Gruppo Gastroenterologia" contract no. 86.00506.04. A. T. and A. I. are grateful to the Association for International Cancer Research, London for financial support.

References

Albano E, Tomasi A (1987) Spin trapping of free radical intermediates produced during the metabolism of isoniazid and iproniazid in isolated hepatocytes. Biochem Pharmacol 36:2913-2920

Albano E, Lott KAK, Slater TF, Stier A, Symons MCR, Tomasi A (1982) Spin-trapping studies on the free radical products formed by metabolic activation of carbon tetrachloride in rat liver microsomal fractions, isolated hepatocytes and in vivo in the rat. Biochem J 204:593-603

Albano E, Poli G, Chiarpotto E, Biasi F, Dianzani MU (1984) Paracetamol-stimulated lipid peroxidation in isolated rat and mouse hepatocytes. Chem Biol Interact 47:249-263

Albano E, Tomasi An, Vannini V, Dianzani MU (1985) Detection of free radical intermediates during isoniazid and iproniazid metabolism by isolated hepatocytes. Biochem Pharmacol 34:381-382

Augusto O, Ortiz de Montellano PR, Quintanilha A (1981) Spin-trapping of free radicals formed during microsomal metabolism of ethylhydrazine and acetylhydrazine. Biochem Biophys Res Commun 101:1324-1330

Augusto O, du Plessis LR, Weingrill CLV (1985) Spin-trapping of methyl radical in the oxidative metabolism of 1,2 dimethylhydrazine. Biochem Biophys Res Commun 126:853-858

Back CK, Thomas AA (1970) Aerospace pharmacology and toxicology. Annu Rev Pharmacol 10:395-412

Battioni P, Mahy IP, Delaforge M, Mansuy D (1983) Reaction of monosubstituted hydrazines and diazenes with rat liver cytochrome P-450. Eur J Biochem 134:241-248

Bodell MA, Lewis JG, Billings KC, Swendberg JA (1982) Cell specificity in hepatocarcinogenesis: preferential accumulation of O^6-methylguanine in target cell DNA during continuous exposure of rats to 1,2 dimethylhydrazine. Cancer Res 42:3079-3083

Fiala ES (1977) Investigation into the metabolism and mode of action of the colon carcinogens 1,2 dimethylhydrazine and azomethane. Cancer 40:2436-2445

Goodman A, Gilman LS (1985) The pharmacological basis of therapautics. McMillan, New York, pp 532-536

Hines RN, Prough RA (1980) The characterization of an inhibitory complex formed with cytochrome P-450 and a metabolite of 1,1 di- substituted hydrazines. J Pharmacol Exp Ther 214:80-86

International Agency for Research on Cancer (1974) Some aromatic amines, hydrazines and related substances, N-nitroso compounds and miscellaneous alkylating agents. IARC Monogr 4:137-170

Kalyanaraman B, Sinha BK (1985) Free radical mediated activation of hydrazine derivatives. Environ Health Perspect 64:179-184

Kalyanaraman B, Sivarjah K, Eling TE, Mason RP (1983) A free radical mediated cooxidation of tetramethylhydrazine by prostaglandin hydroperoxidase. Carcinogenesis 4:1341-1343

Laurier C, Tatematsu M, Rao PM, Rajalakshmi S, Sarma DSR (1984) Promotion by orotic acid of liver carcinogenesis in rats initiated by 1,2 dimethylhydrazine. Cancer Res 44:2186-2191

Moloney SJ, Prough RA (1983) Biochemical toxicology of hydrazines. Biochem Toxicol 5:313-348

Moloney SJ, Snider BJ, Prough RA (1984) The interaction of hydrazine derivatives with rat-hepatic cytochrome P_{450}. Xenobiotica 14:803-814

Nelson SF (1978) Early intermediates in hydrazine oxidation: hydrazine cation radicals hydrazyl and diazenium cations. ACS Symp Ser 69:309-320

Ortiz de Montellano PR, Augusto O, Viola F, Kunze KL (1983) Carbon radicals in the metabolism of alkylhydrazines. J Biol Chem 258:8623-8629

Prough RA, Moloney SJ (1985) Hydrazines. In: MW Anders (ed) Bioactivation of foreign compounds. Academic, New York, pp 433-449

Prough RA, Wittkop JA, Reed DJ (1979) Evidence for the hepatic metabolism of some monoalkylhydrazines. Arch Biochem Biophys 131:369-373

Prough RA, Freeman PC, Hines RN (1981) The oxidation of hydrazine derivatives catalyzed by the purified liver microsomal FAD-containing monoxygenase. J Biol Chem 256:4178-4184

Prough RA, Brown MI, Moloney SJ, Wiebkin P, Cummings SW, Searman ME, Guengerich FP (1985) The activation of hydrazines to reactive intermediates. In: Boobis AR, Caldwell J, de Matteis F, Elcombe CR (eds) Microsomes and drug oxidation. Taylor and Francis, London, pp 330-339

Sinha BK (1983) Enzymatic activation of hydrazine derivatives. A spin trapping study. J Biol Chem 258:796-801

Sinha BK (1984) Metabolic activation of procarbazine. Evidence for carbon-centered free radical intermediates. Biochem Pharmacol 33:2777-2781

Toth B (1975) Synthetic and naturally occurring hydrazines as possible cancer causative agents. Cancer Res 3693-3697

Toth B (1975) Synthetic and naturally occurring hydrazines as possible cancer causative ring in nature and in the environment. Cancer 40:2427-2431

Zimmerman HJ (1978) Hepatotoxicity. Appleton-Century-Crofts, New York, pp 307-209

Biochemical Aspects
of Gynaecological and Liver Cancer

Studies on Cancer of the Human Cervix in Relation to Thiol-Groups

C. Benedetto[1], S. Cianfano[1], M. Zonca[1], G. Nöhammer[2], E. Schauenstein[2], W. Rojanapo[3] and T. F. Slater[4]

Despite the encouraging decrease in the incidence of invasive cancer of the cervix and in the associated mortality rate observed over the last 20 years in western countries, cancer of the cervix is still a major and distressing disease especially because it frequently affects young women in the third and fourth decades of life (Ayiomamitis and Math 1987; Bain and Crocker 1983; Carmichael et al. 1984; Dunn and Schweitzer 1981). Table 1 gives a good example of the increase in the incidence of positive smears in women under 30 years either previously screened or not screened over 14 years (Wolfendale et al. 1983). It can be seen that, although the number of positive smears/1000 smears examined was always higher in women not previously screened than previously screened, the incidence of positive smears increased approximately three times from 1970 to 1979 in both groups.

An additional worrying aspect of the overall problem of cancer of the cervix is the possible effect sexually transmitted diseases such as Herpes simplex type 2 and Papilloma virus infections have on its incidence, especially in young persons (de Brux et al. 1985; McCance et al. 1985; Mazur and Cloud 1984; Prakash et al. 1985; Scholl et al. 1985).

Therefore, the requirement for a reliable, relatively inexpensive mass screening method that can recognise such cancer at an early stage is evident. The Papanicolaou smear test fulfils most requirements in these respects, but the pub-

Table 1. Positive smears in women aged under 30, either previously screened or not screened

Period	Previously screened	Not previously screened
1965–1969	0.5	3.1
1970–1974	1.4	3.3
1975–1979	3.7	9.2

Figures are numbers of positive smears/1000 smears examined.
(From Wolfendale et al. 1983)

[1] Istituto di Ginecologia e Ostetricia, Cattedra A, Universita di Torino, 10126 Torino, Italiy
[2] Institut für Biochemie, Karl-Franzens-Universität, 8010 Graz, Austria
[3] National Cancer Institute, Rama VI Road, Bangkok 10400, Thailand
[4] Department of Biology and Biochemistry, Brunel University, Uxbridge, Middlesex UB8 3PH, UK

Nigam et al. (Eds.), Eicosanoids,
Lipid Peroxidation and Cancer
© Springer-Verlag Berlin Heidelberg 1988

lished data show a somewhat disturbing variation in the proportion of false ne-
gatives and false positives in groups of patients from different clinics (Gay et al.
1985; Yobs et al. 1985). A number of other approaches to routine screening have
been tried over the last 20 years or so in the attempt to find specific biochemical
changes in gynaecological cancer cells that could be used in automatic screening
methods (Davina et al. 1985a, b; Koprowska et al. 1986; Moncrieff et al. 1984a,
b; Rees et al. 1970; Sasson et al. 1985; Tay et al. 1986; Valkova and Laurence,
1985).

In this context Slater and Cook in 1970 applied electron spin resonance (ESR)
techniques to samples of normal and malignant cervix and found a striking
quantitative difference between normal samples, which gave an unexpectedly
strong ESR signal, and cancer samples which gave a much diminished or un-
dectable response. This original report was the incentive for a more extensive
and detailed study showing that (a) normal cervix gives a strong ESR signal both
in intact and powdered form (Benedetto et al. 1981) and (b) the free radical
species present in frozen powders has features characteristic of a peroxyl radical
(Tomasi et al. 1984). This type of free radical has oxidising properties, and
would have disturbing effects on the redox balance, including the ratio of thiol/
disulphide groups, especially around its locus of formation.

Therefore we decided to investigate the content of protein-bound thiol groups
in samples of human uterine cervix. Protein-SH groups can be quantitatively
measured cytochemically in fixed cells stained with dihydroxy-dinaphthyl-disul-
phide (DDD). DDD reacts which protein thiols. The main product of this reaction, a protein bound β-naphthol, is subsequently coupled with the bivalent azo-
coupling reagent, Fast blue B. After azo-coupling, two products are obtained: a
red monoazocoupling product and a blue diazocoupling product, whose molar
absorption has been determined by Esterbauer (1972, 1973).

The protein-bound sulphydryl (PSH)-groups of cysteine occur at various sites
on polypeptide chains, therefore the chemical reactivity of the cysteine residues
is affected by their location within the chain. Some PSH groups are situated in
the inner parts of the macromolecular structures and may be inaccessible to SH-
reagents. These so-called "masked" PSH-groups may show, therefore, relatively
low reactivities with such reagents. In contrast, some PSH-groups are located in
the external regions of protein molecules and react more rapidly with SH-rea-
gents. Such PSH-groups, the so-called "reactive" PSH-groups, were the special
target of our investigations because, generally speaking, they are responsible for
the specificities and catalytic activities of a great number of important enzymes
and the functional properties of other proteins (Schauenstein 1984). An additional factor influencing the reactivity of SH-groups is the pk_a of the thiols: if the
pk_a increases, "reactive" PSH-groups decrease (Friedman 1973). It follows that a
method for the quantitative determination of protein-bound sulphydryl groups
will be more valuable if it can discriminate between unreactive or masked PSH
and reactive PSH-groups. Nöhammer (1982) was successful in modifying the original method of Barrnett and Seligman (1952), using DDD and Fast blue B,
thereby enabling its application to kinetic studies.

Some results obtained with Ehrlich ascites tumour cells showed that the kinetic of the reaction of DDD with the protein-bound sulphur results from the

Fig. 1. The kinetics of the reaction of DDD with the protein bound sulphur in Ehrlich ascites tumour cells. *Ordinate,* percentage of total extinction measured at 560 nm after 14 days interaction and subsequent coupling with Fast blue B; *abscissa,* days of reaction time. *PSHr,* reactive protein SH-groups; *TRPS,* total reactive protein sulphur

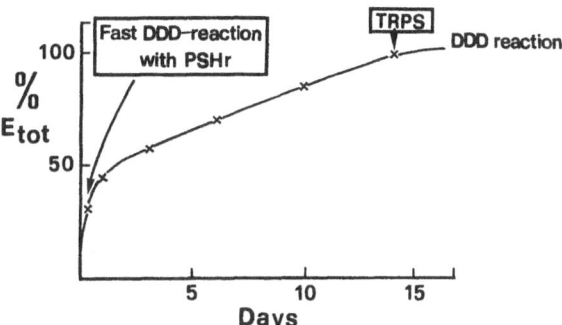

superimposition of a fast reaction, which comes to an equilibrium after approximately 7 hours, on to a much slower reaction that reaches an equilibrium only after approximately 14 days (Fig. 1). Nöhammer (1982) ascribed the first phase of the overall reaction to the reaction of reactive PSH-groups (PSHr) with DDD and the latter phase to the reaction of the so-called "total" reactive protein sulphur (TRPS), that is of masked PSH-groups together with slow reacitve PSH-groups plus some labile disulphide bonds opened by DDD.

We applied the method developed by Nöhammer (1982) for single cells to tissue slices and studied 56 samples of normal cervix obtained from patients undergoing hysterectomy for fibroleiomyoma of the uterine corpus, 34 samples of cervical intraepithelial neoplasia (CIN) grade I and II, 36 samples of CIN III and 29 samples of invasive carcinomas of the cervix. Samples of CIN and invasive carcinomas were taken during operations for cone biopsy of hysterectomy. In all cases three adjacent cryostat sections were prepared from the uterine cervix: one section was stained with haematoxylin-eosin for histopathological evaluation, one for reactive protein-SH groups and the last one for total reactive protein sulphur.

The frozen sections were then stained and quantitatively measured as shown in Fig. 2. In order to eliminate the variation due to section thickness, we calculated the ratio of the value for reactive protein thiol groups found in the epithelium to that in the adjacent stroma in each section. This ratio was expressed as a quotient, defined Q_{PSHr}, whereas the ratio of the value for total reactive protein-sulphur found in the epithelium to that in the adjacent stroma was expressed as Q_{TRPS}.

Table 2 shows the results obtained with the quantitative measure of PSHr (Slater et al. 1985). Reactive protein thiols are more concentrated in epithelium than in stroma both in normal and in pathological conditions. However, in CIN and invasive cancer their concentration decreases in the epithelium and increases in the stroma compared to the controls. Consequently, the ratio of reactive protein thiol groups in epithelium to stroma, the Q_{PSHr}, is significantly higher in sections prepared from normal patients compared to those prepared from patients with CIN or invasive carcinoma. However, the method does not allow unequivocal recognition in individual sections of the pathological lesions studied because of the considerable degree of overlap of the different popula-

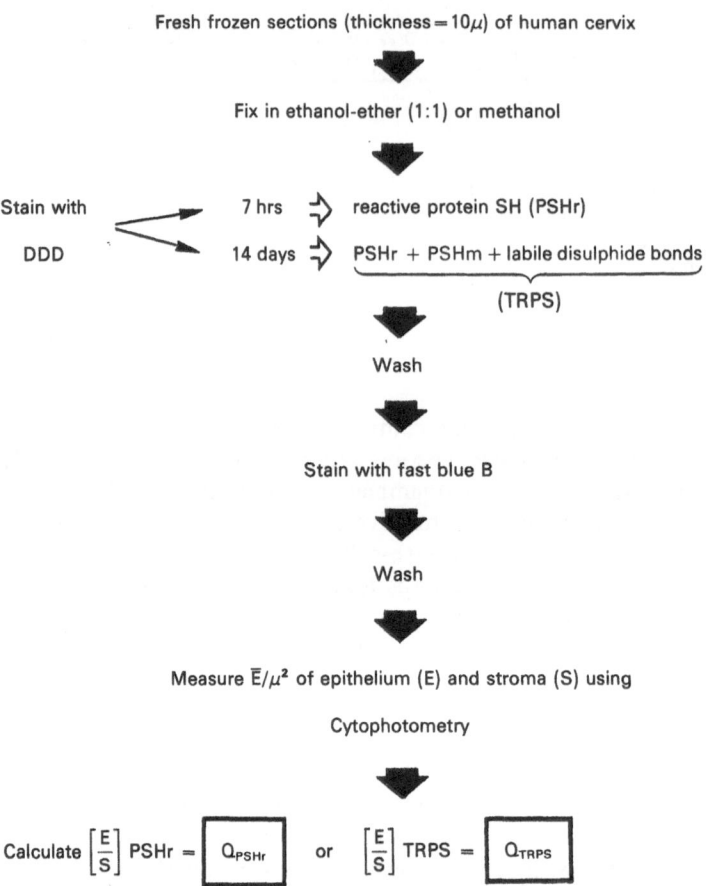

Fig. 2. Scheme of the methodology used to measure reactive protein SH-groups *(PSHr)* and total reactive protein sulphur *(TRPS)* in fresh frozen sections of human uterine cervix. The average extinction *(E)* per unit area (μ^2) was calculated after scanning areas of epithelium and adjacent stroma with an incident light of 560 nm wavelength

tions of values (Fig. 3), probably due to variation in section morphology. To eliminate the variation due to differences in section morphology we calculated in epithelium (EP) and in stroma (ST) the ratio between PSHr-groups, stained in one section, and TRPS, stained in the immediately adjacent serial section (Fig. 4), since the difference in morphology between two adjacent serial sections is very small. We then made the epithelium to stroma ratio of thiols to TRPS in order to decrease the variation due to sections thickness, as stated above. The distribution of the single values of this double ratio, defined as $Q_{PSHr/TRPS}$, shows only limited overlap between the normal group and CIN or invasive carcinoma (Fig. 5).

In other words, the $Q_{PSHr/TRPS}$ emphasises the disturbed biochemical events that have occurred in CIN compared to normal and seems to be a discriminatory function.

Table 2. Mean extinction values of reactive protein thiols per unit area of sections from normal and neoplastic human uterine cervix

Group	Reactive protein thiols ($E/\mu m^2$)		
	Epithelium	Stroma	Epithelium/ Stroma
Normal	0.34 ± 0.02	0.13 ± 0.01	2.73 ± 0.10
CIN I-II	0.27 ± 0.02	0.18 ± 0.02^b	1.61 ± 0.09^c
CIN III	0.25 ± 0.03^b	0.16 ± 0.02^a	1.51 ± 0.06^c
Invasive Ca.	0.22 ± 0.01^b	0.15 ± 0.01	1.55 ± 0.08^c

Significance with respect to the corresponding normal group.
[a] $p < 0.05$
[b] $p < 0.01$
[c] $p < 0.001$
(From Slater et al. 1985)

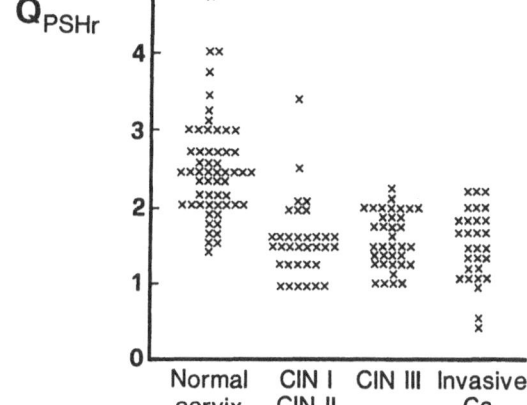

Fig. 3. Distribution of the Q_{PSHr} values obtained from samples of normal cervix, CIN and invasive carcinomas of the cervix

A further interesting observation which is discussed in detail by Nöhammer et al. (see accompanying contribution, this volume), is that similar changes in $Q_{PSHr/TRPS}$ values are observed in the apparently normal epithelium and stroma, according to histological criteria, in the neighbourhood of CIN and invasive cancer. Such a field-like effect or diffuse change would have important biological implications and is under further investigation.

Acknowledgements. This work was supported by the Association for International Cancer Research.

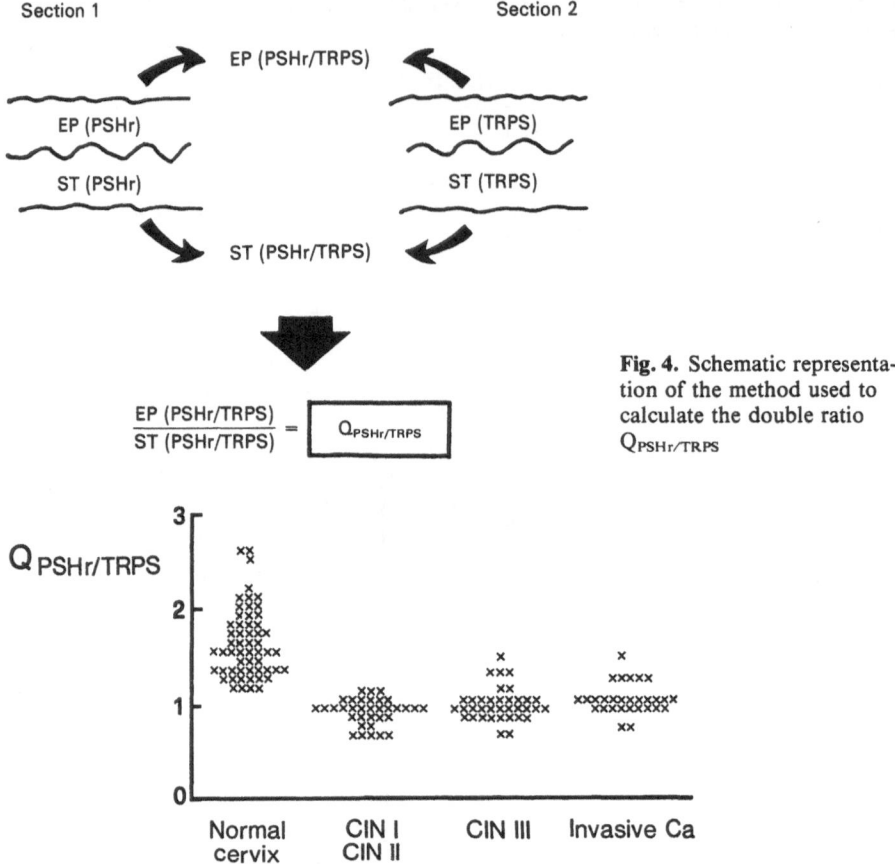

Fig. 4. Schematic representation of the method used to calculate the double ratio $Q_{PSHr/TRPS}$

Fig. 5. Distribution of the $Q_{PSHr/TRPS}$ values obtained from samples of normal cervix, CIN and invasive carcinomas of the cervix

References

Ayiomamitis A, Math B (1987) The epidemiology of cancer of the uterine cervix in Canada: 1931 to 1984. Am J Obstet Gynecol 156:1075–1080

Bain RW, Crocker W (1983) Rapid onset of cervical cancer in an upper socioeconomic group. Am J Obstet Gynecol 146:366–371

Barrnett RJ, Seligman AM (1952) Histochemical demonstration of protein bound sulfhydryl groups. Science 116:323–327

Benedetto C, Bocci A, Dianzani MU, Ghiringhello B, Slater TF, Tomasi A, Vannini V (1981) Electron spin resonance studies on normal human uterus and cervix and on benign and malignant uterine tumors. Cancer Res 41:2936–2942

Carmichael JA, Heffrey JF, Steele HD, Ohlke ID (1984) The cytologic history of 245 patients developing invasive cervical carcinoma. Am J Obstet Gynecol 148:685–690

Davina JH, Stadhauders AM, van Haelst UJ, Lamers JE, Kenemans P (1985a) Concanavalin A-peroxidase labeling in cervical exfoliative cytopathology. I. Labeling of normal squamous cells and the detection of cancer. Gynecol Oncol 22:212–223

Davina JH, Stadhauders AM, van Haelst UJ, de Graaf R, Kenemans P (1985b) Concanavalin A-peroxidase labeling in cervical exfoliative cytopathology. II. Routine assessment of labeling results Gynecol Oncol 22:224–232

Brux J, Ionesco M, Barasso R (1985) Papillomavirus infection and cervical intraepithelial neoplasia (CIN), study of CIN II in 2 series of conizations and hysterectomies (1957–1968 and 1981–1983). Bull Cancer 72:290–297

Dunn JE jr, Schweitzer V (1981) The relationship of cervical cytology to the incidence of invasive cervical cancer and mortality in Alameda County, California, 1960 to 1974. Am J Obstet Gynecol 139:868–876.

Esterbauer H (1972) Beitrag zum quantitativen histochemischen Nachweis von Sulfhydrylgruppen mit der DDD-Färbung. I. Untersuchung der Farbstoffe. Acta Histochem (Jena) 42:351–355

Esterbauer H (1973) Beitrag zum quantitativen histochemischen Nachweis von Sulfhydrylgruppen mit der DDD-Färbung. II. Bestimmung von SH-Gruppen in unlöslichen Proteinen. Acta Histochem (Jena) 47:94–105

Friedman M (1973) In: The chemistry and biochemistry of the sulfhydryl group in amino acids peptides and proteins. Pergamon Oxford, pp 1–24

Gay JD, Donaldson LD, Goellner JR (1985) False negative results in cervical cytologic studies. Acta Cytol (Baltimore) 29:1043–1046

Koprowska I, Zipfel S, Ross AH, Herlyng M (1986) Development of monoclonal antibodies that recognize antigens associated with human cervical carcinoma. Acta Cytol (Baltimore) 30:207–213

Mazur MT, Cloud GA (1984) The koilocyte and cervical intraepithelial neoplasia: time-trend analysis of a recent decade. Am J Obstet Gynecol 150:354–358

McCance DJ, Campion MJ, Clarkson PK, Chesters PM, Jenkins D, Singer A (1985) Prevalence of human papillomavirus type 16 DNA sequences in cervical intraepithelial neoplasia and invasive carcinoma of the cervix. Br J Obstet Gynecol 92:1101–1105

Moncrieff D, Ormerod MG, Coleman DV (1984a) Immunocytochemical staining of cervical smears for the diagnosis of cervical intraepithelial neoplasia. Anal Quant Cytol 6:201–205

Moncrieff D, Ormerod MG, Coleman DV (1984b) Tumor marker studies of cervical smears. Potential for automation. Acta Cytol (Baltimore) 28, 407–410.

Nöhammer G (1982) Quantitative microspectrophotometrical determination of protein thiols and disulfides with 2,2,-dihydroxy-6,6'-dinaphthyldisulfide (DDD). Histochemistry 75:219–250

Prakash SS, Reeves WC, Sisson GR, Brenes M, Godoy J, Bacchetti S, de Britton RC, Rawls WE (1985) Herpes simplex virus type 2 and human papillomavirus type 16 in cervicitis, dysplasia and invasive cervical carcinoma. Int J Cancer 35:51–57

Rees KR, Slater TF, Gibbs DF, Stagg BH (1970) A modified assay for 6-phosphogluconate dehydrogenase in samples of vaginal fluid from women with and without gynecologic cancer. Am J Obstet Gynecol 107:857–864

Sasson AF, Said JW, Nash G, Shintaku IP, Banks-Schlegel S (1985) Involucrin in intraepithelial and invasive squamous cell carcinomas of the cervix: an immunohistochemical study. Hum Pathol 16:447–470

Schauenstein E (1984) In: McBrien DCH, Slater TF (eds) Cancer of the uterine cervix. Biochemical and clinical aspects. Academic, London, pp 197–204

Scholl SM, Pillers EM, Robinson RE, Farrell PJ (1985) Prevalence of human papillomavirus type 16 DNA in cervical carcinoma samples in East Anglia. Int J Cancer 35:215–218

Slater TF, Cook JWR (1970) In: Evans DMD (ed) Cytology automation. Livingstone, Edinburgh, pp 108–120

Slater TF, Bajardi F, Benedetto C, Bussolati G, Cianfano S, Dianzani MU, Ghiringhello B, et al (1985) Protein thiols in normal and neoplastic human uterine cervix. FEBS 2:267–271

Tay SK, Singer A, Griffin JFA, Wickens DG, Dormandy TL (1986) Recognition of cervical neoplasia by the estimation of a free-radical reaction product (octadeca-9, 11-dienoic acids) in exfoliated cells. Free Radic Res Commun 3:27–31

Tomasi A, Benedetto C, Nilges M, Slater TF, Swarts HM, Symons MCR (1984) Studies on human uterine cervix and rat uterus using S-, X- and Q-band electron-spin-resonance spectroscopy. Biochem J 224:431–436

Valkova B, Laurence DJ (1985) Automated screening of cervical smears using immunocytochemical staining: a possible approach. J Clin Pathol 38:886–892

Wolfendale MR, King S, Usherwood MM (1983) Abnormal cervical smears: are we in for an
epidemic? Br Med J 287:526–528

Yobs AR, Swanson RA, Lamotte LC Jr (1985) Laboratory reliability of the Papanicolaou smear.
Obstet Gynecol 65:235–244

Microphotometric Determination of Protein Thiols and Disulphides in Tissue Samples from the Human Uterine Cervix and the Skin Reveal a "Field Effect" in the Surroundings of Benign and Malignant Tumours

G. Nöhammer[1], F. Bajardi[2], C. Benedetto[3], H. Kresbach[4],
W. Rojanapo[5], E. Schauenstein[1] and T. F. Slater[6]

1 Introduction

Cytophotometry enables the investigation of the "biochemistry of the whole cell" (Chayen 1984). It allows the measurement of properties of different kinds of cells in their natural locations in tissues. For example, cytophotometry can be used to study the properties of tumour cells and of disturbances in apparently normal cells in the presence of a tumour; the latter disturbances can be called tumour-associated changes.

During the last four years there has been an increasing number of studies directed at tumour-associated changes. These studies can be divided into four main categories:

1. Tumour-associated changes of the immune system, of both the cellular and the humoral defense system (Bashford and Gough 1983; Schauenstein et al. 1984, 1986a,b, 1987; Berndt et al. 1984; Fumita et al. 1984; Lichtenstein et al. 1985; Nemoto et al. 1985; Onsrud 1985).
2. Tumour-associated changes of apparently normal tissue in the neighbourhood of tumours (Hou-Jensen 1972; Sultatos and Vesell 1980; Benedetto et al. 1981; Jensen et al. 1982; Millet et al. 1982; Nöhammer et al. 1984; Davina et al. 1984; Ireland 1985; Lathrop et al. 1985; Slater et al. 1985).
3. Tumour-associated changes of tissues eithout any close spatial connection with the tumour (Schrempel and Kürchner 1985).
4. Tumour-associated viral infections (Aurelian et al. 1981; Fu et al. 1983; Crum et al. 1984; Sillman et al. 1984).

The present work is concerned preferentially with tumour-associated changes of apparently normal tissue in the *immediate* neighbourhood of tumours. These in-

---

[1] Institut für Biochemie, Karl-Franzens-Universität, 8010 Graz, Austria.
[2] Zytologisches Institut, 8010 Graz, Austria.
[3] Istituto di Ginocologia e Ostetricia, Cattedra A, Universita di Torino, 10126 Torino, Italy
[4] Universitätsklinik für Dermatologie und Veneralogie, 8010 Graz, Austria.
[5] National Cancer Institute, Rama VI Road, Bangkok 10400, Thailand
[6] Department of Biology and Biochemistry, Brunel University, Uxbridge, Middlesex UB8 3PH, UK

Nigam et al. (Eds.), Eicosanoids,
Lipid Peroxidation and Cancer
© Springer-Verlag Berlin Heidelberg 1988

vestigations were stimulated by the findings of Benedetto et al. (1981), who detected very strong esr signals in normal samples of the human uterine cervix, whereas the signal was much reduced or absent in samples of the tumour. In the studies with cervix tumours the piece of frozen tissue taken for esr measurements was much bigger than the tumour. Therefore, it appeared possible that the tumour was affecting neighbouring normal tissue (Benedetto et al. 1981).

The free radical species that is present in frozen powders of normal human uterine cervix has features characteristic of a peroxyl radival (Benedetto 1982; Tomasi et al. 1984). This type of radical has oxidizing properties, and could have disturbing effects on the redox-balance of the tissues involved. Such changes of the redox-balance could be reflected also in the ratio of thiol: disulphide groups, one of the reasons for the quantitative cyto- and histo-photometric determination of these parameters.

Tumour-associated changes of apparently normal tissue furthermore are interesting with respect to tumour diagnosis. Thus, one of the main reasons for the investigation of apparently normal tissue adjoining the tumours has been to use possible tumour-associated changes for a new approach to tumour diagnosis by measurements on apparently normal cells and tissues.

2 Histophotometry and Histochemical Parameters

Compared with cytophotometry, histophotometry allows the study of groups of similar cells, for example epithelial cells, relative to other types of cell or extracellular matrix. Thus, histochemical parameters from cells and extracellular substances within their natrual environment can be measured, which allows the interactions between neighbouring cell groups to be studied. There is one considerable disadvantage in histophotometry, however. Microphotometry principally yields extinction-values per area measured. In cytophotometry the mean extinction determined from a cell is multiplied by the cell-area thus yielding an integrated extinction which is proportional to the mass of the absorbing substance within the cell. Histophotometry only yields mean extinctions of measured areas, reflecting the concentration of the stained parameter. Normally there are no single isolated cells and mostly the cells are cut irregularly. Although there is the advantage of the knowledge of the cell-type within a distinct measured area, it should be evident that each value obtained histophotometrically strongly depends on both the thickness and the morphological content and orientation of a given section. In consequence if possible differences between normal and cancerous tissue are to be investigated, then it will be necessary to make measurements of a considerable number of tissues from different patients. It is practically impossible to cut sections with equal thickness and furthermore the morphologic content and orientation of the tissue samples is different from different donors. Thus, any naturally occurring biochemical variation will be enhanced by the differing thickness and morphology of the sections. Therefore, it should be evident that histophotometrically determined values of single parameters will exhibit a strong variation.

3 Methods

Frozen sections were prepared from samples of human uterine cervix as described previously (Nöhammer et al. 1986); frozen sections of abdominal skin were prepared in a similar manner. The parameters discussed in the present study are: (a) reactive protein thiols (PSHr), (b) total protein thiols (PSH$_{tot}$), (c) total reactive protein sulfur (TRPS) which comprises PSH$_{tot}$ and the reactive disulphides (PSSX), and (d) total proteins (BROT). These were demonstrated and measured by quantitative histochemical staining methods. PSHr and TRPS were measured after 7h and 14d reaction with 2,2'-dihydroxy-6,6'dinaphthyldisulfide (DDD) respectively, followed by coupling with Fast blue B (Nöhammer 1982). PSH$_{tot}$ was stained and measured using the Mercurochrom-cyanide-method (Nöhammer et al. 1981); the total protein content of the tissues was determined using a modified amidoblack staining (Nöhammer 1984).

4 Results and Discussion

Results obtained for PSHr, TRPS and protein in sections of uterine cervix will be described first. The data obtained with 62 normal cases have been compared with those from 34 cases with CIN 1–2 (DYS), and with 36 cases with CIN 3 (CIS) carcinoma in situ. As shown in Tables 1 and 2 the sections from patients with CIN 3 are characterized by a significant decrease of PSHr and of protein density compared with squamous epithelium of normals. The results for PSHr are similar to our previous studies (Schauenstein et al. 1983; Bajardi et al. 1983a, 1983b; Slater et al. 1985; Nöhammer et al. 1986). In contrast, sections from CIN 1–2 (DYS) did not show decreased protein densities. Thus CIN 1–2 can be discriminated from CIN 3 by the lower thiol content of the epithelial proteins.

Whereas the values for PSHr and protein were significantly lower in CIN 3 compared with normal, the value for PSHr in stroma underlying CIN 3 was significantly increased whilst the protein content tended to be lower although this was not statistically significant. consequence, the ratio of PSHr to protein is much increased in the stroma underlying CIN 3 than in normal samples.

4.1. The "Field Effect" of Tumours

We call tumour-associated changes of apparently normal tissue immediately adjacent to tumours the "field effect" (Nöhammer et al. 1984).

Table 1. Histophotometry of reactive protein thiols (PSH_r) and total protein thiols (PSH_{tot}) in human uterine cervix

Group	Sample	PSH_r					PSH_{tot}				
		n	Epithelium		Stroma		n	Epithelium		Stroma	
			Mean	SD	Mean	SD		Mean	SD	Mean	SD
1	Normal										
	NS-Epi+ST	62	0.32	0.14	0.12	0.05	43	0.44	0.23	0.15	0.07
2	Dysplasia (CIN 1–2)										
	a) ANS-Epi+ST	26	0.25	0.14	0.14	0.09	20	0.31	0.17	0.15	0.09
	b) DYS-Epi+ST	34	0.27	0.17	0.17	0.12	23	0.27	0.13	0.16	0.07
3	CIS (CIN 3)										
	a) ANS-Epi+ST	29	0.22	0.11	0.11	0.06	27	0.33	0.09	0.16	0.05
	b) CIS+ST	36	0.25	0.14	0.16	0.09	28	0.36	0.17	0.25	0.14
t-test, groups:											
2a versus 1			$p<0.05$		NS			$p<0.05$		NS	
2b versus 1			NS		$p<0.01$			$p<0.01$		NS	
3a versus 1			$p<0.01$		NS			$p<0.05$		NS	
3b versus 1			$p<0.05$		$p<0.01$			NS		$p<0.01$	
2b versus 3b			NS		NS			$p<0.05$		$p<0.01$	

Normal, patients suffering either from chronic cervicitis or from fibroleimyoma of the uterine body with no evidence of significant pathological disturbances of the uterine cervix; NS-Epi, normal squamous epithelium of normal patients; St, stroma; dysplasia, cases with mild or moderate dysplasia (CIN 1–2); ANS-Epi, apparently normal squamous epithelium immediately adjacent to cervical neoplasias; CIS, carcinoma in situ (CIN 3); n, number of individual cases; mean, mean value of the extinction/unit area – average extinction values ($\bar{E}/\mu m^2$) calculated from the data obtained by scanning small areas (~ 0.3 mm^2) of epithelium or stroma of stained sections; SD, standard deviation of $\bar{E}/\mu m^2$ values.

The reactive protein thiols (PSHr) of apparently normal epithelium (ANS-Epi) adjoining CIN 1–2 or CIN 3 were highly significantly decreased compared with normal epithelium of healthy patients (NS-Epi) Tables 1 and 2). The protein densities of ANS-Epi were significantly decreased in the neighbourhood of CIN 3 but not CIN 1–2.

These data show that there are similarities in the behavious of PSHr and protein in apparently normal epithelium close to a CIN 3 lesion, when compared with normal controls. However, such changes in single parameters are not sufficiently large to allow clear discrimination between the different sample groups examined. This lack of discrimination may result from the inherent variability of measurements of single histophotometric parameters as mentioned earlier.

In principle there are two different ways out of this dilemma. The most convenient way in our case, since we have determined protein thiols and reactive protein disulphides, would have been the relation to the respective protein mass. In practice, however, this can be realised only with double-stained tissues, and no quantitative histochemical double staining methods are yet available.

Table 2. Histophotometry of total reactive protein sulphur (TRPS) and total protein (PRO-TEIN) in human uterine cervix

Group	Sample	TRPS					PROTEIN				
		n	Epithelium		Stroma		n	Epithelium		Stroma	
			Mean	SD	Mean	SD		Mean	SD	Mean	SD
1	Normal NS-Epi+ST	62	0.69	0.26	0.43	0.23	29	0.17	0.12	0.11	0.10
2	Dysplasia (CIN 1–2)										
	a) ANS-Epi+ST	26	0.59	0.18	0.31	0.11	13	0.18	0.08	0.09	0.05
	b) DYS-Epi+ST	34	0.62	0.17	0.37	0.14	15	0.18	0.10	0.12	0.08
3	CIS (CIN 3)										
	a) ANS-Epi+ST	29	0.50	0.18	0.28	0.09	23	0.11	0.07	0.07	0.04
	b) CIS+ST	36	0.61	0.25	0.39	0.17	21	0.13	0.08	0.08	0.05
t-test, groups:											
2a versus 1			NS		$p<0.05$			NS		NS	
2b versus 1			NS		NS			NS		NS	
3a versus 1			$p<0.01$		$p<0.01$			$p<0.05$		$p<0.05$	
3b versus 1			NS		NS			NS		NS	
2a versus 3a			NS		NS			$p<0.01$		NS	
3b versus 4b			NS		NS			$p<0.05$		NS	

Abbreviations, see Table 1.

Based on the observations of Bajardi et al. (1983a), that changes of the epithelium regularly are accompanied by corresponding changes of the adjacent con-

Table 3. Epithelium:stroma ratios (Q) of PSH_r, TRPS and PSH_r/TRPS (uterine cervix)

Group	Sample	n	Q_{PSH_r}		Q_{TRPS}		$Q_{PSH_r/TRPS}$	
			Mean	SD	Mean	SD	Mean	SD
1	Normal NS-Epi+ST	62	2.64	0.71	1.75	0.49	1.56	0.37
2	Dysplasia (CIN 1–2)							
	a) ANS-Epi+ST	26	1.87	0.53	1.96	0.63	0.98	0.18
	b) DYS-Epi+ST	34	1.60	0.49	1.74	0.47	0.93	0.15
3	CIS (CIN 3)							
	a) ANS-Epi+ST	29	2.09	0.55	2.02	1.20	1.15	0.32
	b) CIS+ST	36	1.58	0.36	1.58	0.40	1.01	0.18
t-test, groups:								
2a versus 1			$p<0.001$		NS		$p<0.001$	
2b versus 1			$p<0.001$		NS		$p<0.001$	
3a versus 1			$p<0.01$		NS		$p<0.001$	
3b versus 1			$p<0.001$		NS		$p<0.001$	

Mean, mean epithelium:stroma ratio; other abbreviations, see Table 1.

nective tissue, our basic hypothesis has been the existence of a distinct epithelium-stroma ratio for PSHr (QPSHr) as well as for TRPS (QTRPS). As shown previously (Bajardi et al. 1983b; Nöhammer et al. 1984, 1986; Slater et al. 1985) the use of QPSHr gives a better discrimination between normal cases and CIN 1–3 than measurements on epithelium or stroma alone. However, the variation (Table 3) in QPSHr is still too large to be used for cancer diagnosis. This is due to the fact that the relation of the epithelial to stromal PSHr eliminates the influence of varying thickness of sections but is unable to eliminate the morphological problems referred to earlier. Two serial thin sections should have a very similar morphological content and orientation. If now two serial sections are stained, one for PSHr, the other for TRPS, both parameters can be related to each other. The new ratio PSHr:TRPS, measured from equivalent serial sections, is in some way a measure of the SH:SS ratio in that part of tissue measured. The double ratio provides a good discrimination between normal control sections and CIN 1–3, and to some extent also between normal control sections and apparently normal tissue adjacent to CIN 1–3. The latter relationship is remarkable since mild dysplasias (CIN 1) are included together with CIN 2. Attention should be paid to the marked decrease in the standard deviation of the QPSHr/TRPS values compared with the corresponding values of QPSHr and QTRPS. Figure 1 illustrates the distribution of the single QPSHr/TRPS values measured. A sometimes considerable overlap exists between the QPSHr/TRPS-values of the normals compared with the respective values obtained from CIN 1–3, and much more from apparently normal tissue adjoining the tumours. Therefore, the QPSHr/TRPS value in the present state cannot be used for cancer diagnosis.

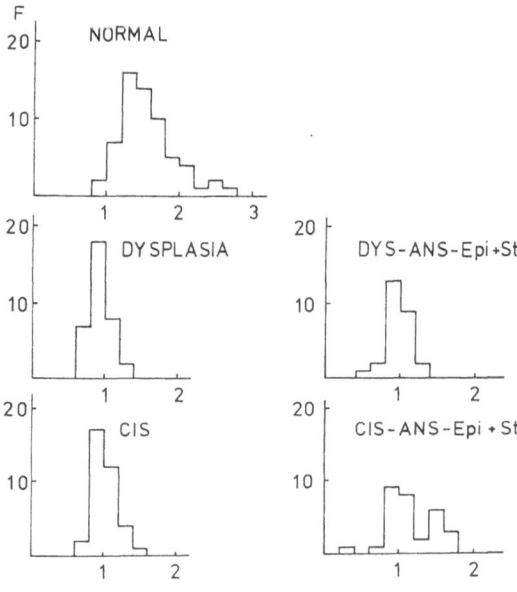

Fig. 1. Distribution of the PSH$_r$/TRPS ratios for uterine cervix. *Abcissa,* PSH$_r$/TRPS; *ordinate,* absolute frequency (*F*)

4.2 A Possible "Extended Field Effect" of Tumours

As demonstrated, small but obvious changes can be detected in apparently normal tissue in the immediate neighbourhood of CIN 1-3. The question arises: how far away from the lesion can such changes still be observed? In preliminary studies we have investigated abdominal skin from "healthy" volunteers (chronic cervicitis, myoma uteri, liver cirrhosis) as well as from patients with CIN 3 or invasive cancer of the uterine cervix, or those suffering from liver cancer or cancer of the ovaries. We have used for these preliminary investigations the double ratio (QPSHr/TRPS). Figure 2 illustrates that there is no significant difference between the Q(PSHr/TRPS) obtained from normal skin of healthy volunteers taken from different parts of the human body and the Q(PSHr/TRPS) of exclusively obdominal skin. It should be mentioned that the first group of Fig. 2 represent skin from normal healthy Austrian volunteeres, whereas the second group of Fig. 2 represents normal abdominal skin of patients from Thailand; QPSHr/TRPS seems to be quite independent from the degree of pigmentation. The QPSHr/TRPS obtained from abdominal skin of patients suffering either from CIN 3 or from different invasive tumours are highly significantly different from the respective values of healthy volunteers.

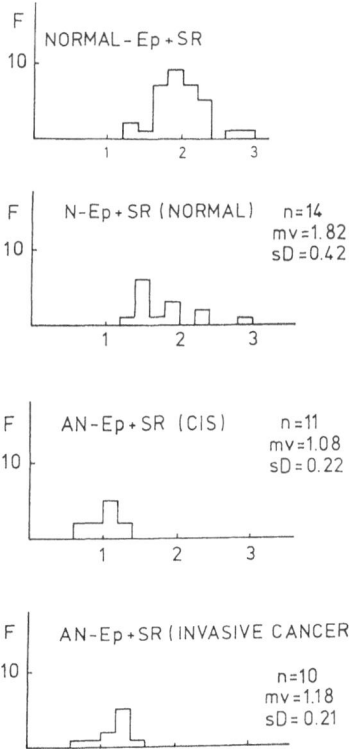

Fig. 2. Distribution of the $PSH_r/TRPS$ ratios for abdominal skin. *Abcissa*, $PSH_r/TRPS$ ratio; ordinate, absolute frequency (F)

Our preliminary results indicate the existence of widespread tumour-associated changes, at least of abdominal skin and underline the usability of the QPSHr/TRPS to detect these changes.

Acknowledgement. We are grateful to the National Foundation for Cancer Research and the Association for International Cancer Research for financial support and enabling this collaborative study.

References

Aurelian T, Kessler II, Rosenshein NB, Barbour G (1981) Viruses and gynecological cancers: herpes virus protein (ICP10/AG-4), a cervical tumour antigen that fulfils the criteria for a marker of cancerogenicity. Cancer 48:455–471

Bajardi F, Jüttner F, Smolle J (1983a) Korrespondierende Verhaltensweisen von Epithel und Stroma der Cervix uteri. Zentralbl Gynäkol 105:257–264

Bajardi F, Benedetto C, Nöhammer G, Schauenstein E, Slater TF (1983b) Histophotometrical investigation on the content of protein and protein thiols in the epithelium and stroma of the human uterine cervix. II. Intraepithelial neoplasias. Histochemistry 78:95–100

Bashford J, Gough IR (1983) Depression of high-affinity rosette formation in dysplasia and carcinoma in situ of the uterine cervix: Mediation by serum factors. Cancer Res 43:3959–3962

Benedetto C (1982) Biochemical studies on human uterine cancer. In: McBrien DCH, Slater TF (eds) Free radicals, lipid peroxidation and cancer. Academic, London, pp 27–54

Benedetto C, Bocci A, Dianzani MU, Ghiringhello B, Slater TF, Tomasi A, Vannini V (1981) Electron spin resonance studies on normal human uterus and cervix, and on benign and malignant uterine tumours. Cancer Res 41:2936–2942

Berndt T, Jankovic D, Stojanovski A, Bolanca M, Vecek N (1984) Immunoglobulins in women with cancer of the uterine body (before treatment). Jugosl Ginekol Opstet 24:10–15

Chayen J (1984) Quantitative cytochemistry: a precise form of cellular biochemistry. Biochem Soc Trans 12:887–898

Crum CC, Ikenberg H, Richart RM (1984) Human papilloma virus type 16 and early cervical neoplasia. N Engl J Med 310:880

Davina IHM, Lamers GEM, van Haelst UJGM, Kenemans P, Stadhouders AM (1984) Tannic acid binding of cell surfaces in normal, premalignant, and malignant sequamous epithelium of the human uterine cervix. Ultrastruct Pathol 6:275–284

Fu YS, Reagan JW, Richart RM (1983) Precursors of cervical cancer. Cancer Surv 2:359

Fumita Y, Tanaka F, Soji F, Nakamuro K (1984) Immunosuppressive factors in ascites fluids from ovarian cancer patients. Am J Reprod Immunol 6:175–178

Hou-Jensen K (1972) Simultaneous epidermoid carcinoma in situ of the portio/cervix and the endometrium of the uterus. Acta Pathol Microbiol Scand [A] 80:1–4

Ireland D (1985) Metaphase-arrest technique applied to human cervical epithelium. II. Cell production rates in normal and pathological cervical epithelium. Cell Tissue Kinet 18:321–331

Jensen HM, Chen I, Devault MR, Lewis AE (1982) Angiogenesis induced by "normal" human breast tissue: a probable marker of precancer. Science 218:293–295

Lathrop JC, Ree JH, Mc Duff HC (1985) Intraepithelial neoplasia of the neovagina. Obstet Gynecol [Suppl] 65:915–945

Lichtenstein AK, Berek J, Zighelboim J (1985) Natural killer inhibitory substance produced by the peritoneal cells of patients with ovarian cancer. INCI 74:349–355

Millett JA, Husain OAN, Bitensky L, Chayen J (1982) Feulgen hydrolysis profiles in cells exfoliated from the cervix uteri: a potential aid in the diagnosis of malignancy. J Clin Pathol 35:345–349

Nemoto S, Koiso K, Aoyagi K, Tojo S (1985) Elastase - does this enzyme play any role in bladder cancer invasion? J Urol 134:996-998

Nöhammer G (1982) Quantitative microspectophotometrical determination of protein thiols and disulfides with 2,2'-dihydroxy-6,6'-dinaphthyl-disulfide (DDD). The variety of DDD-staining methods demonstrated on Ehrlich ascites tumour cells. Histochemistry 75:219-250

Nöhammer G (1984) A modification of the amidoblack-TCA-staining for quantitative micro-spectrometrical determination of proteins in tissue sections. Histochemistry 80:395-400

Nöhammer G, Desoye G, Khoschsorur G (1981) Quantitative cytospectrophotometrical determination of the total protein thiols with "Mercurochrom". Optimization and Calibration of the histochemical reaction. Histochemistry 71:291-300

Nöhammer G, Bajardi F, Benedetto C, Schauenstein E, Slater TF (1984) Studies on the relationship between epithelium and stroma in sections of human uterine cervix in different pathological conditions. In: McBrien DCH, Slater TF (eds) Cancer of the uterine cervix. Academic, London, pp 205-224

Nöhammer G, Bajardi F, Benedetto C, Schauenstein E, Slater TF (1986) Quantitative cytospectrophotometric studies on protein thiols and reactive protein disulphides in samples of normal human uterine cervix and on samples obtained from patients with dysplasia or carcinoma-in-situ. Br J Cancer 53:217-222

Onsrud M (1985) Serum-mediated immunosuppression: A possible tumor marker in patients with ovarian carcinoma. Gynecol Oncol 21:94-100

Schauenstein E, Bajardi F, Benedetto C, Nöhammer E, Slater TF (1983) Histophotometrical investigations on the contents of protein and protein thiols of the epithelim and stroma of the human cervix. Histochemistry 77:465-472

Schauenstein E, Dachs F, Reiter M, Gombotz H, List W (1984) Labile Disulfidbrücken in der IgG$_1$- und freie SH-Gruppen in der IgG$_2$-Subklasse im Blutserum benigne und maligne erkrankter Menschen. Z Physiol Chem 365:944-945

Schauenstein E, Reiter M, Gombotz H, List W (1986a) Labile disulfide bonds and free thiol groups in human IgG. II. Characteristic changes in malignant diseases corresponding to shifts of IgG$_1$ and IgG$_2$ subclasses. Int Arch Allergy Appl Immunol 80:180-184

Schauenstein E, Reiter M, Leitsberger A, Gombotz H, List W (1986b) Verschiebungen in den IgG-Subklassen G$_1$ und G$_2$ bei malignen Erkrankungen. Naturwissenschaften 73:505-508

Schauenstein E, Lahousen M, Reiter M (1987) Labile Disulfidbedingungen und freie Thiolgruppen im Human-IgG. III. Der "Es-Wert" bei gynäkologischen Malignomen. Wien Klin Wochenschr 99:276-279

Schrempel A, Kürschner M (1985) Zur Bedeutung der N-Azetylneuraminsäure (NANA) im Serum von Patientinnen mit gynäkologischen Karzinomen. Zentralbl Gynäkol 107:175-178

Silburn PA, Khoo SK, Hill R, Daunter B, Mackay EV (1984) Demonstration of tumor-associated immunoglobulin G isolated from immune complexes in ascitic fluid of ovarian cancer. Diagn Immunol 2:30-35

Sillman F, Stanek A, Sedlis A, Rosenthal J, Lanks KW, Buchhagen D, Nicastri A, Boyce J (1984) The relationship between human papillomavirus and lower genital intraepithelial neoplasia in immunosuppressed women. Am J Obstet Gynecol 150:300-308

Slater TF, Bajardi F, Benedetto C, Bussolati G, Cianfano S, Dianzani MU, Ghiringhello B (1985) Protein thiols in normal and neoplastic human uterine cervix. FEBS Lett 187:267-271

Sultatos LE, Vesell ES (1980) Enhanced drug-metabolizing capacity within liver adjacent to human and rat liver tumors. Proc Natl Acad Sci USA 77:600-603

· Tomasi A, Benedetto C, Nilges M, Slater TF, Swartz HM, Symons MCR (1984) Studies on human uterine cervix and rat uterus using S-, X- and Q-band electron-spin-resonance spectroscopy. Biochem J 224:431-436

Protein-Thiol Levels in Relation to Deviation from Normality in Rat Liver Cell Lines

P. Principe[1], P. A. Riley[1] and T. F. Slater[2]

1 Introduction

Protein-thiol levels in cells and tissues may be altered in connection with disturbed cellular growth and in certain types of neoplasia.

Several studies have reported a higher content of cellular thiols in tumour tissues and tumour cells than in the corresponding normal cells. Barrnett (1955) investigated the protein-thiol levels in a series of human tumours; he found that benign tumours usually contained a smaller quantity of protein thiol groups than did malignant tumours, of which those of epithelial origin had the highest protein thiol concentrations. Bahr and Moberger (1958), in a study on carcinoma of the breast and of the uterine cervix, also observed differences in the protein thiol content.

Quantitative cytochemical measurements on mouse ascitic tumour cells revealed an elevated protein-thiol content (Caspersson and Révész 1963). In an interesting cytological study of exfoliated bronchial cells Lars-Gösta (1964) showed that malignant cells had a higher concentration of protein-thiol groups and that the measurement of cellular protein-thiol content could be a useful parameter in cytological diagnosis of primary bronchogenic carcinoma. In certain cases he found that the method used for measuring protein thiols was more accurate in the prediction of malignancy than the conventional staining techniques such as the Papanicolaou method. Grimaldi and Caprio (1981) have reported altered serum protein thiol levels in patients with malignant lymphoma. Knock (1981) has published evidence of a higher content of protein-thiol groups in cancers than in the corresponding normal tissues, with the more malignant and anaplastic cancers exhibiting the highest values. Taś et al. (1984) observed an altered nucleo-cytoskeletal structure in leukaemic cells , involving SH-containing proteins. The results of Nöhammer et al. (1986), on sections of cervical biopsy material, indicate that the relative distributions of protein-thiols in the epithelium and stroma are altered in cervical dysplasia and in carcinoma in situ (CIS), when compared with normal cervix; normal tissue contained a higher content of

[1] Department of Chemical Pathology, Academic Unit, University College and Middlesex School of Medicine, Windeyer Building, Cleveland Street, London W1P 6DB, UK
[2] Department of Biology and Biochemistry, Brunel University, Uxbridge UB8 3PH, Middlesex UK

Nigam et al. (Eds.), Eicosanoids,
Lipid Peroxidation and Cancer
© Springer-Verlag Berlin Heidelberg 1988

protein-thiols than samples taken from CIS lesions. The same authors (Bajardi et al. 1977) also reported that the nuclear protein-thiol content seemed to be a useful parameter for quantitative measurements of cellular proliferative activity. They found that the ratio of nuclear protein-thiols versus cytoplasmic protein-thiols was greater for all types of the examined pathological cells than for normal cells. Cancer cells had distinctly higher quotients than dysplastic cells: the maximal ratio was found in undifferentiated malignant cells.

The aim of the present study was to gain further information on the possible relationship between protein-thiol levels and malignancy. Quantitative estimations of cellular protein-thiols were carried on three rat liver cell lines possessing different growth characteristics and degrees of tumorigenicity. In this paper we present the results obtained using computerised microdensitometry on single cells stained by the DDD-FBB histochemical method (Barrnett and Seligman 1952; Nöhammer 1982). In addition, the cellular thiol equilibrium has been modified by inhibition of glutathione synthesis, in order to study the influence of cellular levels of reduced glutathione on reactive protein thiols (PSHr) and total reactive protein sulphur (TRPS).

Table 1. IAR rat liver cell lines: general characteristics

	Cell line		
	IAR	IAR 6.1	IAR 6.1 RT7
Origin	Rat liver BDVI strain	Rat liver BDVI strain	From a tumour induced by IAR 6.1
Treatment	None	DMN[c]	None
Tumorigenicity[a]	None	Epithelial tumours at site of injection	Metastasizing epithelial tumours
Modal cell volume[b] (fl)	731	645	560
Population doubling times (hours)	32.0	16.8	21.0

[a] Via i.p. to newborn syngeneic rats.
[b] Modal cell volumes in femtolitres (fl; Coulter Counter analysis).
[c] Dimethylnitrosamine (100 µg/ml, twice a week for 1 week).

2 Materials and Methods

2.1 Cell Lines

The cell lines used in this study were rat hepatocytes designated IAR cells. These IAR cell lines were developed in the laboratory of Dr. R. Montesano, at the International Agency for Cancer Research, Lyon, France. The cell lines originated from primary cultures of normal BD VI rat livers: IAR 20 originated from a 10-days-old rat liver; IAR 6-1 was derived from a line similar to IAR 20, called IAR 6, that had received treatment with the chemical carcinogen dimethylnitrosamine (DMN); IAR 6.1 RT7 was derived from a tumour produced by IAR 6.1 injected into a syngeneic animal (Montesano et al. 1975, 1977). The general characteristics of these rat liver cell lines are summarized in Table 1. The population doubling times were calculated from the growth profiles, according to the method of Sussman (1964). The modal cell volumes of each cell line were calculated from the cell size distributions obtained using a Coulter Counter (Coulter Electronics Ltd., U.K.). The cell cultures were routinely maintained in William's medium supplemented with 2 mM L-glutamine, penicillin (100 U/ml), streptomycin (100 mg/ml) and 10% foetal bovine serum (Flow Labs). The cells were grown at 37°C in a humidified atmosphere containing 5% CO_2.

2.2 Cytochemical Staining

Cells growing exponentially were detached by trypsinization and seeded onto 18×18 mm sterile coverslips placed in 20 mm diameter tissue culture dishes at concentrations of 5×10^4 cells/ml of growth medium. The cultures were then incubated at 37°C for 48 hours. At the end of this incubation time, the cells were washed with phosphate-buffered saline (PBS) and fixed in 70% ethanol for 30 minutes before staining. The determination of reactive protein-thiols (PSHr) and total reactive protein sulphur (TRPS), i.e. the sum of reduced and oxidised protein-thiol groups and mixed disulphides, was carried out following the method originally described by Barrnett and Seligman (1952) and modified by Nöhammer et al. (1977) and Nöhammer (1982). The reaction between thiol and disulphide groups with 2-2,-dihydroxy-6-6'-dinaphthyl-disulphide (DDD) yields a coloured insoluble product after coupling with the azo-dye Fast Blue B (FBB). As published by Esterbauer (1972, 1973), the coloured product has an absorption maximum at 560 nm wavelength. The molar extinction coefficient of the chromophore was reported to be $\varepsilon = 1.9 \times 10^4$ M^{-1} cm^{-1} (Esterbauer 1973). This value is in agreement with that obtained by calibration using standard specimens of RNAase in agarose gels stained with the DDD-FBB method.

2.3 Quantitative Computerized Video-Microdensitometry

Cell densitometry was performed using an image processing software routine called 'CYDENS', written in Fortran and developed by Dr. D.J.Spargo. A description of this program is given by Spargo and Riley (this volume). The modal cell volumes of each rat liver cell line (Table 1) were used to calculate the path length from the measured cell area in order to derive the concentration of the chromophore from its absorbance. There is a systematic bias in this procedure since it overestimates the path length for those cells that are smaller than the modal volume and underestimates it for those larger than the modal volume. The extent of the error is non-linear and dependent on the relative degree of cell flattening. For well-flattened cells the over- and under- estimates are unlikely to exceed 12% and 1%, respectively. Thus, the apparent range of concentrations will be extended and this must be borne in mind when analysing the data.

Each analysis included the measurement of at least 100 individual cells. At the end of each count, the CYDENS program displays the cumulative data in tabular form with the statistical calculations of the relative absorbances transformed into millimolar (mM) concentrations per cell. The frequency distributions of the concentrations of protein-thiol groups were plotted using LOTUS 1–2–3 software on an IBM PC-XT computer.

2.4 Depletion of Intracellular Glutathione

The cellular thiol equilibrium was modified by inhibition of the synthesis of intracellular reduced glutathione (GSH). Rat liver cells were treated for different incubation times, with 0.02 mM buthionine-sulphoximine (BSO), an irreversible inhibitor of the enzyme γ-glutamyl-cysteine-synthetase (Griffith and Meister 1979). The treatment times and the concentration of BSO were chosen to give maximal depletion of GSH without producing significant toxicity. The fluorimetric method developed by Hissin and Hilf (1976) was used to measure reduced glutathione concentrations in control and BSO treated cells. Cellular PSHr and TRPS were measured in rat liver cells that at the time of the analysis had been exposed to BSO for a period of time necessary to result in a 70% depletion of glutathione i.e. 24 hours for IAR 20 and 12 hours for IAR 6.1 and IAR 6.1 RT7. Quantitative computerised microdensitometric estimations were carried out on glutathione depleted cells as described for the control cells.

3 Results

IAR 20 cells in culture behave differently from the two tumorigenic cell lines. As summarised in Table 1, these cells were derived from a primary culture of a normal BD VI rat liver, and still retain characteristics of 'normal' cells such as well-defined epithelial morphology, contact-inhibition, inability to grow in agar and failure to induce tumours in syngeneic animals. IAR 20 cells are larger and have a slower population doubling time compared with IAR 6.1 and IAR 6.1 RT7.

Reactive protein-thiols (PSHr) are defined as the protein thiol groups available for relatively rapid reaction with a thiol-specific reagent. The DDD reaction with cellular protein thiols is a function of time: an arbitrary reaction time of 7 hours has been suggested (Nöhammer 1982) as the time necessary for DDD to react with PSHr before a significant contribution to the staining is due to cleavage of S-S bonds and reaction with all the potential SH groups present in the proteins (total reactive protein sulphur, TRPS). The appearance of rat liver cells from each cell line incubated with DDD for 7 hours and 14 days and then stained with FBB is shown in Fig. 1.

In Table 2 are reported the mean millimolar concentrations of PSHr as estimated by computerised microdensitometry, using the CYDENS program. These values were derived from 4 separate analyses and refer to measurements on a total of 900 cells for each cell line. IAR 6.1 and IAR 6.1 RT7 have greater amounts of PSHr compared to IAR 20: the difference of the mean cellular PSHr concentrations between each tumorigenic line and IAR 20 is statistically significant.

In Fig. 2 are shown the frequency distributions of PSHr in each of the three populations of rat liver cells: the mM concentrations calculated for single cells are plotted as classes of concentrations (band width 2 mM) versus percentage of cells per class. The two tumorigenic cell lines show a higher variability in the distribution of PSHr compared with IAR 20.

Total reactive protein sulphur, TRPS, is considered to represent the sum of PSHr, masked PSHr, protein disulphide and mixed disulphide groups. An arbitrary reaction time of 14 days was selected as the time necessary for the DDD reagent to combine with all the thiol groups and reactive disulphides present in proteins (Nöhammer 1982). The appearance of rat liver cells incubated with DDD for 14 days and stained with FBB is illustrated in Fig. 1.

In Table 2 are reported the mean millimolar concentrations of TRPS estimated by computerised microdensitometry using the CYDENS program. These values were derived from 4 separate analyses and refer to measurements on a total of 900 cells, for each cell line. The statistical analysis reveals a highly significant difference between the values for each tumorigenic cell line in comparison with IAR 20.

The frequency distributions of TRPS for the three rat liver cell lines were derived and plotted as described for PSHr values. As is evident from Fig. 3, the variability in total reactive protein sulphur content of the two tumorigenic cell lines is higher than the value indicated for PSHr.

PSHr and TRPS values were estimated in glutathione depleted rat liver cells in the same way as described for the control cells. Initial GSH levels, measured by fluorimetric assay, were as follows: IAR 20, 25.5 nmoles/mg protein; IAR 6.1, 37.6 nmoles/mg protein; IAR 6.1 RT7, 17.2 nmoles/mg protein. These values were derived from 4 separate analyses, for each cell line.

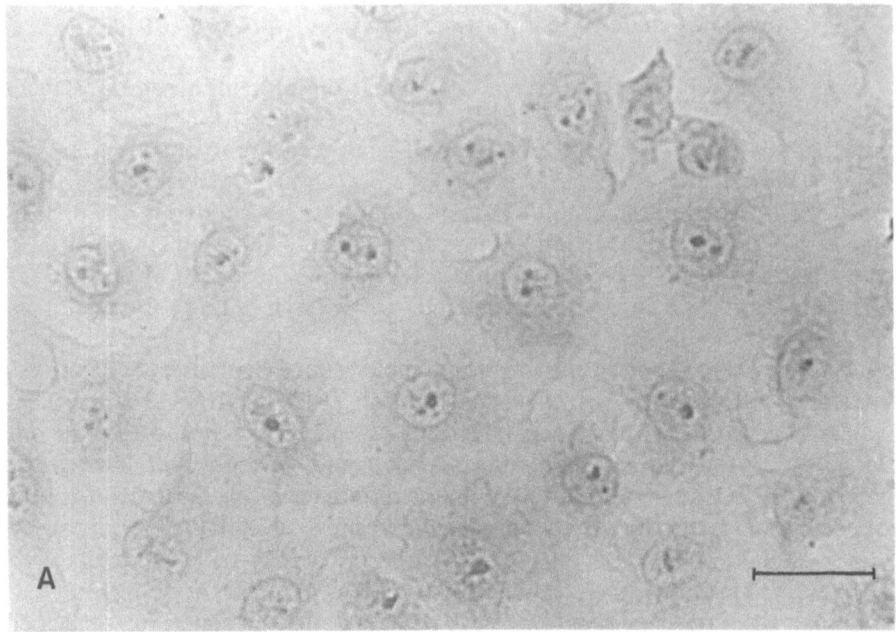

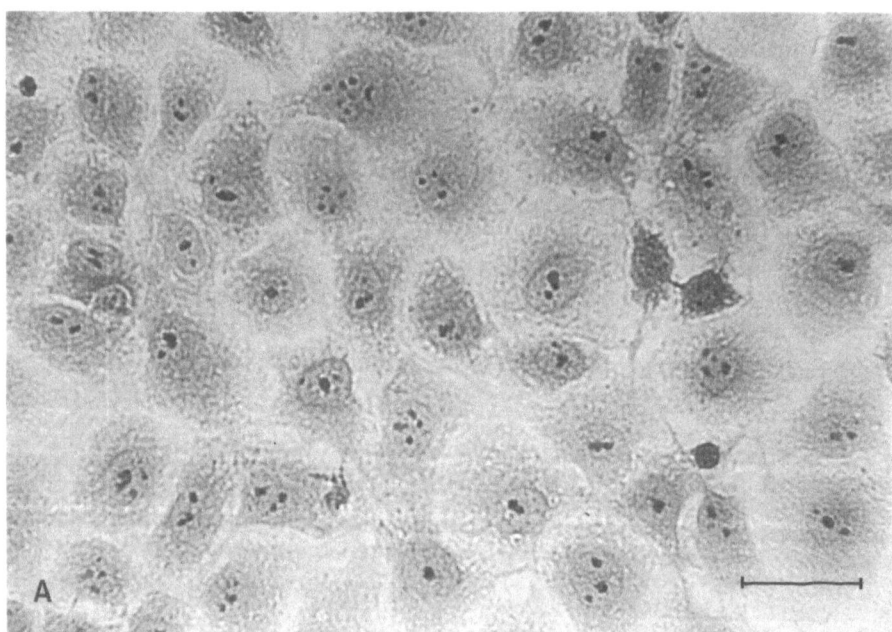

Fig. 1. Appearance of IAR rat liver cells incubated 7 h (above) and 14 days (below) with DDD and stained with FBB. The *bar* is equivalent to 30 μm. **A** IAR 20. **B** IAR 6.1. **C** IAR 6.1 RT7. *Above,* 7 h, *below,* 14 days

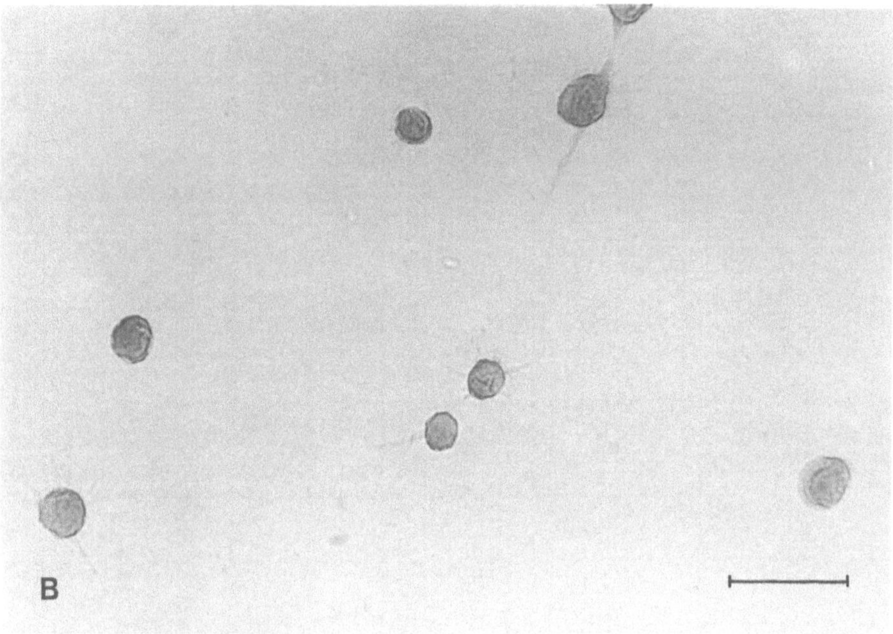

B

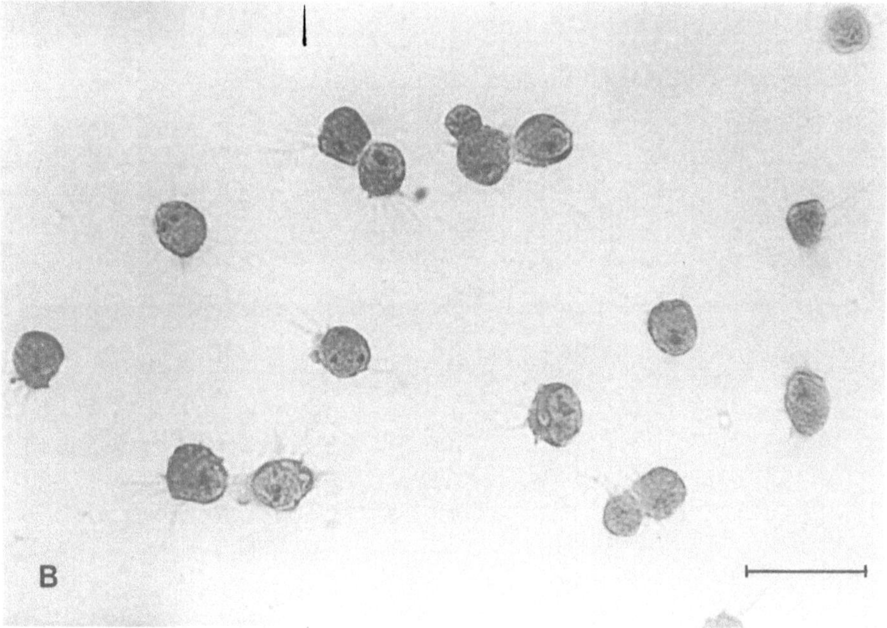

B

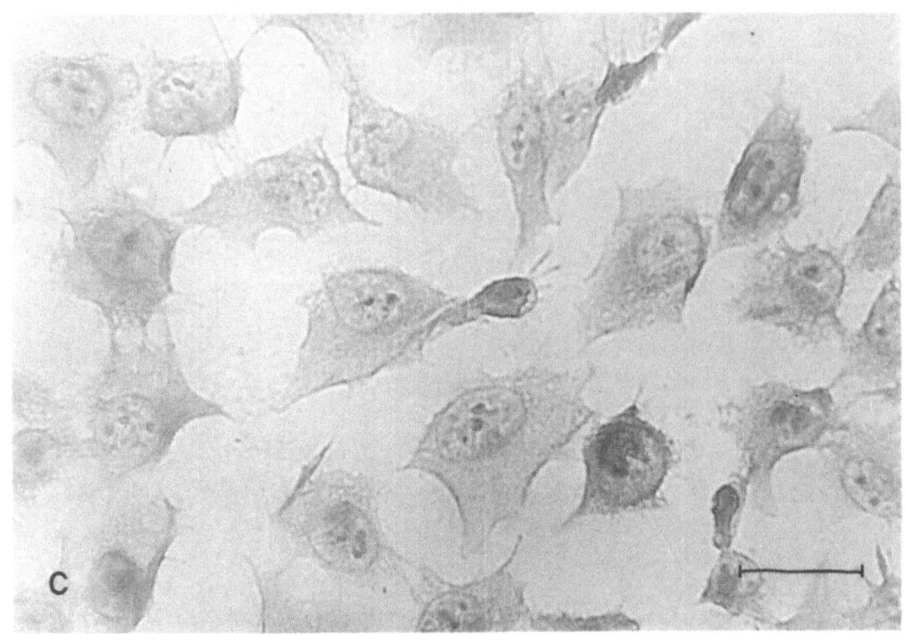

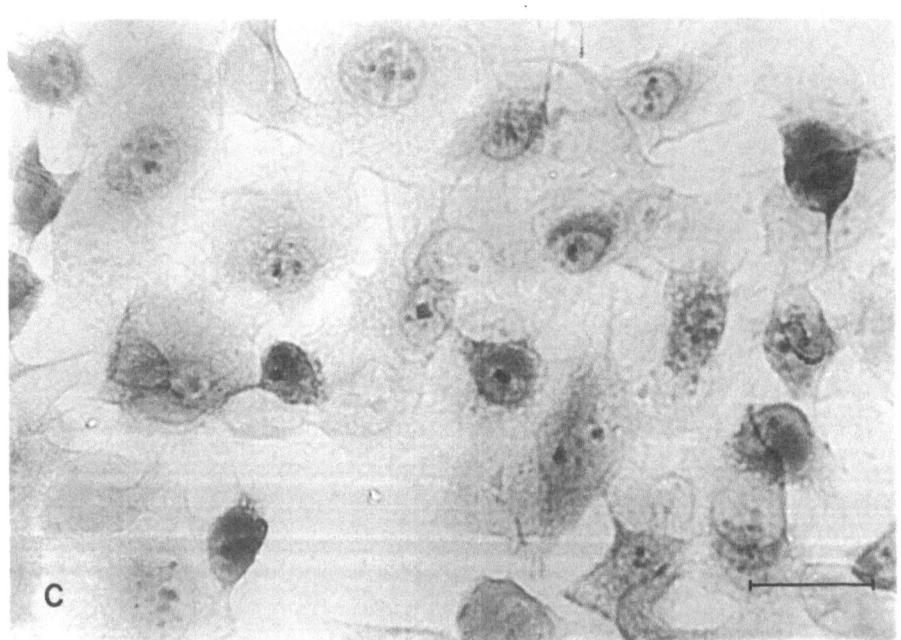

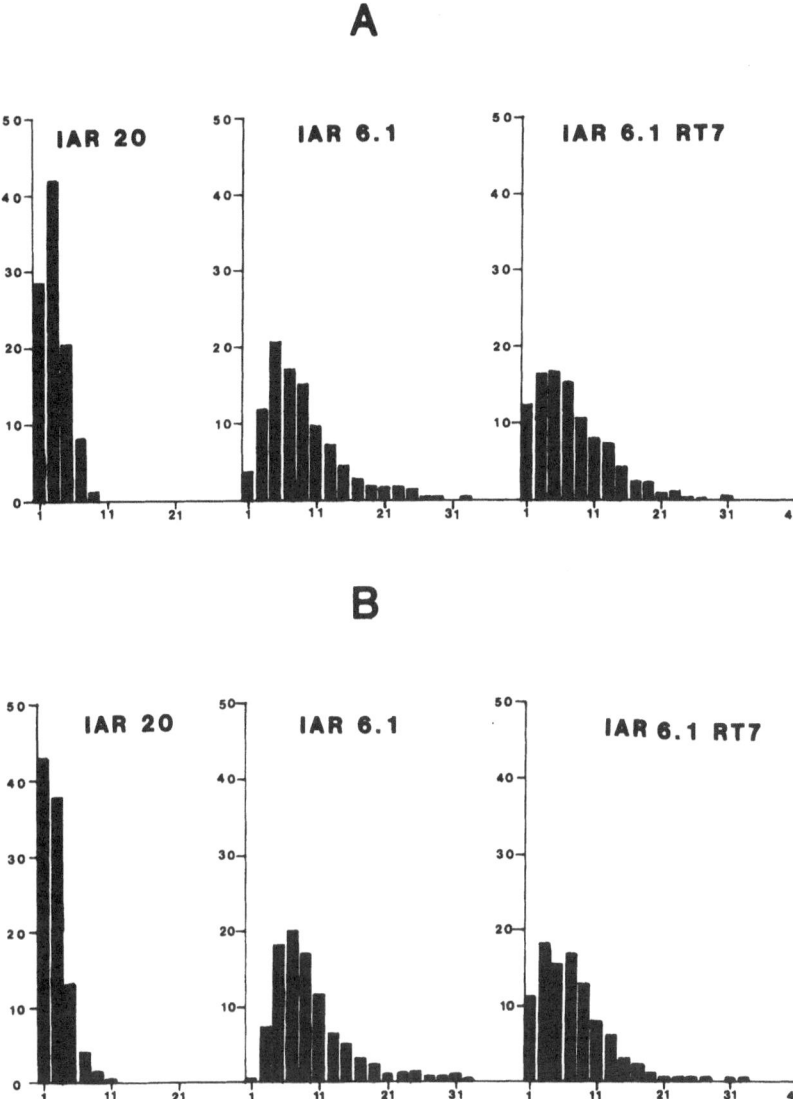

Fig. 2. Frequency distribution of PSH$_r$ (mM concentrations) estimated in IAR cell lines by computerised microdensitometry. *Abscissa,* classes of concentrations (class interval: 2 mM); *ordinate,* frequency (%). **A** Control cells. **B** Glutathione-depleted cells

IAR 6.1 was more sensitive to glutathione depletion: after 12 hours of BSO treatment these cells loose their ability to attach to the plastic surface and start to float into the medium.

The values obtained by cytodensitometry for PSHr and TRPS content of cells depleted of glutathione are reported in Table 2. These values are the mean mM

Table 2. Reactive (PSH$_r$) and total reactive protein sulphur (TRPS) concentrations in control and glutathione-depleted IAR rat liver cells, estimated by computerised microdensitometry

Cell line	Treatment	PSH$_r$	TRPS
IAR 20		3.2±0.06	24.1±0.3
IAR 20	0.02 mM BSO 24 h inc.	2.6±0.07	25.6±0.3
IAR 6.1		8.9±0.2	29.8±0.6
IAR 6.1	0.02 mM BSO 12 h inc.	9.9±0.2	32.4±0.6
IAR 6.1 RT7		7.9±0.2	46.6±0.9
IAR 6.1 RT7	0.02 mM BSO 12 h inc.	7.3±0.2	51.1±1.1

Values for PSH$_r$ and TRPS are the mean mM concentrations ± SEM, calculated from 4 separate analyses for the control cells and 3 separate analyses for the glutathione-depleted cells (BSO treated cells). Probability values of comparison of the means by Student's t-test are indicated as follows: a, $p < 0.5$; b, $p < 0.1$; c, $p < .0001$.

concentrations per cell calculated from 3 separate analyses and refer to a total of 800 cells per cell line.

These values show that there is a slight increase in the mean values which is significant at the 5% level in a two-tailed Student's t test (Table 2).

Glutathione depletion had no obvious effect on the pattern of variability of PSHr and TRPS levels in the three rat liver cell lines. The frequency distributions of PSHr and TRPS in glutathione-depleted cells are illustrated in Fig. 2 and Fig. 3, respectively. These distributions were derived and plotted as described for the control cells.

4 Discussion

As previously reported by Nöhammer (1982), the cytochemical reaction of thiol and disulphide groups with DDD, followed by the azo-coupling with FBB, represents a suitable and reproducible method for the estimation of total reactive protein sulphur content, defined as the sum of reactive thiols (PSHr), slow-reacting and masked thiol groups, and reactive disulphides. Because of the existence of thiol groups differing in their reactivities, the conditions of the histochemical staining procedure must be carefully controlled. The DDD reaction with protein

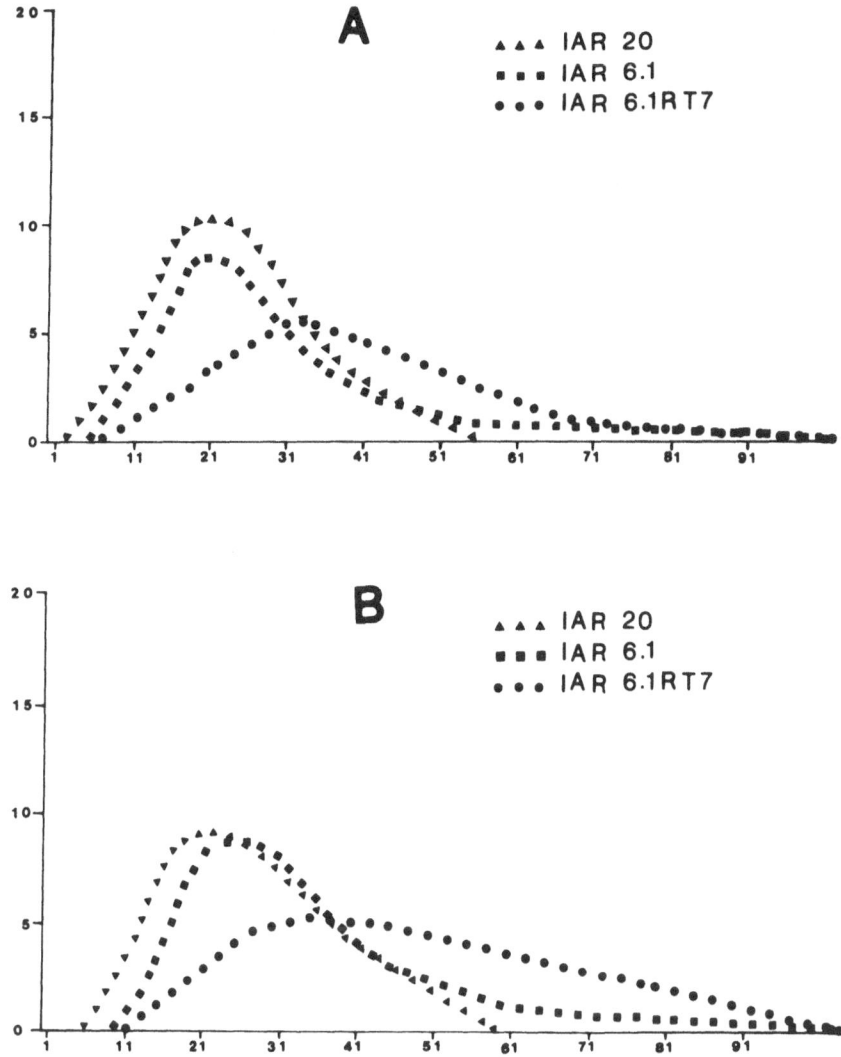

Fig. 3. Frequency distribution of TRPS (mM concentrations) estimated in IAR cell lines by computerised microdensitometry. *Abscissa,* classes of concentrations (class interval: 2 mM); *ordinate,* frequency (%). **A** Control cells. **B** Glutathione-depleted cells

thiols is a function of pH and time (Nöhammer 1982). At pH 8.6, exposed thiol groups react relatively rapidly with DDD. Masked -SH groups react more slowly and disulphide exchange with the DDD reagent also takes place (Barrnett and Seligman 1952). These slow reactions exhibit a quasi-linear time dependence after the first 7 hours of incubation (Nöhammer 1982). On the basis of the reaction kinetics there is, therefore, a case for distinguishing between rapidly reacting thiols (PSHr) and those that require longer incubation periods to be demonstrated. The sum of the two comprises the total reactive protein sulphur (TRPS).

Clearly, the kinetic distinction between the two classes of macromolecular cellular thiols and disulphides will vary according to the tissue and the conditions of incubation: the periods of 7 hours and 14 days employed in the procedures adopted from Nöhammer (1982) are somewhat arbitrary. Some indication that this method does measure different classes of cellular thiols comes from the fact that the PSHr is not a constant proportion of the TRPS in the different cell lines examined in this study. This could be explained either by the existence of different types of proteins in the cell lines with different thiol content or by differences in the redox status of the cellular protein thiol groups. To test this latter possibility, we examined the effect of depleting the cells of glutathione on the PSHr and TRPS levels in the cell lines, since glutathione depletion would be expected to modify the redox status of intracellular thiols and should be reflected in an alteration in the PSHr levels. However, our results demonstrate that severe depletion of glutathione does not diminish the PSHr. Indeed, the data indicate that there is a slight increase both in PSHr and TRPS in all the cell lines, perhaps due to diversion of additional cysteine to protein synthesis consequent on inhibition of GSH synthetase. This evidence favours the view that the cytochemical method does not measure the redox status of cellular macromolecular thiols but that the PSHr and TRPS values reflect the abundance of different species of proteins. This conclusion is supported by parallel studies using biochemical methods (Principe et al. 1988a). In the biochemical estimation of cellular protein thiols with the 5,5'-dithiobis-2-nitrobenzoic reagent (DTNB; Ellman 1959), the difference between reactive protein thiols and total reactive protein sulphur levels depends on the availability of the macromolecular thiol groups for reaction with the chromogenic agent. Similar values have been found for the mean total protein sulphur content estimated by the DTNB method and by microdensitometry (Principe et al. 1988a). The correspondence of values is closer for TRPS estimations than for PSHr where differences in the preparative procedure may account for the discrepancy. The biochemical method uses acid precipitation which may expose more cryptic thiol sites by denaturation of the proteins than are revealed by the alcohol fixation used in the cytochemical procedure.

Cytofluorimetric studies, using the probe o-phthalaldehyde (OPT; Cohn and Lyle 1966) to estimate macromolecular thiols in alcohol-fixed cells of the same IAR cell populations, have confirmed the differences in protein-thiol content found in the cell lines and these values are unresponsive to glutathione depletion (Principe P. et al. 1988b).

Our data clearly show that there are dissimilarities between the three rat liver cell lines (IAR 20, IAR 6.1 and IAR 6.1 RT7) which differ in total reactive protein sulphur content, in the proportion of reactive protein thiols and in the variability of their distribution in the population of cells examined. All these parameters exhibit a correlation with the degree of abnormality of the cell lines. Thus, the most tumorigenic line, IAR 6.1 RT7, possesses the highest TRPS value and the greatest variability in the cell population, whilst the 'normal' cell line, IAR 20, is at the other extreme of the range with IAR 6.1 occupying an intermediate position. Therefore, whilst the cellular macromolecular thiols as measured by the DDD-reaction, are apparently insensitive to alteration in reduced glutathione content, the differences in the protein thiol content and their variability in the cell lines examined imply inequalities in the type and/or distribution of structural proteins in the cell. This suggests the possibility that some element of the biochemical commitment to cell transformation may be reflected by altered macromolecular thiol content.

5 Summary

Protein-thiol levels may be altered in disturbed cellular growth and in certain types of neoplasia. To investigate the possible relationship between protein-thiol levels and cell transformation, three rat liver cell lines which possess different degrees of tumorigenicity have been studied. One cell line (IAR 6.1 RT7) was isolated from an epithelial tumour and cultivated in-vitro; a second (IAR 6.1) was transformed in vitro with a chemical carcinogen: this cell line induces epithelial tumours at the site of injection in syngeneic rats. The third rat liver cell line (IAR 20) has retained many characteristics of normal cells and is not tumorigenic. Computerised quantitative cytochemical investigations have been carried out to estimate and compare the protein-thiol content of cells of each line. The results demonstrate that three rat liver cell lines differ in their protein-thiol content, and there appear to be differences in the distribution of protein thiols in the cellular population within each cell line. These characteristics are correlated with the tumorigenic potential of the cell lines: the highest mean value of protein-thiol content and the greatest variability was measured in the tumour cell line (IAR 6.1 RT7). It appears that an abnormal macromolecular thiol content may be associated with cell transformation and could be regarded as an aspect of the biochemical commitment to abnormal cell growth.

Acknowledgements. P. Principe thanks Dr. R. Montesano for the IAR cell lines and Mr. P. Nicolas for helping in the preparation of the frequency distributions. The investigations reported here were supported by the Association for International Cancer Research.

References

Bahr GF, Moberger G (1958) Histochemical methods for the demonstration of SH groups in normal tissues and malignant tumours. Acta Pathol Microbiol Scand 42:109–132

Bajardi F, Schauenstein E, Nöhammer G, Unger-Ullmann C (1977) Quantification of protein-thiols in normal and pathological cells of the epithelium of the protio uteri. Acta Cytol (Baltimore) 21(4):573–577

Barrnett RJ (1955) SH and S-S groups of proteins. Tex Rep Biol Med 13:611–622

Barrnett JR, Seligman AM (1952) Histochemical demonstration of protein-bound sulphydryl groups. Science 116:323–327

Caspersson O, Révész L (1963) Cytochemical measurement of protein thiols in cell lines of different radiosensitivity. Nature 199:153–155

Cohn VH, Lyle J (1966) A fluorometric assay for glutathione. Anal Biochem 14:434–440

Ellman GL (1959) Tissue sulphydryl groups. Arch Biochem Biophys 82:70–77

Esterbauer H (1972) Beitrag zum quantitativen histochemischen Nachweis von sulphydrylgruppen mit der DDD-Färbung. I. Acta Histochem (Jena) 42:351–355

Esterbauer H (1973) II. Mitteilung: Bestimmung von SH-gruppen in unlöslichen Proteinen. Acta Histochem (Jena) 47:94–105

Esterbauer H, Nöhammer G, Schauenstein E, Weber P (1973) III. Mitteilung: quantitative cytospektrometrische bestimmungen an Ehrlich-Ascites Tumorzellen. Acta Histochem (Jena) 47:106–114

Griffith OW, Meister A (1979) Potent and specific inhibition of glutathione synthesis by buthionine-sulphoximine. J Biol Chem 254(16):7558–7560

Grimaldi HG, Caprio G (1981) Serum thiol concentrations in patients with malignant lymphoma. Tumori 67:411–413

Hissin PJ, Hilf R (1976) A fluorimetric method for determination of oxidised and reduced glutathione in tissues. Anal Biochem 74:214–226

Knock FE (1981) Possible role of accessibility of protein-thiol groups to the cancer state. Arch Geschwulstforsch 51(1):75–79

Lars-Gösta W (1964) A study on protein-bound SH groups in pulmonary cytodiagnosis. Almquist and Wiksell, Stockholm

Montesano R, Saint-Vincent L, Drevon D, Tomatis L (1975) Production of epithelial and mesenchymal tumours with rat liver cells transformed in vitro. Int J Cancer 16:550–558

Montesano R, Drevon C, Kuroki T, Saint-Vincent L, Handleman S, Sanford KK, DeFeo D, Weinstein IB (1977) Test for malignant transformation of rat liver cells in culture: cytology, growth in soft agar and production of plasminogen activator. INC'I 59(6):1651–1658

Nöhammer G (1982) Quantitative microspectrophotometrical determination of protein thiols and disulfides with 2-2'-dihyroxy-6-6'dinapthyldisulfide (DDD). Histochemistry 75:219–250

Nöhammer G, Schauenstein E, Bajardi F, Unger-Ullmann C (1977) Microphotometrical quantification of protein-thiols in morphologically intact cells of the cervical epithelium. I. Acta Cytol (Baltimore) 21(2):341–344

Nöhammer G, Bajardi F, Benedetto C, Schauenstein E, Slater TF (1986) Quantitative cytospectrophotometric studies on protein thiols and reactive protein disulphides in samples of normal uterine cervix and on samples obtained from patients with dysplasia or carcinoma in situ. Br J Cancer 53:217–222

Principe P, Wilson GD, Riley PA, Slater TF (1988a) Flow cytometric analysis of protein-thiol groups and their distribution in rat hepatocytes in relation to the cell cycle and intracellular glutathione levels, (in preparation)

Principe P, Riley PA, Slater TF (1988b) Biochemical estimations of protein-thiol contents in rat hepatocytes depleted of glutathione, (in preparation)

Sussman M (1964) Growth and development. Prentice-Hall, Englewood cliffs

Tas S, Rodriguez LV, Drewinko B, Trujillo JM (1984) Evidence for extensive intermolecular disulfide bonds of the proteins of non-ionic detergent high-salt-resistant skeletons of normal lymphocytes and the altered structure in leukaemia. Int J Cancer 34:329–333

Computerized Digital Video Microdensitometry Applied to the Quantitation of Intracellular Macromolecules or Their Functional Groups

D. J. Spargo and P. A. Riley

Department of Chemical Pathology, University College and Middlesex School of Medicine, Cleveland St., London W1P 6DB, UK

1 Introduction

The use of specific histochemical stains to demonstrate features of cell metabolism or the state of cellular differentiation has provided much qualitative data about the biochemical status of cells and tissues. A quantitative estimate of a particular biochemical phenomenon may be obtained by using specific histochemical stains and microspectrophotometry (microdensitometry). In this way a measure of the variability in concentration of a substance in a cell population or within a tissue may be obtained, which could not be provided by biochemical assays which involve cellular disruption.

Measurement of absorbance, and thereby concentration of a substance essentially involves the comparison of light intensities in a reference sample and a sample absorbing light at a particular wavelength. The greater the concentration of absorbing material the greater the absorbance in accordance with the Beer-Lambert equation i.e.

$$\text{Transmittance } (T) = \frac{\text{Transmitted light}}{\text{Incident light}} = \frac{I_e}{Io} \tag{1}$$

$$\text{Absorbance } (A) = \log_{10} (1/T) \tag{2}$$

$$\text{but also: } A = e.C.l \tag{3}$$

where: e is an extinction coefficient for a particular chromophore at a specific wavelength; C is the concentration of the chromophore (e.g. DDD Fast Blue B); and l is the pathlength of material absorbing light at a specific wavelength.

There are, however, several factors to be taken into account when undertaking measurements using digitized images with a video camera as the detector. These are: (a) the resolution of the digital conversion and electronic characteristics of the camera; (b) spectral response of the detector; (c) optical effects such as those produced by scattered light and changes in numerical aperture, of the objectives and condenser lenses; and (d) temporal variation in intensity due to change in camera stability or fluctuations in the light source.

Nigam et al. (Eds.), Eicosanoids,
Lipid Peroxidation and Cancer
© Springer-Verlag Berlin Heidelberg 1988

In this paper we present details of a densitometric program CYDENS which uses computer assisted image analysis on video images of cells for the relatively rapid measurement of substances within single cells grown as a monolayer culture or prepared as smears, and present data that enable the contribution of the above factors in the measurement of absorbance of substances in cells and tissues to be assessed.

2 Materials and Methods

2.1 Microscopy

Objects were viewed with a Zeiss photomicroscope using transmitted light adjusted for Köhler illumination. Depending upon the absorption characteristics of the stain involved different narrow bandwidth filters (Glen Creston, UK) were employed. In the case of the DDD – FBB stain for protein thiols a 540/25 nm filter was used. In experiments to test the spectral response of the system, broad bandwidth filters (Ilford, UK) were employed (see Results). Absorbance measurements for calibration were made using neutral density filters (Carl Zeiss, Oberköchen, FDR) with specific transmittance characteristics.

The video image was formed by a VKM98E monochrome MOS-TV camera (Hitachi – Denshi, UK) connected to an Intellect 200 image analyser (Quantel Ltd, UK), interfaced to a PDP11/23 + computer (Digital Equipment Corporation, USA). The output from the camera was split and the operator could view an immediate image (i.e. that falling on the MOS HE998224 sensor) on a black and white monitor as well as the computer controlled image which was displayed on a colour monitor (Electronic Visuals Ltd, UK) (see Fig. 1).

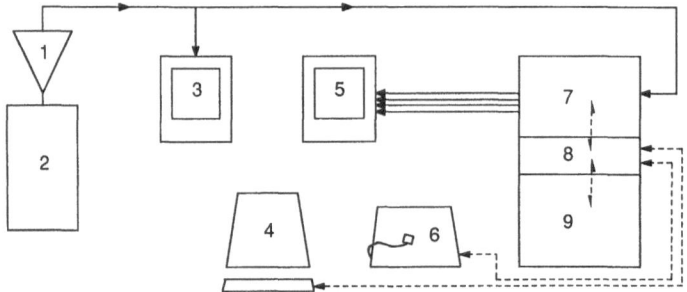

Fig. 1. A schematic representation of the apparatus used in videomicrodensitometry. *1,* Hitachi VKM98E MOS-TV camera; *2,* Zeiss photomicroscope; *3,* monochrome monitor; *4,* visual display unit and keyboard; *5,* RGB colour monitor; *6,* summagraphics 'Bit Pad One' digitizing tablet; *7,* Quantel Intellect 200 image analyser; *8,* Digital Equipment Corporation PDP11/23 + computer; *9,* Magnetic storage devices (discs and tape). *Solid lines* indicate video data signals; *broken lines* indicate computer control/data lines

Table 1. Microscope objective resolution

Objective	Microns/pixel	
	Horizontal	Vertical
6×	2.00	1.43
10×	1.32	0.93
25×	0.53	0.37
40×	0.32	0.23
63× (oil)	0.21	0.16
100× (oil)	0.14	0.09

The lenses of the microscope were calibrated using a microscope stage micrometer slide, (Carl Zeiss, Oberköchen, FDR) so that measurements of length in pixels could be converted to standard units. The aspect ratio (3/2) of the television system results in a difference in the horizontal and vertical resolutions. Table 1 shows the resolution obtained with different lenses.

2.2 Image Analysis

2.2.1 Hardware

The analogue video signal from the TV camera is digitised into 8 bits (256 grey levels) by the Intellect 200, and the image can be directed to any one of four framestores, which are software addressable microchip memories representing an array of 512 × 512 picture elements (pixels), each pixel having a grey value (luminosity) between 0 and 255. Video output from the Framestore is to a colour monitor via a colour processor which converts the monochrome image into a pseudocolour image by altering the individual red, green and blue elements of each pixel using a software loaded look-up table to control the output processor function.

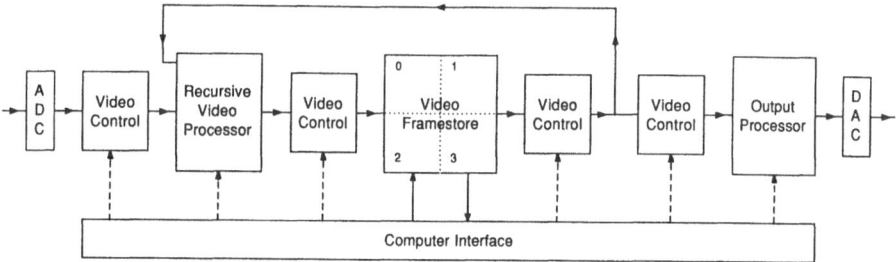

Fig. 2. A schematic representation of the basic hardware in the Intellect 200, showing video data paths *(solid lines)* and computer control paths *(broken lines)*. *ADC,* analogue to digital converter. *DAC.* digital to analogue converter

Regions of the framestore can be demarcated using a digitizing tablet (Summagraphics, USA). The system is configured so that the addresses of the divisions in the tablet correspond to addresses of pixels in the framestore(s). In this way it is possible to 'point' to a pixel address quickly and accurately, or demarcate a feature in the image by corresponding movements of the cursor over the surface of the digitising tablet (Fig. 2). The individual pixels in each framestore can be accessed via a data highway (computer interface in Fig. 2) which permits communication between the PDP11/23+ computer and the Intellect. In this way measurements of individual grey values in each pixel may be obtained by the computer, or a grey value inserted in a pixel locus (or loci) in any framestore (for image processing).

The use of multiple framestores to store separate (background and object) images allows for the correction of variation in illumination across the field of view and any differences in the sensitivity of the regions of the MOS sensor without laborious computation since pixels in equivalent positions in the two images will be subject to equivalent influences of sensitivity and shading. Inhomogeneities in areas of the specimen background are compensated for using recursive video integration of 8 reference images from various areas of the slide. This is accomplished by the recursive video processor; a device that located between the ADC and the framestore (see Fig. 2) which adds, or substracts (in real time) the grey values from the incoming video signal and the value from the corresponding location in a selected framestore according to the formula:

$$FS = K \cdot ADC + (1 - K) \cdot FS \tag{4}$$

where FS = framestore data; ADC = analogue to digital converter input (incoming video); K = shift factor or noise reduction constant and which can have values between $1/2^1$ and $1/2^8$, i.e. 1/2 to 1/256.

$$FS = (1/2^N) \cdot ADC + (1 - 1/2^N) \cdot FS \tag{5}$$

A weighted average is derived where n has successive values between 1 and 8 and the process is repeated eight times (see Fig. 3) since this is less subject to transient variations in illumination of the ADC component at the time of image addition.

2.2.2 Software

The CYDENS densitometric program is depicted as a flow chart (see Fig. 4) showing the essential steps involved in densitometric measurement using computerised digital video images. The design of the program however permits considerable scope for embellishment and development of the algorithm to suit individual needs and problems. CYDENS begins by interrogating the user for information such as experiment code, cell type, stain, extinction coefficient etc. The information is written as a heading to the results file that is subsequently generated. The next task for the user is to select different reference areas of the slide and capture the image by a single keystroke from the keyboard. Once a reference image is captured and integrated by recursive video processing (see Fig. 3) the program returns to live video until the cycle is completed. The final reference image is stored in framestore 1. The slide is then searched by the op-

1. Load initial image(s) into Framestore

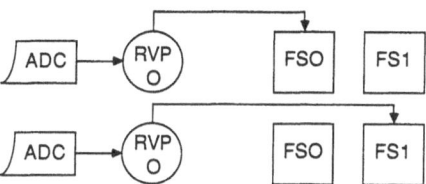

2. Add subsequent images with recursive-video

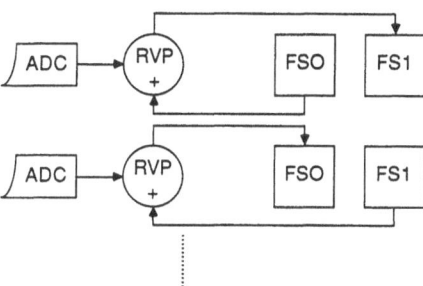

Fig. 3. A schematic representation of the recursive video averaging process for the formation of the reference image. The initial image is loaded into the framestore with the recursive video processor off *(RVP O)*; the output from the framestore is then put through the recursive video processor and added to the incoming video signal with the recursive video processor on *(RVP +)*. The proportion of incoming to stored data is software controlled. The addition is repeated until the cycle is complete

erator for a field of cells for measurement, and a suitable image stored in framestore 0.

The operator then selects areas of the image (10 blocks of 10·10 pixels) where there are no cells, these areas are compared with the corresponding areas of the reference image. If the luminosities in these areas match those in the background the user is permitted to move to the next stage of the program, if there is no match the object image is recaptured. Once a suitable object image is obtained the user indicates the cells for measurement by pointing to them with the bit pad cursor, and depressing the key on top of the cursor to mark them. The cells are then enclosed in a box (drawn as an overlay on the colour monitor) defining the area of the framestores in which the pixel luminosities are to be measured. These are then read from the corresponding regions of framestores 0 and 1 and the mean transmittance per pixel computed. From this the absorbance and concentration of the substance in each selected cell is calculated according to equations (1), (2) and (3). The computation of the pathlength (l in equation (3)) is achieved from the mean volume of the particular cell type, obtained from measurements on a Coulter Counter (Coulter Electronics Ltd, Luton, UK) and the maximal area enclosing the cell determined by the program CYDENS. Hence:

$$\text{Mean Pathlength } (l) = \frac{\text{Mean cell volume}}{\text{Cell area}} \tag{6}$$

The extinction coefficient in equation (3) is provided by the user as data from values obtained by calibration. In the case of DDD – Fast blue B it was $1.90 \cdot 10^2$ a value consistent with published data (Nöhammer 1982).

When measurement of all the selected cells in one image is complete the operator repeats the selection of another image and goes through the background

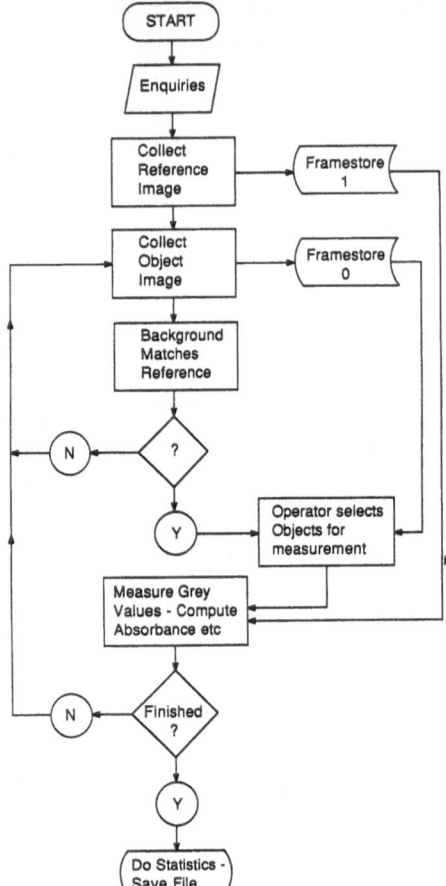

Fig. 4. Flow chart for the program CYD-ENS. (See text for details.)

checking routine before proceeding with further measurements. At the end of each set of measurements a data file, and statistical calculations are performed.

3 Results

3.1 The Effect of Digital Resolution

The analogue to digital converter (ADC) converts the incoming video signal voltage into a binary digit of up to 8 bits which is then stored in the framestore. Fluctuations in the voltage are due to proportional fluctuations in the light intensity and camera performance. At low light intensities the degree of uncertainty is high since the alteration of the least significant bit (LSB) in a 3 bit number rep-

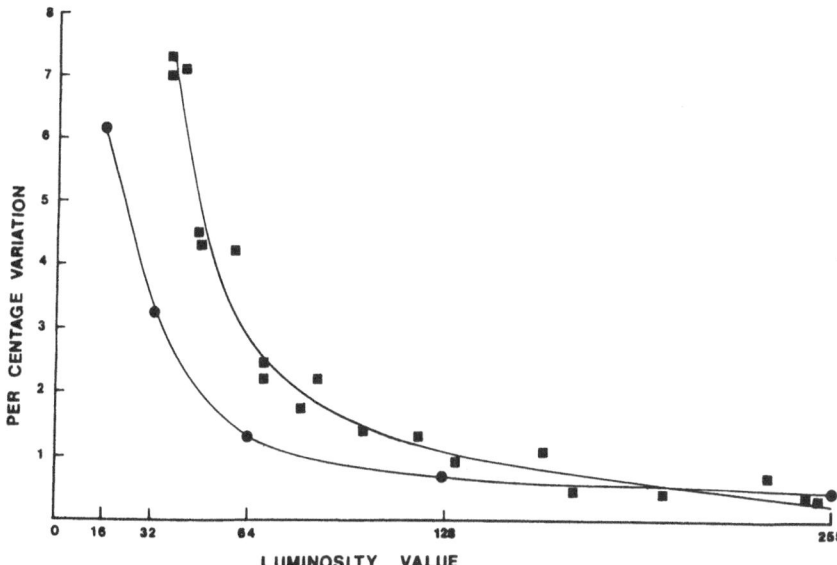

Fig. 5. Variation in the pixel luminosity values. The line with *squares* shows the variation obtained in 100 different measurements of the central pixel, at varying luminosities. The line with *circles* shows the theoretical variation due single (least significant) bit fluctuations. Measurements were made with the ADC tuned to account for the camera dark current, i.e. the luminosity value obtained when no light is falling on the detector

resents a greater proportion of the number than the LSB of an 8 bit number. Thus the system response is most accurate in the higher luminosity values. This is illustrated in Fig. 5 where the percentage variation for a luminosity value was derived from sampling 100 values of the central pixel luminosity over a 4 second period, at different light settings. It can be seen that the resolution falls off at luminosities below 128 (4 bits), above this only the LSB appears to be fluctuating, below this value fluctuations begin to occur in the two LSBs.

Video cameras are manufactured to operate at a wide range of light intensities and under a variety of circumstances, for this reason they possess an automatic gain control circuit (AGC). This enables the camera to adjust its sensitivity and output so that should there be a transient variation in illumination (e.g. a cloud temporarily obscuring the sun) a picture can be obtained at diminshed light levels. For analytical purposes this circuitry may be disabled (see Porro et al. 1984).

In the case of the VKM98E diode D30 is disabled which permits manual trimming of the gain with potentiometer RT7). The effect of the AGC on absorbance measurements is to introduce a systematic deviation from the theoretical value. This can be seen in Fig. 6. The deviation is compensated for in the programming.

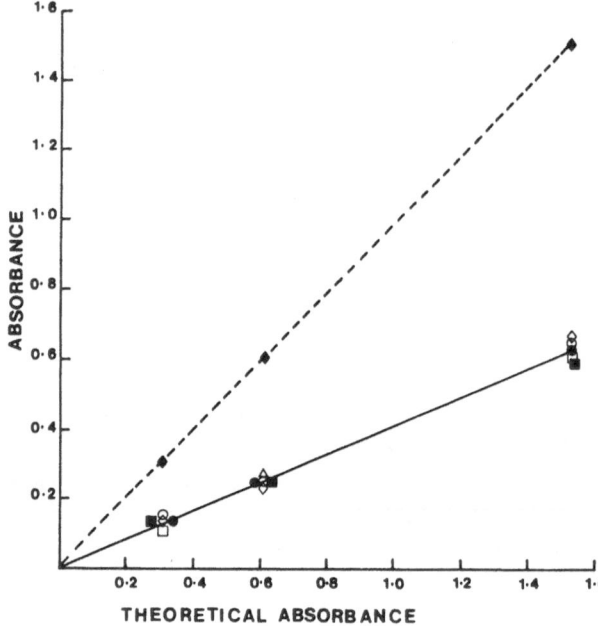

Fig. 6. The function of observed absorbance *(solid line)* against theoretical absorbance *(broken line)*, using lenses with different numerical apertures. □, 10× (N.A. 0.22); ◇, 25× (N.A. 0.65); ○, 40× (N.A. 0.65); ■, 63× (N.A. 1.40); ●, 100× (N.A. 1.30)

3.2 The Effect of Numerical Aperture of the Objective

Numerical aperture (N.A.) is dependent upon the refractive index (n) of the object space, the focal length (f), and the clear aperture diameter of the lens (∅). The relation is given by:

$$N.A. = n.Sin(f/∅) \tag{7}$$

Using neutral density filters with known transmittances, absorbance values for the central pixel were calculated, and the results shown in Fig. 6. It can be seen from this that the numerical aperture has no effect on the absorbance measurements.

3.3 The Effect of Scattered Light on Absorbance

The degree of flare or scattered light falling on the object plane can be altered by altering the field iris diameter. The field iris is normally opened so that the light fills the field of view. The iris diameter in the object plane of a 10× objective was measured with a stage micrometer, and transmittance measurements for the central pixel taken at different iris aperture settings (200 – 700 μm). The values obtained at 560 nm are shown in Table 2. It can be seen from the table that the contribution due to flare is between 8% and 9% of the luminosity value measured.

Table 2. The effect of field iris diameter on central pixel luminosity

	Transmittance		
	100%	50%	25%
Luminosity	17	14	10
(Range)	(198–215)	(147–161)	(113–123)
Median	206.5	154	118
Range/median	8.2%	9.0%	8.5%

3.4 The Effect of Differing Wavelength on Absorbance Measurements

When setting up the microscope and camera for measurements, the light intensity is normally set to give a mean pixel luminosity of a certain value, a background luminosity of 120 was used for measurements involving protein thiols in hepatocytes. The operator always sets the system up in the same way before measurement, i.e. with the same initial mean pixel luminosity. Using broad bandwidth filters to provide illumination in the ranges 400–500 nm, 500–600 nm and 600–700 nm, it can be seen from Fig. 7 that by adjusting the initial conditions of the system to give the same mean luminosity no difference in response is observed in the different spectral regions despite differential sensitivity of the detector element.

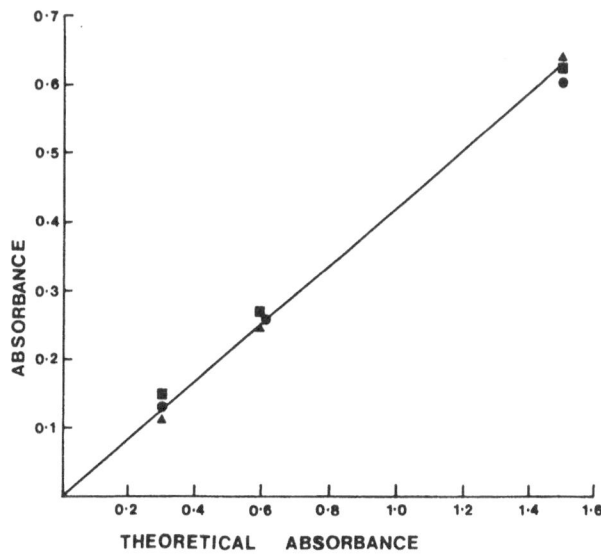

Fig. 7. The effect of different spectral regions on the obsorbance produced by neutral density filters. ▲, 600–700 nm (Ilford filters 502 and 202); □, 500–600 nm (Ilford filter 404); ●, 400–500 nm (Ilford filter 501)

4 Discussion

The quantitation of substances or their reactive groups in cells and tissues by video microdensitometry is subject to a number of problems and practical errors. Our results demonstrate that for transmitted light in the visible range, provided the illumination is correctly adjusted, the precision, sensitivity and reproducibility of the system is adequate and variation is within acceptable range ($< 10\%$). Adjustment of the illumination takes into account changes in the colour temperature of the source and the effect of temporal variations in filament current are monitored during the procedure adopted in our program. We have demonstrated that the system is unaffected by the choice of objective and condenser optics, provided that they are properly aligned and adjusted for Köhler illumination with the field iris opened to just beyond the edge of the observed area of the object. Under these conditions the effect of 'flare' from scattered light is within acceptable limits for most purposes where a relatively homogeneous field is observed. Our results demonstrate that the luminosity response of the detector is not dependent on the spectral response. Adjustments of the zero setting to take into account the camera gain are made initially and do not require routine attention. The shape of the luminosity response curve is modified by the automatic gain control circuitry present in most video cameras and this has either to be disabled or taken into account by the construction of calibration curves.

The procedure used in the CYDENS program obviated the need to adjust for any spatial bias in the detector elements since all the data are constructed from pixel by pixel comparisons of object and reference fields at identical locations. Temporal variations in camera responsiveness is monitored during the sampling procedure.

A technical limitation of the system is the interposition of digital to analogue conversion in the camera which converts an element matrix of 388×477 into a video signal which is then reconverted by the ADC in the Intellect and stored as a 512×512 pixel array. There is therefore some empty magnification of the field. In addition, the frequency range of the ADC limits the digitisation to an 8 bit number for each pixel which leads to the errors that we have noted in the low luminosity range.

The range and degree of variability in the biological samples to be measured far exceeds the errors inherent in any technical limitations of the system, and we are satisfied that video microdensitometry can be usefully applied to absorbance measurements on cells and tissues.

A problem arises where it is desired to estimate the concentration of a substance in cells, either cultivated as a monolayer or dispersed as a smear. In order to convert the absorbance measurements into concentration it is necessary to calculate the path length. This problem does not arise with histological sections where the pathlength is equivalent to the section thickness, although other difficulties regarding the sampling of the specimen complicate the issue. We have solved the pathlength problem for cells by making use of the modal cell volume of the cells to be sampled (estimated from Coulter Counter data). The CYDENS program calculates the projected area of the cell being measured and derives an average pathlength by the procedure set out in the paper. This estimate inevitably introduces bias into the results because for cells smaller than the modal volume the pathlength will be an overestimate and vice versa, and this caveat should borne in mind when interpreting concentration data. Where it has been applied, however, the correspondence with biochemical estimates of the mean concentrations has been satifactory (see Principe et al., this volume).

Our results show that an essentially simple routine set-up operation, followed by the measurement of initial and transmitted luminosities stored as digitized images in framestore memory permits the reproducible measurement of absorbance or concentration of a chromophore. The novel features of the program CYDENS are the recursive video averaging of reference images, the use of a background checking routine to avoid temporal aberrations in illumination and sensitivity of the detector and a pixel by pixel comparision of luminosities in reference and object images to obviate spatial aberrations in sensitivity and illumination.

The flexibility of computerised image analysis allows several approaches to measurement. Our approach has been to measure small areas of the framestores without recourse to image processing, which tends to be demanding on computer time. In our procedure only the significant part of the image is subjected to analysis, and the limiting factor in measurement time is the selection of cells by the operator. CYDENS, therefore, permits the research worker to use the image analyser as an experimental tool rather than as a device designed for automated measurement. Often the nature of the material e.g. cells grown as monolayers or presented as smears, does not lend itself easily to a program that can automatically detect them since the diversity of form, size and luminosity distributions in the objects presents a complex non-uniform image. The discrimination between cell and non-cell is perhaps best left to the user rather than to a program, especially in the case of research applications.

One of the advantages of using computer aided image analysis over conventional microspectrophotometry is that measurements are made from an image, which may be stored, and subsequent or repeated measurements do not require the presence of the specimen, and the stored image is not subject to the same processes of degradation as the original specimen.

Acknowledgements. We are grateful for financial support from the AICR. The Intellect 200 and computer equipment were purchased with a generous grant from the National Foundation for Cancer Research. We thank Mr. T. J. Herklots for designing and building ancillary electronic equipment and computer interfaces.

References

Nöhammer G (1982) Quantitative microspectrophotometrical determination of protein thiols and disulphides with 2-2-dihydroxy-6-6-dinaphthyldisulphide (DDD). Histochemistry 75:219–250

Porro C, Fonda S, Baraldi P, Biral GP, Cavazzuti M (1984) Computer-assisted analyses of [^{14}C]2-DC autoradiographs employing a general purpose image processing system. J Neurosci Methods 11:243–250

Modulation of Hepatocarcinogenicity of Aflatoxin B_1 by the Chlorinated Insecticide DDT

W. Rojanapo[1], A. Tepsuwan[1], P. Kupradinun[2] and S. Chutimataewin[1]

1 Introduction

Aflatoxin B_1 (AFB$_1$), a metabolite of the fungi *Aspergillus flavus* and *Aspergillus paraciticus,* is a widespread contaminant of human food supplies (Busby and Wogan 1984) and has been shown to be a potent hepatocarcinogen for several species of animals, including primates (Busby and Wogan 1984; Newberne and Butler 1969; Sieber et al. 1979). It has also been implicated as one of the etiological factors in the development of acute hepatitis (Serck-Hansen 1970) and primary hepatocellular carcinoma in Africa and Asia including Thailand (Alpert et al. 1971; Peers and Linsell 1973; van Rensburg et al. 1985; Bulatao-Jayme et al. 1982; Shank et al. 1972a–d; Peers et al. 1987).

p,p'-Dichlorodiphenyltrichloroethane (DDT), an organochlorine insecticide, is also found to be present in various kinds of food supplies in Thailand since it is still used to control the spread of malarial disease. DDT has been shown to increase the incidence of liver tumours in certain strains of rats and mice and is, therefore, considered as a liver carcinogen (Fitzhugh and Nelson 1947; Innes et al. 1969; Terracini et al. 1973; Tomatis et al. 1972). However, subsequent experiments have indicated that DDT may not be a carcinogen but might be appropriately classified as a tumour promoter (Grasso and Crampton 1972; Thorpe and Walker 1973; Peraino et al. 1975). It has been shown to enhance the effect of 2-acetyl-aminofluorene, diethylnitrosamine and 3'-methyl-4-(dimethylamino)-azobenzene in inducing liver tumours in rats and mice when given after these carcinogens (Peraino et al. 1975; Nishizumi 1979; Kitagawa et al. 1984; Williams and Numoto 1984). Moreover, DDT has also been reported to increase the incidence of mammary gland tumours in male rats treated with 2-acetamino-phenanthrene (Scribner and Mottet 1981).

These effects of DDT were very similar to those of phenobarbital which had been shown previously to enhance the hepatocarcinogenicity of those carcinogens (Peraino et al. 1971, 1973, 1975; Ito et al. 1980; Kitagawa and Sugano 1978). On the other hand, however, phenobarbital has also been reported to inhibit the hepatocarcinogenicity of various carcinogens including AFB$_1$ if given to animals prior to carcinogens (Ishidate et al. 1967; Kunz et al. 1969; Peraino et al. 1971;

Biochemistry and Chemical Carcinogenesis Section[1] and Laboratory Animal Section[2], Research Division, National Cancer Institute, Rama VI Road, Bangkok 10400, Thailand

Nigam et al. (Eds.), Eicosanoids,
Lipid Peroxidation and Cancer
© Springer-Verlag Berlin Heidelberg 1988

McLean and Marshall 1971). To our knowledge, the modulation of DDT on the hepatocarcinogenicity of AFB_1 has never been previously documented. Since these 2 compounds are widely spread in the environment it is of interest to examine the effect of DDT on the carcinogenicity of AFB_1. This paper reports the results of DDT treatment on AFB_1-induced hepatocarcinogenesis in Wistar male rats when given prior to, simultaneously with, or following the carcinogen.

2 Materials and Methods

2.1 Chemicals

Aflatoxin B_1 was purchased from Makor, Israel. DDT and dimethylsulfoxide (DMSO) were from BDH chemicals Ltd, Poole, UK and E. Merck, Darmstadt, FRG, respectively.

2.2 Animals and Diets

Male Wistar rats weighing 70–80 g were obtained from the National Laboratory Animal Center, Mahidol University, Salaya, Nakronpathom, Thailand. All animals were housed in suspended stainless steel mesh cages in an air-conditioned room at 25–27°C under natural light-dark cycle. They were given a standard laboratory chow diet (Pokphand Animal Feed Co. Ltd, Bangkok, Thailand) and water ad libitum.

2.3 Schedule for Chemical Administration

A total of 140 rats were randomly divided into 6 groups and treated with various chemicals as summarized in Table. AFB_1 dissolved in DMSO (500 µg/ml) was administered to rats by gastric intubation for a total of 1.9 mg per rat in a period of 14 weeks according to the protocol described by Kalengayi and Desmet (1975). DDT was dissolved in safflower oil (75 mg/ml) and given intragastically to rats for 15 weeks starting at various times, i.e. 1 week prior to AFB_1, simultaneously with AFB_1 and after the completion of AFB_1 doses. Five animals were housed together in a $58 \times 28 \times 18$ cm stainless steel cage.

Animals were weighed and examined for liver tumours once a week, and they were necropsied when they died or became moribund. All surviving animals were sacrificed 52 weeks after the first treatment and various organs including liver, spleen, lung, kidney and intestine were removed, weighed and then fixed in formalin-sodium acetate solution. Routine histological examination of these tissues was performed after they were embedded in paraffin, sectioned and stained with hematoxylin and eosin. Special staining methods were also used in some

Table 1. Experimental design

Group	Number of animals	Body weight (g)	Treatment	Duration (weeks)
1	20	70–80	DMSO	
			0.10 ml, 2 times/week	4
			0.15 ml, 2 times/week	10
2	20	70–80	DDT	
			75 mg/kg, 2 times/week	15
3	25	70–80	AFB_1	
			50 µg, 2 times/week	4
			75 µg, 2 times/week	10
4	25	70–80	DDT (as in group 2)	15
			AFB_1 (as in group 3)	14
			DDT was given 1 week prior to AFB_1	
5	25	70–80	DDT (as in group 2)	15
			AFB_1 (as in group 3)	14
			DDT was given simultaneously with AFB_1	
6	25	70–80	DDT (as in group 2)	15
			AFB_1 (as in group 3)	14
			DDT was given 15 weeks after AFB_1	

instances for differentiation between hepatocellular carcinoma (HCC) and cholangiocellular carcinoma (carcinoma of the bile duct, CCC).

The data on the tumour incidence was analyzed using the statistical techniques described by Peto et al. (1980) while others were analyzed using Student's t-test.

3 Results

AFB_1 and DDT whether given alone or in combination with each other had no effect on the growth of the rats throughout the experiment (Fig. 1). Table 2 shows the survivals, body and liver weights of animals at time of killing. A total of 13 rats from all groups died of pulmonary pneumonia within 10 weeks of the first treatment. There was no remarkable lesion in the liver, thus they were excluded from the experiment. Body weights of animals in any treated groups were not significantly different either from each other or from those in the control group. However, liver weights of animals treated with AFB_1 alone, but not with DDT alone, were significantly higher than those of control rats. DDT treatment affected the liver weights of animals receiving AFB_1 but differently depending upon the sequence of exposure to AFB_1 and DDT. It was found that DDT pretreatment (group 4) significantly inhibited the increase in liver weights while it

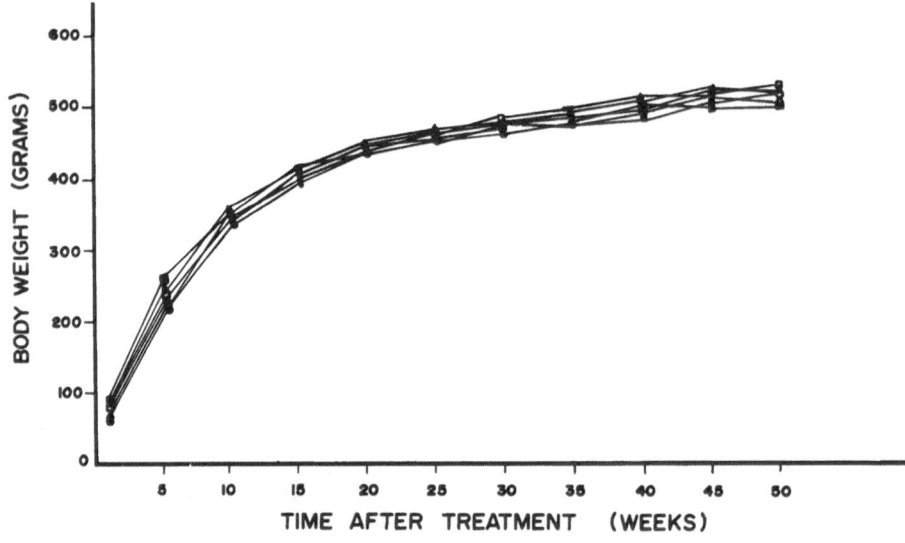

Fig. 1. Growth of male rats receiving AFB₁ (△—△), DDT (●—●), DDT prior to AFB₁ (■—■), DDT simultaneously with AFB₁ (▲—▲) and DDT after ABF₁ (□—□). Animals receiving DMSO served as controms (○—○)

strongly promoted such increase if given after the complete doses of AFB₁ (group 6). DDT also slightly inhibited, although not significantly, the increase in liver weights when given simultanously with AFB₁. The increase in liver weights of animals treated with AFB₁ followed by DDT was due largely to the presence of neoplasms.

Table 3 summarizes the incidence and histological findings of liver tumours observed in each group of animals. There was no remarkable change in the liver of rats treated with DMSO which served as control animals. However, 2 rats in this group were found to develop cancers of stomach and kidney which may be regarded as spontaneous cancers. In rats treated with DDT alone there was also no cellular alteration or tumours observed in the liver except for fatty changes in some cases, which is a general effect of DDT on rat liver (Ortega 1986).

AFB₁, when given alone (group 3), induced a moderate degree of foci of cellular alteration in all rats. These included foci of acidophilic (Fig. 2a), basophilic (Fig. 2b) and clear cells (Fig. 2c) and most animals developed all of these foci. A high incidence of hyperplastic nodules (83%, 20/24 rats) was also observed in this group of animals with a total of 100 nodules per group and 4.2 nodules per rats. These nodules were not very big, mostly smaller than 1 cm in diameter (Fig. 3). Liver carcinomas were developed in 8 out of 24 rats (33.3%). These included both HCC (Fig. 4) and CCC (Fig. 5) which were equally observed (4 rats each). HCC observed was mainly the well-differentiated or trabecular type. However, there was no distant metastasis observed in any of these animals. One animal was found to develop renal cell carcinoma.

Table 2. Body and liver weights of male rats treated with AFB_1 and DDT

Group	Treatment	Number of animals		Body weight	Liver weight	Liver weight per 100 g body weight
		Initial	Effective[a]			
1	DMSO	20	18	492.2 ± 101.5	17.8 ± 4.2	3.3 ± 0.46
2	DDT	20	19	485.0 ± 69.5	15.9 ± 3.3	3.4 ± 1.01
3	AFB_1	25	24	504.3 ± 61.7	21.0 ± 6.9	4.2 ± 1.50^c
4	DDT then AFB_1	25	23	514.4 ± 85.7	16.6 ± 3.8^e	3.2 ± 0.25^d
5	DDT simultaneously with AFB_1	25	19	513.5 ± 66.6	18.2 ± 4.7	3.5 ± 0.80
6	AFB_1 then DDT	25	23	474.8 ± 88.2	26.1 ± 13.1^c	$5.5 \pm 2.6^{b,f}$

Data are means $\pm$ SD.

[a] Number of rats surviving until the end of experiment, i.e. 52 weeks.
[b] Significantly different from group 1 at $p < 0.001$.
[c] Significantly different from group 1 at $p < 0.025$.
[d] Significantly different from group 3 at $p < 0.005$.
[e] Significantly different from group 3 at $p < 0.01$.
[f] Significantly different from group 3 at $p < 0.05$.

Table 3. Histopathological findings in livers of rats treated with AFB$_1$ and DDT

Group	No. of animals[a]	Foci of cellular alteration		Hyperplastic nodules			Carcinoma			
		No. of animals (%)	Degree[b]	No. of animals (%)	No. of nodules Total	per animal (M±SD)	No. of animals (%)	HCC	CCC	Mixed
1	18	0 (0)	−	0	0	0	0	0	0	0
2	19	0 (0)	−	0	0	0	0	0	0	0
3	24	24 (100)	++	20 (83.3)	100	4.2±2.7	8 (33.3)	4	4	0
4	23	18 (78.3)[f]	+	5 (21.7)[c]	23[d]	1.0±1.7[c]	0 (0)[c]	0	4	0
5	19	17 (89.4)	++	9 (47.4)[f]	30[d]	1.6±2.5[e]	6 (31.6)[g]	2	3	2
6	23	23 (100)	++++	23 (100)[f]	127	5.5±2.8	20 (87.0)[d,g]	13	9	2

HCC, hepatocellular carcinoma; CCC, cholangiocellular carcinoma; mixed; hepato-cholangiocellular carcinoma.

[a] A total of 13 rats that died during 10 weeks after AFB$_1$ treatment were excluded from the experiment.
[b] Degree of altered foci: + + + +, highly pronounced; + + +, pronounced; + +, moderate; +, slight; −, absent.
[c] Significantly different from group 3 at $p<0.001$.
[d] Significantly different from group 3 at $p<0.002$.
[e] Significantly different from group 3 at $p<0.005$.
[f] Significantly different from group 3 at $p<0.05$.
[g] Some rats developed both hepatocellular carcinoma and cholangiocarcinoma but in different lobes.

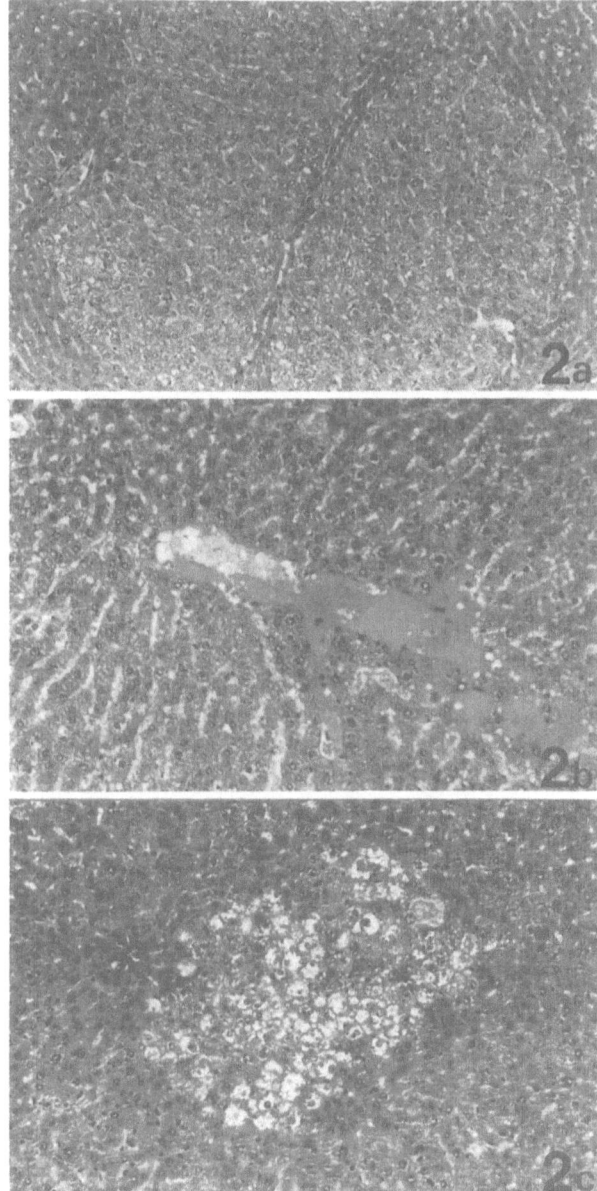

Fig. 2a–c. Micrographs of focal hyperplasia of liver cells. **a** Acidophilic cell focus consisting of parenchymal cells with large nuclei and abundant eosinophilic cytoplasm ($\times 31$). **b** Basophilic cell focus showing small cell with a strong basophilia of cytoplasm ($\times 108$). **c** Clear cell focus showing plant-cell-like hepatocyte. Note the extremely balloned cytoplasm due to glycogen accumulation ($\times 108$)

DDT treatment, which started 1 week prior to AFB$_1$ and continued until the completion of AFB$_1$ doses, resulted in a marked reduction in the incidences of all types of liver lesions (grous 4). Foci of cellular alteration were observed in 78.3% of animals and they were less frequent than in those in animals receiving AFB$_1$ alone. Five out of 23 rats (21.7%) were found to develop a total of 23 hyperplastic nodules which were equivalent to 1.0 nodule per animals. More

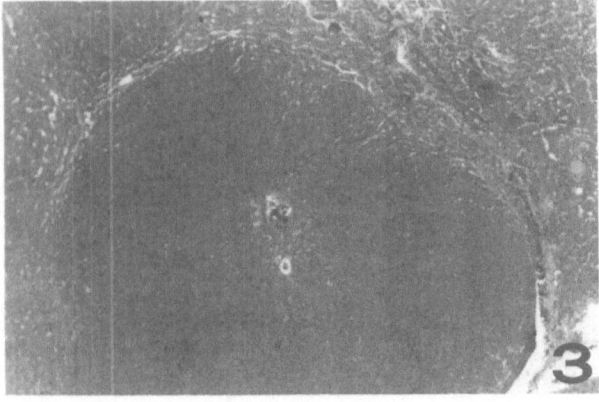

Fig. 3. Micrograph of hyperplastic nodule showing slight compression of the surrounding liver parenchyma. ($\times$31)

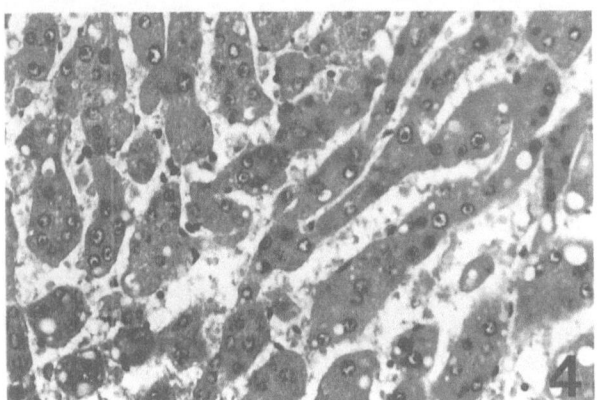

Fig. 4. Micrograph of well-differentiated hepatocellular carcinoma demonstrating small trabeculae of tumour cells separated by dilated vascular channels. Note the nuclei uniform with a single prominent nucleous and fatty changes in the cytoplasm of some cells. ($\times$218)

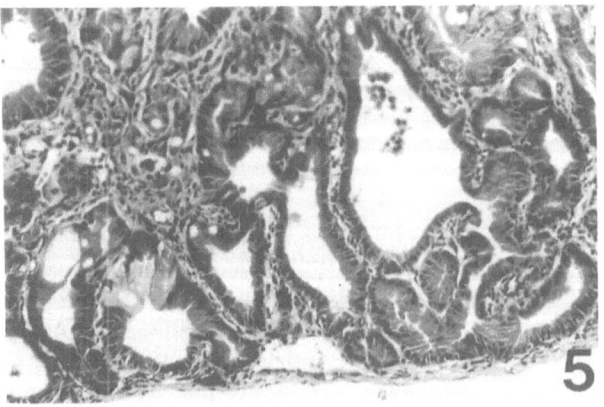

Fig. 5. Micrograph of cholangiocellular carcinoma showing glands lined by cylindrical neoplastic epithelial cells. Some cells exhibit deep basophilic cytoplasm. Abundant connective tissue surrounds the glands. ($\times$108)

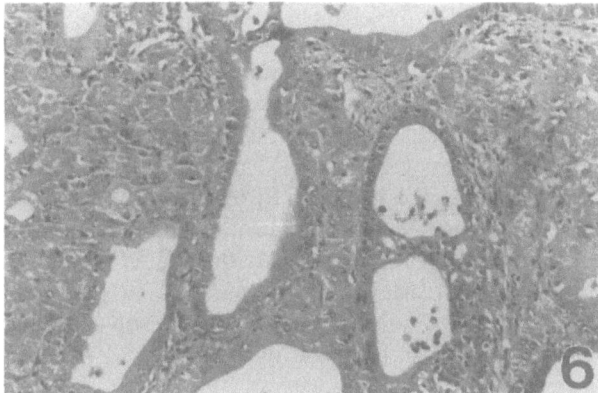

Fig. 6. Micrograph of mixed hepato-cholangiocellular carcinoma. Both tumour components are intimately intermingled.

importantly, no liver carcinoma of either type was observed in any animal of this group.

When DDT was given simultaneously with AFB$_1$ (group 5), 89.4% of animals developed a moderate degree of altered foci and this incidence was not significantly different from that in animals given AFB$_1$ alone. Hyperplastic nodules were observed in 47.4% of rats with the average of 1.6 nodules per rat. These incidences were significantly lower than those observed in animals in group 3. Liver carcinomas were also observed in these animals with the incidence of 31.6% (6 out of 19 rats). These carcinomas included HCC and CCC and 2 rats were found to develop mixed hepato-cholangiocellular carcinoma in the same region (Fig. 6). However, this incidence of liver carcinoma was not significantly different from that in animals receiving AFB$_1$ alone.

In the group receiving AFB$_1$ followed by DDT (group 6), all rats were found to develop a highly pronounced degree of altered foci and hyperplastic nodules. Both the total number of nodules (127 nodules per group) and the number of nodules per rat (5.5 ± 2.8) were higher than those in group receiving AFB$_1$ alone. However, the difference was not statistically significant. Liver carcinomas were observed in 87% of rats (20 out of 23 rats) and this incidence was about 2.6 times higher than that in Group 3 animals. The incidence of HCC observed in these animals was higher than that of CCC and there were 4 rats developed both HCC and CCC but in different lobes. In addition, 2 rats were found to develop mixed hepato-cholangiocellular carcinoma in the same region. Thus, DDT administration following AFB$_1$ enhanced the incidence of both HCC and CCC.

4 Discussion

Results in the present study clearly demonstrate that pre-treatment of rats with DDT significantly reduces the hepatocarcinogenicity of AFB$_1$ and that, conversely, feeding of this insecticide to animals after exposure to AFB$_1$ produces a substantial enhancement of AFB_1 hepatocarcinogenicity. Simultaneous feeding of DDT with AFB_1 also reduced the carcinogenicity of AFB_1, but only slightly and the inhibitory effect was much less than that of DDT pretreatment. Only the incidence of preneoplastic lesions, not of carcinomas, was significantly decreased compared to those in animals receiving AFB_1 alone (Table 3).

The inhibitory effect of DDT pretreatment as well as co-administration, on the hepatocarcinogenicity of AFB_1 observed here was similar to those of phenobarbital and α-benzene hexachloride which have been shown to inhibit AFB_1-induced hepatocarcinogenesis either when given before or simultaneously with the carcinogen (McLean and Marshall 1971; Swenson et al. 1977; Angsubhakorn et al. 1981). DDT, phenobarbital and α-benzene hexachloride have been shown to increase the level of cyt. P_{450} and microsomal drug metabolizing enzymes in the liver (Remmer and Merker 1965; Sell and Davidson 1973; Mgbodile et al. 1975; McLean and Driver 1977). In addition, our recent results also show that DDT is highly effective in inducing the in vitro metabolism of AFB_1 (unpublished results). This evidence therefore suggests that DDT inhibits the hepatocarcinogenicity of AFB_1 presumably by enhancing carcinogen detoxification resulting from the induction of both microsomal mono-oxygenase activity and the major conjugation enzyme systems. This mechanism has been suggested previously for the inhibitory effect of phenobarbital (Wattenberg 1985). The effect of DDT on the activities of conjugating enzymes and the binding of AFB_1 to DNA is being investigated in our laboratory.

The low inhibition of AFB_1 hepatocarcinogenesis by simultaneous feeding of DDT observed in this study may be due to the low level of AFB_1 used as compared to that used by Swenson et al. (1977) and Augsubhakorn et al. (1981). The differences in the effect of DDT and phenobarbital on the different doses of carcinogen have been previously discussed by Kitagawa et al. (1984).

Another interesting effect of DDT is its promotion of hepatocarcinogenesis induced by AFB_1 if given following the carcinogen (Table 3). This result is in agreement with previous findings that DDT enhances the carcinogenicity of various carcinogens, as mentioned in the Introduction. The molecular basis of tumour promotion is not yet clear. However, the interaction with membrane and subsequent inhibition of intercellular communication (metabolic co-operation) maø explain in part the promoting activity of many compounds including DDT (Wiliams et al. 1981; Trosko et al. 1982; Enomoto and Yamasaki 1985).

From our results reported here, it can be concluded that DDT modulates the hepatocarcinogenicity of AFB_1 in at least 2 different ways depending upon the sequence of exposure to these 2 chemicals. DDT exerts its inhibitory effect if given prior to AFB_1 whereas it exhibits promoting activity if given following the carcinogen.

Acknowledgements. The authors would like to express their gratitude to Prof. Tinnarat Satitnimankarn and Withaya Thamavit for their generous help with histological examination. We also thank Miss Chintana Soodprasert and Miss Anong Chunit for excellent technical assistance and Miss Panporn Srimanop for typing the manuscript. This work was partly supported by the Association for International Cancer Research.

References

Alpert ME, Hutt MSR, Wogan GN, Davidson CS (1971) Association between aflatoxin content of food and hepatoma frequency in Uganda. Cancer 28:253–260

Angsubhakorn S, Bhamarapravati N, Romruen K, Sahaphong S, Thamavit W, Miyamoto M (1981) Further study of α-benzene hexachloride inhibition of aflatoxin B_1 hepatocarcinogenesis in rats. Br J Cancer 43:881–883

Bulatao-Jayme J, Almero EM, Castro MACA, Jardeleza MATR, Salamat LA (1982) A case-control dietary study of primary liver cancer risk from aflatoxin exposure. Int J Epidemiol 11:113–119

Busby WF, Wogan GN (1984) Aflatoxins. ACS Monogr 182:945–1136

Enomoto T, Yamasaki H (1985) Phorbol ester-mediated inhibition of intercellular communication in BALB/c 3T3 cells: Relationship to enhancement of cell transformation. Cancer Res 45:2681–2688

Fitzhugh OG, Nelson AA (1947) The chronic oral toxicity of DDT (2,2-Bis (p-chlorophenyl-1.1.1-Trichloroethane). J Pharmacol Exp Ther 89:18–30

Grasso P, Crampton RF (1972) The value of the mouse in carcinogenicity testing. Food Cosmet Toxicol 10:418–426

Innes JRM, Ulland BM., Valerio MG, Petrucelli L, Fishbein L, Hart ER, Pallota AJ et al. (1969) Bioassay of pesticides and industrial chemicals for tumorigenicity in mice: A preliminary note. ING'I 42:1101–1114

Ishidate M, Watanabe M, Odashima S (1967) Effect of barbital on carcinogenic action and metabolism of 4-dimethylaminoazo benzene. Gann 58:267–281

Ito N, Tatematsu M, Nakanishi K, Hasegawa R, Takano T, Imaida K, Ogiso T (1980) The effects of various chemicals on the development of hyperplastic liver nodules in hepatectomized rat treated with N-nitrosodiethylamine or N-2-fluorenylacetamide. Gann 71:832–842

Kalengayi MM, Desmet V (1975) Sequential histological and histochemical study of the rat liver during aflatoxin B_1-induced carcinogenesis. Cancer Res 35:2845–2852

Kitagawa R, Sugano H (1978) Enhancing effect of phenobarbital on development of enzyme-altered islands and hepatocellular carcinomas initiated by 3'-methyl-4-(dimethylamino) azobenzene or diethylnitrosamine. Gann 69:679–687

Kitagawa R, Hino O, Nomura K, Sugano H (1984) Dose-response studies on promoting and anticarcinogenic effects of phenobarbital and DDT in the rat hepatocarcinogenesis. Carcinogenesis 5:1653–1656

Kunz W, Schaude G, Thomas C (1969) The effect of phenobarbital and halogenated hydrocarbons on nitrosamine carcinogenesis. Z Krebsforsch 72:291–304

McLean AEM, Driver HE (1977) Combined effects of low doses of DDT and phenobarbital on cytochrome P_{450} and amidopyrine demethylation. Biochem Pharmacol 26:1299–1302

McLean AEM, Marshall A (1971) Reduced carcinogenic effects of aflatoxin in rats given phenobarbitone. Br J Exp Pathol 52:322–329

Mgbodile MUK, Holscher M, Neal R (1975) A possible protection role for reduced glutathione in aflatoxin B_1 toxicity: effect of pretreatment of rats with phenobarbital and 3-methylcholanthrene on aflatoxin toxicity. Toxicol Appl Pharmacol 34:128–142

Newberne PM, Butler WH (1969) Acute and chronic effects of aflatoxin on the liver of domestic and laboratory animals: a review. Cancer Res 29:236–250

Nishizumi M (1979) Effect of phenobarbital, dichlorodiphenyltrichloroethane und polychlorinated biphenyls on diethylnitrosamine induced hepatocarcinogenesis. Gann 70:835–837

Ortega P (1966) Light and electron microscopy of dichlorodiphenyl-trichloroethane (DDT) poisoning in the rat liver. Lab Invest 15:657–679

Peers FG, Linsell CA (1973) Dietary aflatoxins and liver cancer – a population based study in Kenya. Br J Cancer 27:437–484

Peers FG, Bosch S, Kaldor J, Linsell A, Pluumen M (1987) Aflatoxin exposure, hepatitis B virus infection and liver cancer in Swaziland. Int J Cancer 39:545–553

Peraino C, Fry RJM, Staffeldt E (1971) Reduction and enhancement by phenobarbital of hepatocarcinogenesis induced in the rat by 2-acetylaminofluorene. Cancer Res 31:1506–1512

Peraino C, Fry RJM, Staffeldt E, Kisieleski WE (1973) Effects of varying the exposure to phenobarbital on its enhancement of 2-acetylaminofluorene-induced hepatic tumorigenesis in the rat. Cancer Res 33:2701–2705

Peraino C, Fry RJM, Staffeldt E, Christopher JP (1975) Comparative enhancing effects of phenobarbital, amobarbital, diphenylhydantoin and dichlorodiphenyltrichloroethane on 2-acetylaminofluorene induced hepatic tumorigenesis in the rat. Cancer Res 35:2884–2890

Peto R, Pike MC, Day NE, Gray RG, Lee RN, Parish S, Peto J, Richards S, Wahrendorf J (1980) Guidelines for simple, sensitive significance tests for carcinogenic effects in long-term animal experiments. IARC Monogr [Suppl] 2:311–426

Remmer H, Merker HJ (1965) Effect of drugs on formation of smooth endoplasmic reticulum and drug metabolizing enzymes. Ann NY Acad Sci 123:79

Scribner JD, Mottet NK (1981) DDT acceleration of mammary gland tumors induced in the male Sprague-Dawley rat by 2-acetamidophenanthrene. Carcinogenesis 2:1235–1239

Sell JL, Davidson KL (1973) Changes in the activities of hepatic microsomal enzymes caused by DDT and dieldrin. Fed Proc 32:2003–2009

Serck-Hansen A (1970) Aflatoxin-induced fatal hepatitis? Arch Environ Health 20:729–731

Shank RC, Bhamarapravati N, Gordon JE, Wogan GN (1972a) Dietary aflatoxins and human liver cancer. IV. Incidence of primary liver cancer in two municipal populations of Thailand. Food Cosmet Toxicol 10:171–179

Shank RC, Gordon JE, Wogan GN, Nondasuta A, Subhamani B (1972b) Dietary aflatoxins and human liver cancer. III. Field survey of rural Thai families for ingested aflatoxins. Food Cosmet Toxicol 10:71–84

Shank RC, Wogan GN, Gibson JB (1972e) Dietary aflatoxins and human liver cancer. I. Toxigenic moulds in food and foodstuffs of tropical South-East Asia. Food Cosmet Toxicol 10:51–60

Shank RC, Wogan GN, Gibson JB, Nondasuta A (1972d) Dietary aflatoxins and human liver cancer. II. Aflatoxins in market foods and foodstuffs of Thailand and Hong Kong. Food Cosmet Toxicol 10:61–69

Sieber SM, Correa P, Dalgaro DW, Adamson RH (1979) Induction of oestrogenic sarcomas and tumors of the hepatobidiary system in nonhuman primates with aflatoxin B_1. Cancer Res 39:4545–4554

Swenson DH, Lin JK, Miller EC, Miller JA (1977) Aflatoxin B_1-2,3-oxide as a probable intermediate in the covalent binding of aflatoxin B_1 and B_2 to rat liver DNA and ribosomal RNA in vivo. Cancer Res 37:172–181

Terracini B, Testa MC, Cabral TR, Day N (1973) The effects of long-term feeding of DDT to BALB/c mice. Int J Cancer 11:747–764

Thorpe E, Walker AIT (1973) The toxicology of dieldrin (HEOD). II. Comparative long-term oral toxicity studies in mice with dieldrin, DDT, phenobarbitone, B-BHC. Food Cosmet Toxicol 11:433–442

Tomatis L, Turusov V, Day N, Charles RT (1972) The effect of long-term exposure to DDT on CF-1 mice. Int J Cancer 10:489–506

Trosko JE, Yotti CP, Warren ST, Tsushimoto G, Chang CC (1982) Inhibition of cell-cell communication by tumor promoters. Carcinogenesis 7:565–585

Van Rensburg SJ, Cook-Mozaffari P, van Schalkwyk DJ, van der Watt JJ, Vincent TJ, Purchase IF (1985) Hepatocellular carcinoma and dietary aflatoxin in Mozambique and Transkei. Br J Cancer 51:713–726

Wattenberg LW (1985) Chemoprevention of cancer. Cancer Res 45:1–8

Williams GM, Numoto S (1984) Promotion of mouse liver neoplasms by the organochlorine pesticides chlordane and heptachor in comparison to dichlorodiphenyltrichloroethane. Carcinogenesis 5:1689–1696

Williams GM, Telang S, Tong C (1981) Inhibition of intercellular communication between liver cells by the liver tumor promoter 1.1.1-trichloro 2,2-bis (p-chlorophenyl)ethane. Cancer Lett 11:339–329

Intratumoral Fibrin Stabilization

L. Muszbek and R. Ádány

Department of Clinical Chemistry, University School of Medicine, POB 40, Debrecen, 4012, Hungary

1 Introduction

Fibrin deposition in the interstitial space of the stroma of malignant tumours is a frequent finding (see reviews by Rickles and Edwards 1983; Dvorak et al. 1983) and can be regularly observed, among others, in lymph nodes with Hodgkin's disease (Harris et al. 1982). Hypotheses based on supporting experimental and clinical data implicate intratumoral fibrin in the promotion of tumour growth and metastasis formation (Laki 1974; Dvorak et al. 1983). Fibrin may serve as a barrier making tumour cells relatively inaccessible for the host's immune response, may play an important role in tumour angiogenesis and desmoplasia and might support the implantation of tumour cells at metastatic sites.

The appearance of fibrin is the result of the extravasal activation of clotting pathway initiated by malignant cells and/or tumour associated macrophages (TAMs). The ways in which these cells can induce fibrin formation have been extensively studied in the last years and the mechanisms elucidated involve the expression of tissue factor, and the production of proteases directly activating Factor X and/or prothrombin (Gordon et al. 1975; Curatolo et al. 1979; Gordon 1981; Edwards et al. 1981; Key 1983; Lorenzet et al. 1983; Rickles and Edwards 1983; Guarini et al. 1984; VanDeWater et al. 1985). In the light of the abundance of studies on the activation of the clotting system in malignancies it is rather suprising that practically no information is available on the nature of intratumoral fibrin deposits.

In normal haemostasis as the last step of the coagulation cascade γ and α chains of fibrin polymers become crosslinked by activated Factor XIII (FXIIIa) and a covalently crossbound fibrin network is formed (see review by Muszbek and Laki 1984). In addition, FXIIIa covalently attaches α_2-antiplasmin (α_2-AP), the main physiological inhibitor of fibrinolysis, to fibrin chains making the clot resistant to a prompt fibrinolytic degradation (Sakata and Aoki 1980, 1982; Aoki and Harpel 1984; Jansen et al. 1987). As many types of malignant cells are capable of producing and secreting plasminogen activator (Nicolson 1982; Duffy and O'Grady 1984; Dano et al. 1985; Layer et al. 1987) that induces fibrinolysis the stabilized or unstabilized nature of extravascular fibrin deposits in the tumour stroma is a question of high biological significance. In the present study it is demonstrated that FXIII is present and crosslinks fibrin in the extravascu-

Nigam et al. (Eds.), Eicosanoids,
Lipid Peroxidation and Cancer
© Springer-Verlag Berlin Heidelberg 1988

lar space of lymph nodes with Hodgkin's disease. It attaches α_2-AP to the fibrin network which then, indeed, inactivates plasmin.

2 Materials and Methods

2.1 Specimens

Lymph node biopsies were obtained from 12 patients with Hodgkin's disease of the nodular sclerosing type. Specimens used for immunomorphological investigations were divided into two parts at the time of surgical biopsy. One part was fixed in 3.5% paraformaldehyde fixative (4h, room temperature) then vacuum embedded in paraffin and sectioned into 6 μm slides, while the other part was snap-frozen and cut in a cryostat. Sections from non-neoplastic, reactive lymph nodes served as controls. For immunoblotting specimens were placed into ice cold extracting buffer (50 mM Tris HCl, pH 7.4 containing 10 g/l Triton X-100, 5 mmol/l EGTA, 0.5 mmol/l PMSF, 10 U/ml heparin, 1 U/ml hirudin, 0.1 mol/l ε aminocaproic acid, 10 U/ml Trasylol, 2 mmol/l iodoacetamide, 2 mmol/l N-ethylmaleimide. They were minced, homogenized using a tight fitting teflon-glass homogeniser, then sonicated for 3 × 30 sec. The homogenate was stirred for 1 hour than centrifuged at 30000 g for 20 min. The pellet was suspended in extracting buffer, stirred for 20 min and again centrifuged. The whole extraction procedure was carried out at +4°C. The final residues was taken up in a denaturing solution (Laemmli 1970) containing 6 mol/l urea, the samples were boiled for 5 min and the protein content was determined (BCA Protein assay reagent, Pierce, Oud-beijerland, the Netherlands).

2.2 Immunomorphological Studies

Immunoperoxidase staining for the enzymatically active a subunit of FXIII (FXIII A) was carried out as described (Ádány et al. 1987a). Rabbit antiserum against FXIII A (Behringwerke, Marburg, FRG) was used as primary antibody and the antigen-antibody reaction was detected by biotinylated anti-rabbit IgG and avidin-biotinylated peroxidase complex (Vactastain ABC kit; Vector Laboratories, Burlingame, California).

Fibrin deposits were detected by direct immunofluorescent staining using goat anti-human fibrinogen IgG, FITC conjugated (1:100 dilution, 1 hour). In certain experiments, prior to the immunodetection of fibrin, sections were extracted in 3 mol/l urea solution for 2 hours at 37°C. In other experiments staining for fibrin was combined with the visulization of one of the following antigens: FXIII A, α_2 AP and α_2 AP-plasmin complex neoantigen (α_2 AP-P-Neo). 1 hour incubation with a 1:100 dilution of the respective primary rabbit antiserum (Behringwerke, Marburg, FRG) was followed by horse biotinylated anti-rabbit immunglobulin (Vector Labs., 1:250 dilution, 30 min) and Texas-Red streptavidin (Amersham, UK, 1:40 dilution, 45 min incubation). The double immunofluorescent labelling for FXIII A and Leu M 3, a surface marker of monocyte/macrophage cell line (Dimitriu-Bona et al. 1983), has been described earlier (Ádány et al. 1987a). In all experiments preincubation with an appropriate serum was used to prevent aspecific IgG binding. In the case of negative control slides identical dilutions of non-immune antiserum from the respective species was substituted for the first antibodies.

Following double immunofluorescent staining sections were mounted by 50% glycerol in PBS and examined under an Opton ultraviolet microscope equipped with an epifluorescence condensor containing selective filters for FITC and Texas Red.

2.3 Immunoblotting Technique

Denatured samples were electrophoresed in a 5%–15% gradient sodium dodecyl sulfate polyacrylamide gel (SDS PAG) (Laemmli 1970). In addition to the lymph node residue, M_r standard (Sigma, SDS-6H), purified human fibrinogen (Test Fibrinogen, Behringwerke), noncrosslinked as well as crosslinked fibrin and human plasma diluted 1:14 in denaturing solution were also applied to the gel. Crosslinked fibrin was made from fibrinogen (10 mg/ml) by clotting it with 10 U/ml thrombin in the presence of 0.1 mg/ml highly purified FXIII prepared in our laboratory and 4 mmol/l $CaCl_2$ (3h incubation at 37°C). In the case of non-crosslinked fibrin the action of FXIII was prevented by adding 1 mmol/l EDTA instead of $CaCl_2$. Two parallel runs were performed on the same gel which was divided following electrophoresis. One part was stained by Coomassie blue and destained. The other was electroblotted to nitrocellulose paper and developed for fibrinogen by the immunoperoxidase technique (Bio Rad Immuno-Blot Assay Kit with the modification that 0.01% DAB and 0.003% H_2O_2 in Tris buffered saline, pH 7.6 were applied as peroxidase substrates). Rabbit anti-human fibrinogen antiserum (Cotimmun Fibrinogen, Behringwerke, 1:500 dilution) was used as first antibody.

3 Results and Discussion

The first question we addressed was whether FXIII is available in the tumour stroma. By the immunoperoxidase method numerous relatively large multipolar mononuclear cells were stained for this antigen in lymph nodes with Hodgkin's disease (Fig. 1). Some extracellular elements seemed also to be labelled by anti-FXIII A. The latter finding was further elaborated by a double immunofluorescent technique (see later). When sections were counterstained with haematoxy-

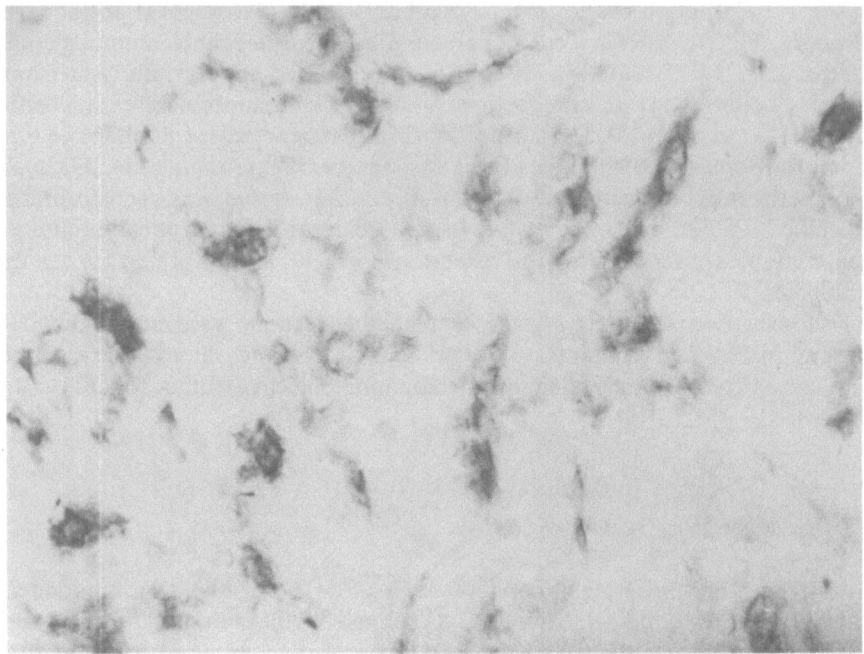

Fig. 1. Immunoperoxidase staining for Factor XIII subunit a on a section of paraformaldehyde-fixed paraffin embedded lymph node with Hodgkin's disease. (× 285).

lin-chromatrop these cells could frequently be localized in the immediate vicinity of malignant Hodgkin's cells and around fibrous material coloured red (not shown). It is to be noted that in reactive lymph nodes, FXIII containing cells were localized almost exclusively in perivascular connective tissue and in sub-capsular or medullary sinuses and only a few positive cells could be detected in the interfollicular areas (Ádány et al. 1987a).

It has been established that monocytes and various types of macrophages contain FXIII A (Muszbek et al. 1985; Ádány et al. 1985, 1987b, 1988; Henriksson et al. 1985; Kradin et al. 1987) and its detection could be considered as a marker reaction of this cell line (Ádány et al. 1987b, c). Figure 2 demonstrates that cells which contain FXIII in lymph nodes with Hodgkin's disease also belong to this cell population. In a double immunofluorescent system they were labelled for the monocyte macrophage marker Leu M 3, i.e., they should be considered tumour associated macrophages (TAMs). A more detailed characterization of FXIII containing TAMs (Ádány et al. 1987a) revealed that they were positive for α naphthyl acetate esterase but negative for HLA-DR antigen and for ATPase, acid and alkaline phosphatase activities.

Fibrin deposits were observed both in nodular and sclerosing areas although in the former fibrin was more abundant and in addition to a fibrillar pattern amorphous deposits were also detected with high frequency. Using a highly sensitive immunofluorescent technique FXIII A was found both in cellular ele-

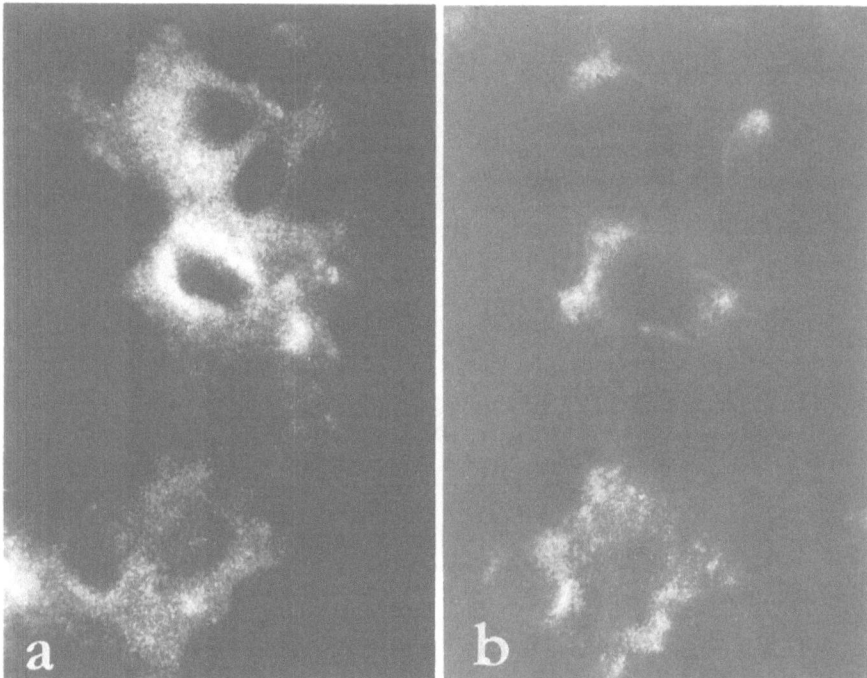

Fig. 2. In a lymph node with Hodgkin's disease identical cells were labelled with immunofluor-escent staining for FXIII subunit a (**a**) and for the monocyte/macrophage surface antigen Leu M3 (**b**). (From Ádány et al. 1987a) ($\times$ 1100)

ments and over fibrin deposits (Fig. 3). As activated FXIII A strongly adsorbs to fibrin (Lorand and Dickenman 1955; Israels et al. 1973) this finding clearly indicates that there is FXIII available for the crosslinking reaction in the interstitial space of tumour stroma. The origin of extracellular FXIII is not known. It may leak out from the plasma through vessel walls of increased permeability and/or may be secreted by intact or released by damaged TAMs. The close association of TAMs to fibrin strands demonstrates the role of these cells in extravascular clotting. The association of macrophages activated by tumour cells with fibrin has also been revealed in an experimental system (Dvorak et al. 1978).

The question whether intratumoral fibrin was crosslinked by FXIII was first approached by testing its solubility in a concentrated urea solution. As demos-trated by imunofluorescence staining, extraction with 3 mol/l urea did not re-move fibrin from the sections (Fig. 4). In fact, perhaps because of the dissolution of some surrounding protein structures, the fluorescence became even brighter. This finding strongly indicates the fibrin had been transformed into an urea-insoluble highly crosslinked clot by FXIIIa. A further and more clear-cut proof for the crosslinked nature of intratumoral fibrin was provided by the immuno-blotting technique (Fig. 5). As is well known thrombin transforms fibrinogen to fibrin by removing fibrinopeptides A and B from the α and β chains which man-ifest this in their increased mobility in SDS PAG (Fig. 5A). If activated FXIII is

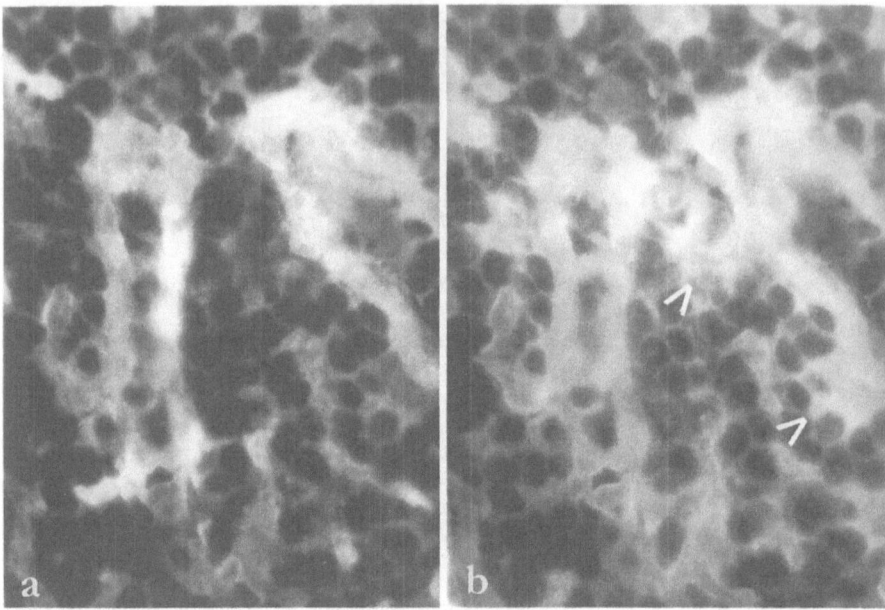

Fig. 3. Double immunofluorescent labelling for fibrin (**a**) and Factor XIII subunit a (**b**) on a frozen section of a lymph node with Hodgkin's disease. *Arrowheads* indicate Factor XIII containing cells. ($\times 570$). Reproduced from Ádány et al., Thrombosis and Haemostasis, (in press) with permission

also present the formation of γ chain dimers and highly crosslinked α chain polymers occurs while β chains do not participate in the crosslinking reaction. The crosslinking of γ chains proceeds faster. However, from the point of view of fibrin stabilization, the relatively slower formation of high M_r α polymers is the event of primary importance. On SDS PAG these events were demonstrated by the disappearance of γ and α chains and the appearance of a new bond corresponding to γ-γ dimers. The high M_r α polymers cannot be seen since they do not enter the concentrating gel. The anti-fibrinogen antibody we used reacted predominantly with the α and β chains. Its specificity is well demonstrated on plasma samples in which only the above two fibrin chains gave a positive reaction by immunoblotting. When a lymph node with Hodgkin's disease was extracted with a buffer that inhibits both artificial clotting and fibrinolysis and dissolves most of the proteins except components of the insoluble extracellular matrix (fibrin, collagen etc.) the β but not α chain of fibrin was well detectable in the residue. As our antibody reacted both with fibrin and fibrinogen it is important to stress that the reactive polypeptide comigrated with the β chain of fibrin but not with the β(B) chain of fibrinogen, i.e., the crossreacting material was fibrin. The absence of α chain indicates the formation of α polymers and proves the highly crosslinked nature of fibrin formed in the stroma of the lymph nodes with Hodgkin's disease. Though, according to our best knowledge, this is the first report that proves the existance of stabilized fibrin in the stroma of a human

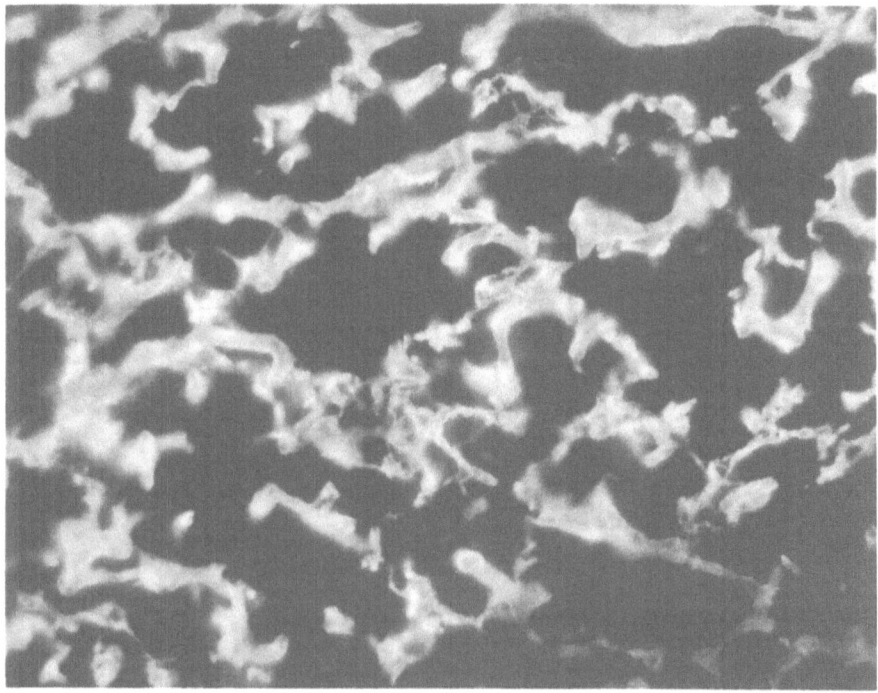

Fig. 4. Immunofluorescent labelling for fibrin on a 3 mol/l urea pretreated section of a lymph node with Hodgkin's disease. ($\times 285$)

tumour. The presence of crosslinked fibrin deposits has also been revealed in guinea pig carcinomas (Dvorak et al. 1984).

As previously mentioned, in addition to crosslinking fibrin chains, another important physiological task of FXIII is to incorporate α_2 AP into α chain polymers during the clotting process. As unbound α_2 AP does not adsorb to fibrin and is ineffective in preventing the rapid dissolution of the clot by the fibrinolytic enzyme, plasmin, this mechanism is of high physiological importance (Sakata and Aoki 1982; Aoki and Harpel 1984; Jansen et al. 1987). The complete coincidence of staining for α_2 AP and fibrin throughout the lymph nodes with Hodgkin's disease (Fig. 6) clearly demonstrates that FXIII activated in the extravascular space fulfilled this task in the tumour stroma, as well. In a number of areas, especially in parts showing sclerosis labelling for fibrin and for α_2 AP-P-Neo were also in co-localization (not shown), i.e. part of α_2 AP had captured native plasmin molecules and existed as inactive complex.

The above results allow us to conclude that FXIII of plasma and/or cellular origin is present and becomes activated in the interstitial space of the tumour stroma. It exerts its normal function, makes a highly crosslinked stabilized clot and attaches α_2 AP to it. In this way it supports the pathological barrier function of fibrin and provides the fibrin network that could promote angiogenesis and cell proliferation (Beck et al. 1961; Ueyama and Urayama 1977; Kasai et al.

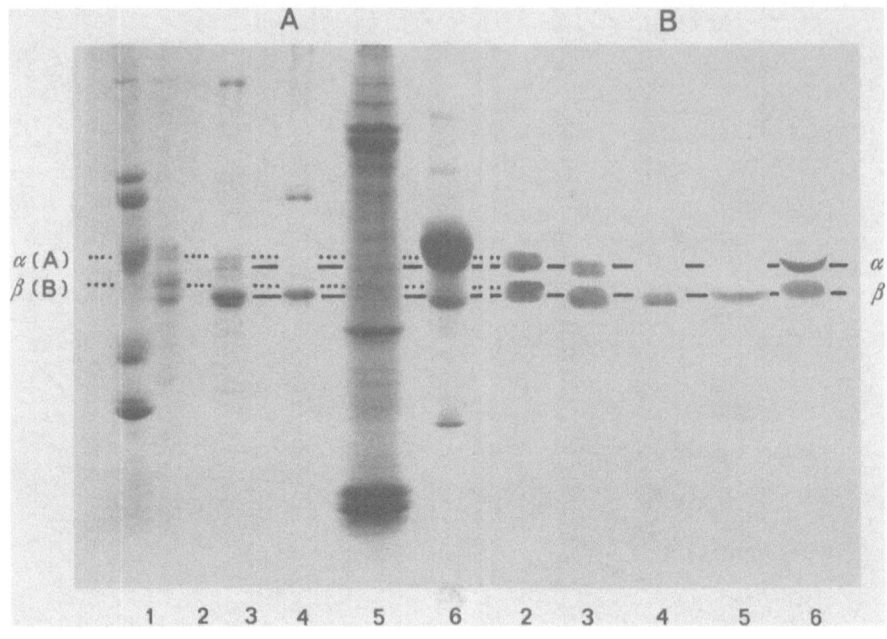

Fig. 5 A, B. Detection of α and β chains of fibrin by immunoblotting in the residue of a lymph node with Hodgkin's disease extracted by 1% Triton X-100 at high ionic strength. **A** SDS PAGE analysis. **B** immunoblotting. *1*, M_r markers, the individual bands correspond to myosin (205 kD), β galactosidase (116 kD), phosphorylase b (97.4 kD), bovine serum albumin (66 kD), egg albumin (45 kD) and carbonic anhydrase (29 kD); *2*, fibrinogen (7.5 μg); *4*, fibrin crosslinked by Factor XIII (7.5 μg); *5*, lymph node residue (210 μg); *6*, human plasma (43 μg); α, β, α(A), β(B), the corresponding chains of fibrin and fibrinogen, respectively. (α chain in this fibrinogen preparation appears as doublet). The γ chain of fibrinogen and fibrin (not marked) that migrates just ahead of the β chain (*A2, 3*), was not detected by immunoblotting. In crosslinked fibrin note the disappearance of the bands corresponding to α and γ chains and the appearance of γ–γ dimer (*A4*). On the immunoblot of lymph node residue the α chain is missing and only a single band which comigrates with the β chain of fibrin could be detected (*B5*)

1983). FXIII might have a direct cell proliferating effect as well (Bruhn 1981; Bruhn and Pohl 1981; Bruhn and Zurborn 1983). A further possibility is that FXIIIa as a transglutaminase could attach host's proteins to the membrane surface of malignant cells and cover up their un-selfness. Unrecognizable malignant cells could then escape elimination by the host's defence system. Clearly, further experiments are needed to elucidate the exact pathological role of fibrin stabilization and FXIII in tumour progression and metastasis formation.

4 Summary

As in many other tumour types fibrin deposition is a regular finding in lymph nodes with Hodgkin's disease. The present study was undertaken to shed light

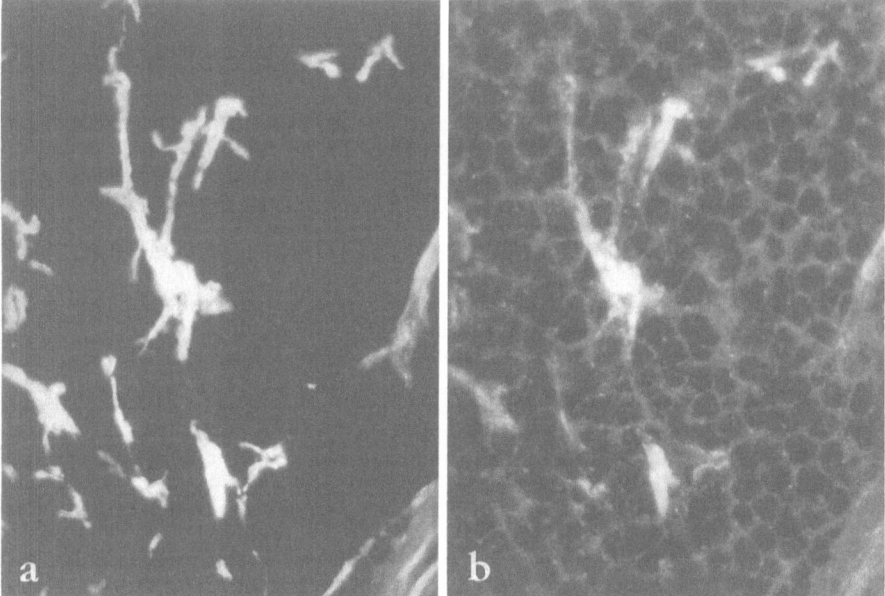

Fig. 6. Double immunofluorescent labelling for fibrin (**a**) and α_2 antiplasmin (**b**) on a frozen section of lymph node with Hodgkin's disease (× 285). Reproduced from Ádány et al., Thrombosis and Haemostasis, (in press) with permission

on the nature of fibrin formed in the extravascular space of this malignant tissue. Factor XIII for blood coagulation, fibrin stabilizing factor, was available in the tumour stroma and was localized both over fibrin strands and in cellular elements by immunomorphological methods. Using established enzyme-cytochemical and immunological marker reactions Factor XIII containing cells were characterized as tumour associated macrophages. The following findings indicate that Factor XIII became activated and exerted its crosslinking effect in the interstitial space: (a) fibrin formed in situ could withstand solubilization in concentrated urea solution, (b) crosslinked fibrin was detected in the residue of Triton X-100 extracted lymph nodes by immunoblotting, (c) α_2 antiplasmin a substrate for Factor XIII was found in co-localization with fibrin by double immunofluorescent labelling. The formation of crosslinked, stabilized fibrin and the binding of α_2 antiplasmin to it by Factor XIII provides a mechanism by which intratumoral fibrin deposits escape a prompt elimination by a powerful fibrinolytic system.

Acknowledgements. The technical assistance of Ms. Ágnes Bana, Ms. Gizella Haramura, Mr. Bálint Katona and the skillful secretarial work of Ms. Maria Kozma are gratefully acknowledged. This study was supportted by the Hungarian Ministry of Health, the Hungarian Academy of Sciences (contract No 1-3-86-298) and the National Science Foundation (contract No 1890).

348   L. Muszbek and R. Adány

References

Ádány R, Belkin A, Vasilevskaya T, Muszbek L (1985) Identification of blood coagulation factor XIII in human peritoneal mcrophages. Eur J Cell Biol 38:171–173

Ádány R, Nemes Z, Muszbek L (1987a) Characterization of factor XIII containing-macrophages in lymph nodes with Hodgkin's disease. Br. J Cancer 55:421–42

Ádány R, Kappelmayer J, Muszbek L (1987b) Expression of FXIII subunit *a* in different types of human macrophages. In: Mauri C, Rizzo SC, Ricevuti G (eds) The biology of phagocytes in health and disease, Pergamon, Oxford

Ádány R, Kiss A, Muszbek L (1987c) Factor XIII:a marker of mono- and megakariocytopoiesis. Br J Haematol 67:167–172

Ádány R, Glukhova MA, Kabakov AY, Muszbek L (1988) Characterization of human connective tissue cells containing factor XIII subunit *a*. J Clin Pathol, (in press)

Aoki N, Harpel PC (1984) Inhibitors of the fibrinolytic enzyme system. Semin Thromb Hemost 10:24–41

Beck E, Duckert F, Ernst M (1961) The influence of fibrin stabilizing factor on the growth of fibroblasts in vitro and wound healing. Thromb Diath Haemorrh 6:485–491

Bruhn HD (1981) Growth regulation of vessel wall cells and of tumor cells by thrombin, factor XIII and fibronectin. Thromb Haemost 46:762

Bruhn HD, Pohl J (1981) Growth regulation of fibroblasts by thrombin, factor XIII and fibronectin. Klin Wochenschr 59:141–146

Bruhn HD, Zurborn KH (1983) Influences of clotting factors (thrombin, factor XIII) and of fibronectin on the growth of tumor cells and leukemic cells in vitro. Blut 46:85–88

Curatolo L, Colucci M, Cambini AL, Poggi A, Morasca L, Donati MB, Semeraro N (1979) Evidence that cells from experimental tumours can activate coagulant factor X. Br J Cancer 40:228–233

Dano K, Andreasen PA, Grondahl-Hansen J, Kristensen P, Nielsen LS, Skirer L (1985) Plasminogen activators, tissue degradation and cancer. Adv Cancer Res 44:139–266

Dimitriu-Bona A, Burmester GR, Waters SJ, Winchester RJ (1983) Human mononuclear phagocyte differentiation antigens. I. Patterns of antigenic expression on the surface of human monocytes and mcrophages defined by monoclonal antibodies. J Immunol 130:145–152

Duffy MJ, O'Grady PO (1984) Plasminogen activator and cancer. Eur J Cancer Clin Oncol 20:577–582

Dvorak AM, Connell AB, Proppe K, Dvorak HF (1978) Immunologic rejection of mammary adenocarcinoma (TA3-St) in C57BL/6 mice: participation of neutrophils and activated macrophages with fibrin formation. J Immunol 120:1240–1248

Dvorak HF, Senger DR, Dvorak AM (1983) Fibrin as a component of the tumor stroma: origins and biological significance. Cancer Metastasis Rev 2:41–73

Dvorak HF, Harvey S, McDonagh J (1984) Quantitation of fibrinogen influx and fibrin deposition and turnover in line 1 and line 10 quinea pig carcinomas. Cancer Res 44:3348–3354

Edwards RL, Rickles FR, Cronlund M (1981) Abnormalities of blood coagulation in patients with cancer: mononuclear cell tissue factor generation. J Lab Clin Med 98:917–928

Gordon ST (1981) A proteolytic procoagulant associated with malignant transformation. J Histochem Cytochem 29:457–463

Gordon ST, Franks JJ, Lewis B (1975) Cancer procoagulant A: a factor X activating procoagulant from malignant tissue. Thromb Res 6:127–137

Guarini A, Acero R, Alessio G, Donati MB, Semeraro N, Mantovani A (1984) Procoagulant activity of macrophages associated with different murine neoplasms. Int J Cancer 34:581–586

Harris NL, Dvorak AM, Smith J, Dvorak HF (1982) Fibrin deposits in Hodgkin's disease. Am J Pathol 108:119–129

Henriksson P, Becker S, Lynch G, McDonagh J (1985) Identification of factor XIII in human monocytes and macrophages. J Clin Invest 76:528–534

Israels ED, Paraskevas F, Israels LG (1973) Immunological studies of coagulation factor XIII. J Clin Invest 52:2398–2403

Jansen JWCM, Haverkate F, Koopman J, Nieuwenhuis HK, Kluft C, Boschman TAC (1987) Influence of Factor XIIIa activity on human whole blood clot lysis in vitro. Thromb Haemost 57:171-175

Kasai S, Kunimoto T, Nitta K (1983) Cross-linking of fibrin by activated factor XIII stimulates attachment, morphological changes and proliferation of fibroblasts. Biomed Res 4:155-160

Key M (1983) Macrophages in cancer metastases and their relevance to metastatic growth. Cancer Metastasis Rev 2:75-88

Kradin RL, Lynch GW, Kurnick JT, Erikson M, Colvin RB, McDonagh J (1987) Factor XIII is synthetized and expressed on the surface of U937 cells and alveolar macrophages. Blood 69:778-785

Laemmli UK (1970) Cleavage of structural proteins during the assembly of the head of bacteriophage T4. Nature 227:680-685

Laki K (1974) Fibrinogen and metastases. J Med 5:32-37

Layer GT, Burnand KG, Gaffney PJ, Cederholm-Williams SA, Mahmoud M, Houlbrook S, Pattison M (1987) Tissue plasminogen activators in breast cancer. Thromb Res 45:601-607

Lorand L, Dickenman RC (1955) Assay method for the "fibrin-stabilizing factor". Proc Soc Exp Biol 89:45-48

Lorenzet R, Peri G, Locati D, Allavena P, Colucci M, Semeraro N, Mantovani A, Donati MB (1983) Generation of procoagulant activity by mononuclear phagocytes: A possible mechanism contributing to blood clotting activation within malignant tissues. Blood 62:271-273

Muszbek L, Laki K (1984) Interaction of thrombin with proteins other than fibrinogen (thrombin susceptible bonds). Activation of FXIII. In: Machovich R (ed) The thrombin. CRC, Boca Raton, pp 321-342

Muszbek L, Ádány R, Szegedi G, Polgár J, Kávai M (1985) Factor XIII of blood coagulation in human monocytes. Thromb Res 37:401-410

Nicolson GN (1982) Cancer metastasis: Organ colonization and the cell surface properties of malignant cells. Biochim Biophys Acta 695:113-176

Rickles FR, Edwards RL (1983) Activation of blood coagulation in cancer: Trousseau's syndrome revisited. Blood 62:14-31

Sakata Y, Aoki N (1980) Cross-linking of α_2-plasmin inhibitor to fibrin by fibrin-stabilizing factor. J Clin Invest 65:290-297

Sakata Y, Aoki N (1982) Significance of cross-linking of α_2-plasmin inhibitor to fibrin in inhibition of fibrinolysis and in hemostasis. J Clin Invest 69:536-542

Ueyama M, Urayama T (1977) The role of factor XIII in fibroblast proliferation. Jpn J Exp Med 48:135-142

VanDeWater L, Tracy PB, Aronson D, Mann KG, Dvorak HF (1985) Tumor cell generation of thrombin via prothrombinase assembly. Cancer Res 45:5521-5525

Subject Index

Stay ahead of an expanding field:

Eicosanoids

Honorary Editor:
John R. Vane, London

a new journal

Managing Editors:
K. Schrör, Düsseldorf; A. M. Lefer, Philadelphia

Research into eicosanoids has expanded during the last decade to become a field that is both established and recognized. Initially, the chemical identification of these natural substances was the predominant research topic. This resulted in the synthesis of structural analogues and the investigation of their biological properties, eventually leading to more selective compounds for experimental and clinical research.

This new journal provides you with the latest research in:
● Fundamental studies of the biology of eicosanoids
● Eicosanoid mechanisms involved in human medicine
● Developments of *in vitro* methods for detection and measurement of eicosanoids
● Characterization of assay methods
● Basic and clinical experience with therapeutic applications of eicosanoids
● Mechanisms underlying the therapeutic action of eicosanoids

In addition to original research articles this new journal also presents reviews, short communications and letters to the editor.

Springer-Verlag
Berlin Heidelberg New York
London Paris Tokyo Hong Kong

Subscription information:
Title No. 156
1989, Volume 2 (4 issues) DM 250,– plus carriage charges: FRG DM 8,99; other countries DM 17,40.

Springer

E. Garaci, Rome; R. Paoletti, Mailand;
M. G. Santoro, Rome (Eds.)

Prostaglandins in Cancer Research

1987. 113 figures. XI, 288 pages.
Hard cover ISBN 3-540-17548-2

Contents: Introductory Lectures. – Carcinogenesis. – Cellular and Molecular Mechanisms of PG Action. – Cell Proliferation and
Differentiation. – Immunomodulation. –
Brief Reports.

The relationship between prostaglandins
and tumor cell growth and function has
recently attracted considerable attention
from researchers in oncology, biochemistry,
pharmacology, and cell and molecular
biology. In this volume international experts
describe the fundamental discoveries as well
as the latest findings on the complex interactions between cancer and prostaglandins
and related compounds. The individual
contributions have been organized under
four main headings: carcinogenesis, cellular
and molecular mechanisms of prostaglandin
action, cell proliferation and differentiation,
and immunomodulation. Researchers and
clinicians will appreciate the comprehensive
coverage of this exciting area of research.

Springer-Verlag
Berlin Heidelberg New York
London Paris Tokyo Hong Kong

Springer